Nuclear Power

Westview Replica Editions

The concept of Westview Replica Editions is a response to the continuing crisis in academic and informational publishing. Library budgets for books have been severely curtailed. Ever larger portions of general library budgets are being diverted from the purchase of books and used for data banks, computers, micromedia, and other methods of information retrieval. Interlibrary loan structures further reduce the edition sizes required to satisfy the needs of the scholarly community. Economic pressures on the university presses and the few private scholarly publishing companies have severely limited the capacity of the industry to properly serve the academic and research communities. As a result, many manuscripts dealing with important subjects, often representing the highest level of scholarship, are no longer economically viable publishing projects--or, if accepted for publication, are typically subject to lead times ranging from one to three years.

Westview Replica Editions are our practical solution to the problem. We accept a manuscript in camera-ready form, typed according to our specifications, and move it immediately into the production process. As always, the selection criteria include the importance of the subject, the work's contribution to scholarship, and its insight, originality of thought, and excellence of exposition. The responsibility for editing and proofreading lies with the author or sponsoring institution. We prepare chapter headings and display pages, file for copyright, and obtain Library of Congress Cataloging in Publication Data. A detailed manual contains simple instructions for preparing the final typescript, and our editorial staff is always available to answer questions.

The end result is a book printed on acid-free paper and bound in sturdy library-quality soft covers. We manufacture these books ourselves using equipment that does not require a lengthy make-ready process and that allows us to publish first editions of 300 to 600 copies and to reprint even smaller quantities as needed. Thus, we can produce Replica Editions quickly and can keep even very specialized books in print as long as there is a demand for them.

About the Book and Editors

Nuclear Power:
Assessing and Managing Hazardous Technology
edited by Martin J. Pasqualetti and K. David Pijawka

Addressing the major issues surrounding the use of nuclear power, twenty-nine social scientists with extensive involvement in the assessment and management of nuclear technology discuss critical areas of concern--problem recognition, risk estimation, and policy formation and implementation. The authors appraise fundamental policy issues and examine the controversies surrounding specific power plants, covering a broad range of topics, among them the acceptability of nuclear risk; spatial aspects of power plant siting and the nuclear fuel cycle; land-use controls and other options to mitigate potential hazards; socioeconomic impacts of nuclear power plants; emergency planning around nuclear facilities; and equity issues concerning waste disposal.

Martin J. Pasqualetti is associate professor of geography at Arizona State University. K. David Pijawka is assistant director of the Center for Environmental Studies and assistant professor at the Center for Public Affairs at Arizona State University.

Nuclear Power
Assessing and Managing Hazardous Technology

edited by
Martin J. Pasqualetti
and K. David Pijawka

with the assistance of
Sara J. Frischknecht

Westview Press / Boulder and London

The paper used in this publication meets the minimum
requirements of the American National Standard for
Permanence of Paper for Printed Library Materials
Z39.48-1984.

A Westview Replica Edition

Published in 1984 in the United States of America by
    Westview Press, Inc.
    5500 Central Avenue
    Boulder, Colorado 80301
    Frederick A. Praeger, President and Publisher

Library of Congress Cataloging in Publication Data
Main entry under title:
Nuclear power.
    1. Atomic power--Addresses, essays, lectures. 2. Technology assessment--
Addresses, essays, lectures. I. Pasqualetti, Martin J., 1945-
II. Pijawka, K. D. (K. David)
TK9155.N77   1984   333.79'24   83-21797
ISBN 0-86531-811-5

Printed and bound in the United States of America

10 9 8 7 6 5 4 3

Contents

Tables and Figures

Figures

Preface

Major technological innovation is accompanied by changes in society. One of the most fundamental technical stimuli is in basic fuel type, and characteristic social institutions have developed along with the acceptance of each one. The weave of social fabric has become more complex with the successive availability of new fuels, and we have found it increasingly demanding to adjust to the quick pace of technological change. In the Twentieth Century we have attempted to move from coal, to oil, to uranium within one human lifetime, but such change has been resisted, most strongly by those concerned with the social adjustments necessary for such a shift.

Without adequate time to study the implications of a change as fundamental as a new energy source, we have been induced to take leaps of faith, trusting that we are clever enough to resolve issues as they arise. This seemed to work adequately with fossil fuels, but the difficulties pursuant to the transition to uranium indicate it may no longer be a viable approach. Now that this leap has been made, however, we are forced to gather together as much information and understanding as possible to help us formulate appropriate policy.

Professionals from virtually all disciplines have attempted to erase some of our deficiencies with regard to technology assessment. They have proceeded in their task by identifying and measuring the direct and indirect as well as the unintended consequences of developing, introducing, or altering technology, and by elucidating management strategies and policies that will enhance the benefits of a technology while minimizing its risks. The Three Mile Island experience demonstrated deficiencies in accident risk assessments, government safety policy, design of training programs and evacuation planning. The objective of this volume of papers is to further the field of technology assessment and hazard management by exploring issues in problem recognition, hazard reduction and impact management, and policy formulation.

There has been numerous studies on nuclear power, and the more recent contributions have dealt with sociopolitical and waste equity issues. The publication of such studies has in itself been tacit recognition that some of the problems encountered by the nuclear industry are not solely technical ones, but rather institutional, a complex integration of human values and reactions and policy development. Such publications have been of several types. First, are the fact books. They focus on what is happening where, but they are lacking in basic principles and concepts. A second type may be labeled crisis-oriented, and are those which respond to a particular apprehension. Often they are responses to "brush fire" topics, and many followed the accident at Three Mile Island, Pennsylvania. A third type are those whose task is to address basic policy questions. Again, these usually are responses to specific, high-visibility issues such as the disposal of high-level radioactive waste. In contrast to these books, the purpose of the present volume is to take a more integrative approach, one which includes all the types of studies mentioned above and examines them during several of the key stages in the life of a nuclear generating station.

The scientific perspective most clearly represented in this book is that of Geography. As the bibliographies in the individual papers illustrate, geographers are well-represented in the field of nuclear studies. Until now, however, these contributions have been scattered in journals outside the discipline, and this history of dispersed publication has obscured the important contribution geographers have made to technology assessment and policy analysis. The purpose of this volume is not, however, to simply bring together papers published previously (most are original) or to circumscribe a "how to" description to technology assessment and management, but rather to explore assessment issues in nuclear power which have come to light since the Three Mile Island accident. What do we know of human evacuation behavior that may help in the preparation of realistic and effective contingency plans? What are the social and institutional impacts of implementing various policy options to manage nuclear waste?

In this regard, Sewell and Coppock (1976) suggested that policy-making consists of four stages: (1) identification of significant problems for which there is no policy or for which policies are inadequate; (2) formulation of policies which attempts to solve the problem; (3) implementation of policy; and (4) monitoring the effects of policy. The papers found in this volume address these policy dimensions in the context of technology assessment of nuclear power.

Arizona State University has supported the effort to edit and complete this volume of papers. Appreciation for encouragement and funding is expressed to Dr. Guido Weigend, Dean, College of Liberal Arts and to Dr. Nicholas Henry, Dean, College of Public Programs; Vice Presidents, Drs. Elmer Gooding and Harold Hunnicutt; Dr. Duncan Patten, Director, Center for Environmental Studies; and Dr. Donald McTaggart, Chair, Department of Geography. Patricia Chase, Administrative Coordinator, Center for Environmental Studies, is owed a special thanks for the administration of the project. Sara J. Frischknecht typed, edited, and helped organize the entire manuscript from original conceptualization to final printing. Without the efforts and encouragement of these individuals and programs, the volume would not have been possible.

REFERENCES

Sewell, W.R.D., and J.T. Coppock. 1976. "Achievements and Prospects." In Spatial Dimensions of Public Policy, J.T. Coppock and W.R.D. Sewell, eds. Oxford: Pergamon Press, pp. 257–262.

Introduction

Wilbur Zelinsky

Writers of introductions such as these inevitably make difficult decisions. Is their role simply that of Master of Ceremonies, someone who ushers the players onto the stage as quickly and gracefully as possible? Or should they also, or alternatively, provide the hurried reader with a synopsis of what lies ahead--and perhaps steal quite a bit of the contributors' thunder as they do so? And how critical or bland should the introducer be? Should he succumb to the temptation to fill in the larger chinks in the subject that were somehow overlooked in the papers to follow? Or, finally, what about the easiest, but most dangerous, strategy: the cosmic overview, or setting the subject within broader perspectives, be they philosophic, scientific, or historical?

The character of the three papers in this section has pretty firmly dictated a choice. By raising, but certainly not exhausting, some very great questions concerning attitudes toward, and the acceptability of, novel technologies having vast social, environmental, and economic consequences and the relationships of such innovations with historical tradition, national character, political and legal institutions, and the dynamics of decision-making in general, our authors have opened certain avenues of inquiry we would do well to explore more deeply. They have approached these issues from rather different angles of vision, while supplying us with useful data on the progress of the nuclear power industry in the United States, Canada, and the United Kingdom. As he ingeniously exploits the concept of scale--industrial, areal, and institutional --as a factor in deciding upon, then designing and siting nuclear power facilities, Wilbanks seems to accept the long-term desirability of nuclear energy and searches for the decision-making format and siting policy that would make it least obnoxious to the greatest number of persons and communities. In his illuminating treatment of the contrasting patterns of nuclear power development in the United States and Canada, Cook has remained remarkably evenhanded as he presents both the positive and negative features of this late Twentieth-Century technology while also interpreting the international differences in terms of political culture and historical heritage. Much less dispassionate is the manner in which Fernie and Openshaw attack the institutional framework whereby nuclear power plants have appeared in the British countryside. It requires no special sensitivity to detect between the lines their fear and loathing of these new arrivals. Clearly, even among the well-informed, nuclear power evokes no simple, unequivocal response. Instead it generates perplexity, strong feelings, and many forms of anxiety, as the various essays in this volume amply demonstrate. But rather than reviewing points best left to the authors in this and other sections of the volume, I would prefer to share some thoughts concerning two substantial themes implied in the contributions of Wilbanks, Cook, and Fernie and Openshaw.

The first of these is the geography of attitudes toward nuclear power. Given the level of interest of many geographers today in matters perceptual and attitudinal, the prospects for useful research would seem to be substantial. At the intranational scale, to date we seem to have little of note, aside from the flurry of studies carried on among persons residing in the vicinity of Three Mile Island. But it would be most profitable, for both scholarly and practical reasons, to learn how the residents of various sections of the United States and of different types of places look upon nuclear power issues. Any such investigation would also do well to measure differences among people classified along all the usual demographic, social, economic, and cultural dimensions; and, ideally, it would be longitudinal in nature, so that the shifting currents of perception could be charted. The results would be helpful to anyone seeking to identify optimal sites for future plants according to social as well as technical criteria--assuming that there are to be any future siting decisions in the United States.

At the broader international scale, such an approach may generate valuable findings for the student of social and cultural, as well as political, geography. Indeed the comparative cross-national strategy, for which I have lobbied in other contexts, is a two-edged device. It informs the student of the energy economy, about present and future prospects for nuclear power development in the countries under analysis and also, quite possibly, by controlled extrapolation, in other lands with comparable characteristics. But even more enlightening are the insights into the collective mind and structure of a national society, as I believe has been nicely illustrated in the American,

Canadian, and British instances by the present essays. The phenomenon of a central government, with or without the intervention of privately owned utilities, confronting the question of whether or not to join the nuclear power club offers us that rarest of opportunities: a quasi-controlled experiment in human history, and at the international level to boot. Unlike nearly all other technological innovations, the yes/no decision concerning nuclear power installations calls for intense deliberations at the loftier levels of the political hierarchy. The more usual course of events for new technologies, including some bringing on the most profound consequences, is to have them arrive casually, even absentmindedly, often gradually, via private or corporate channels, perhaps only with modest public attention. Perhaps the nearest analogy to the nuclear power phenomenon would be the recent history of the SST, the supersonic aircraft, whose apparent impending demise may tell us little about the fate of nuclear power. In any case, the SST seriously involved only three national governments, whereas nuclear power is of interest, currently or prospectively, to dozens of countries.

The decision concerning nuclear power brings into focus numerous forces and factors. In addition to availability of investment capital, cost-benefit ratios, and other economic considerations, there can be little doubt that concerns about national prestige and, if only clandestinely, military side-effects enter the calculations. Of particular interest to the human geographer who views the facts in a variety of national settings is the light cast by such decision-making on the cultural matrix and historical circumstances within which crucial officials play out their roles. And through judicious parings of various national cases we can also learn much about the workings of the political process and the effectiveness of public opinion. Indeed the potential payoffs for the exploiter of the cross-national approach are limited only by his imagination.

In order to pursue such macrogeographic strategy, we need more than the present materials on three countries, valuable though these might be. Studies in greater depth for the U.S., U.K., and Canada are required. We also must generate detailed accounts of nuclear power developments in other nations only briefly noted in passing by Cook--France, Sweden, Japan, and the Soviet Union--while there still remains the analysis of many other present and potential cases, including, inter alia, China, India, Israel, South Africa, Brazil, West Germany, Argentina, Syria, and Pakistan.

The second theme I should like to explore here in introductory fashion overlaps the first: the need for

4

a comprehensive study of the social response to technological innovation, and how the case of nuclear power fits within such a larger framework. It is reassuring to realize that the history of science and technology has become a respectable scholarly pursuit; however much still remains to be learned. But, with a few scattered exceptions, the systematic study of how the general public reacts to novel scientific notions or new technologies, sometimes accepting or rejecting them, or often simply learning to tolerate them without any articulate response, any such organized field of inquiry remains at a rudimentary stage.

But, even in lieu of such background material, it is clear enough that in a number of ways the advent of nuclear power presents us with many features that fail to fit whatever standard pattern covers most other innovations. First, much as its advocates might wish the sequence had been reversed, nuclear power was conceived in sin, and it is indeed the offspring of nuclear weaponry. And no matter how cleverly we might replay history according to hypothetical scenarios, the imagination boggles at a world in which nuclear power plants preceded the atomic bomb. Consequently, now and forever, the stigma of its parentage lingers in the public mind.

The sheer massiveness of the investment needed for research, development, and construction also relegates power plants using fissionable materials to a rather special category. Perhaps the only other projects of comparable cost are the development and operation of spacecraft or, perhaps sometime in the next century, fusion power. (Coincidentally, neither produces any new commodity of consequence, just as in the case of nuclear power.) Creating a system of nuclear power plants, or even in some instances only a single one, can seriously strain the combined financial resources of state and corporation. The scale at which cash and skills have been injected into the nuclear enterprise is connected with the abrupt appearance of a new profession, nuclear engineering, and associated bureaucracies in government and industry. Thus we have a rather sizeable corps of highly trained and committed individuals, many of them genuinely fascinated with the scientific and technological intricacies of their craft, who have wagered great emotional energy and entire careers on the gamble that nuclear power will pay off. Seldom, if ever, has the world witnessed the like: a substantial constituency whose livelihood and professional status await the popular verdict as to the economic, political, and environmental acceptability of their product.

And the nature of this product compounds the problem of the acceptability of nuclear power. For a generation or more, nearly all of the world's people have

become accustomed to having electricity available,
either from batteries, on-site generators, or, most
often and conveniently, via long-distance transmission
lines. When a switch is flicked on, the consumer
generally has no way to determine just where the current is coming from or whether it was generated from
flowing water, coal, gas, oil, wood, thermal wells,
windmills, garbage, or uranium. The energy is anony-
mous. In any event, the substitution of fissionable
material for other energy sources means nothing in
terms of material comfort, social relationships, changing life-styles, or patterns of thought unless, as is
increasingly doubtful, there is a noticeable drop in
the utility bill. Thus the consumption of consumer
power has instigated none of those profound alterations
in social or cultural practice that have been triggered
by a dozen or more other technological revolutions of
the past two centuries. Moreover, unless one is
employed in the industry or lives in proximity to a
nuclear power plant, the questions we have been discussing do not intrude into our daily routine. In fact,
many, perhaps most, Americans and Canadians may live
out their entire lives without once laying eyes upon a
nuclear power plant.

Obviously, the one great virtue of nuclear power
technology is that, by substituting uranium and plutonium for dwindling supplies of petroleum, natural
gas, and, eventually, coal, the era of fossil fuel use
can be extended significantly. For certain countries
and regions, the nuclear option also increases the
chances for autarky in their energy economy. But,
equally obvious is the fact that the adoption of
nuclear power represents only a temporary reprieve, and
does not solve the ultimate question: how to effect a
transition from an economy based overwhelmingly upon
exhaustible minerals to one sustained indefinitely by
renewable energy resources. Without such a successful
transition, one that may require several centuries, it
will not be possible to maintain the sort of complex
civilization offering the material abundance that much
of humankind has achieved and most of the remainder
aspire to. Thus the case for accepting nuclear power
is that, while not making life different or noticeabley
better, it does provide the breathing space for making
critical decisions.

On the debit side, there are the physical perils,
known and unknown, and the perceptions thereof. Despite much research, and the endless spate of polemics
on both sides of the issue, much uncertainty still
surrounds the immediate and long-term impact of the
normal operation of nuclear power plants upon air,
water, and land and the health of living creatures.
The debate is complicated by the fact that such operations form only a single link in a long energy cycle

that reaches from the initial mining of ores to the
ultimate disposal of radioactive wastes and the dismantling of the power plants. It is arguable that, in
the aggregate and under normal circumstances, the environmental insults inflicted by nuclear power generation are no greater than those created by coal-burning
plants, possibly much less.

Nevertheless, doubts and qualms persist, in a
manner paralleling public experience with various
industrial chemicals, drugs, fertilizers, herbicides,
and pesticides that, after the euphoria attending their
introduction, were found to produce harmful effects
after a certain interval upon persons exposed to them.
Incomplete knowledge and the aura of mystery that
hovers about nuclear matters are not conducive to
wholehearted public acceptance.

But the largest, blackest cloud hanging over the
prospects for general acceptance of the new technology
is the possibility of a catastrophic accident. It is a
danger not to be taken lightly, just as we cannot dismiss public apprehensions over such an event.   The
question is not whether it will occur, but just where,
when, and how. The nature of the technology is such
that eventually, perhaps before these words reach print
or perhaps several centuries from now, the inevitable
will happen because of errors in design and construction or human or mechanical failures. In spite of all
the most sophisticated efforts to model such accidents,
we simply cannot predict how large an area would be
affected for how long a period and with what damage to
plant and animal life. The only near-certainty is that
a major accident would be a lethal blow to the nuclear
power establishment at the national, or even the international level.

There is one final, quite remarkable characteristic
of nuclear power we must keep in mind if we are to
assess its eventual acceptability properly. As already
hinted, this technology has neither responded to deep-
seated socioeconomic and cultural change nor ushered in
any meaningful qualitative shifts in our ways of life.
On the contrary, it may be the purest sort of example
of technology for the sake of technology, of hubris on
the part of a professional elite, a manifestation of
forces, other than normal human necessities, driving a
set of powerful decision-makers. (Once again, the case
of the SST comes to mind.) They may indeed have pushed
us beyond our capacity for cultural change.

Consider a number of other innovations that did
truly reshape and transform the world, all of them
responsive to powerful evolutionary societal imperatives, whether clearly diagnosed beforehand or not.
Perhaps none was more inevitable or more profound in
impact than the invention of agriculture in Neolithic
times. But much closer in time we have the alphabet,

the university, the printing press, photography, the railroad, the telephone, the internal combustion engine, the electronic computer, and the silicon chip. It would take a wild leap of the imagination to place nuclear power in the same league with any of these in terms of profundity of cause or effect. But it has spawned one dramatic result. As Wilbanks has noted, on no other issue have the battle lines been drawn so sharply between two world-views, to use the crudest of terms, the technocratic and the environmental. In the clash over the acceptability of nuclear power, they may have found the most appropriate battleground.

1
Scale and the Acceptability of Nuclear Energy

Thomas J. Wilbanks

Among the various energy options for the United States, nuclear energy is unique in the emotionally-charged opposition it provokes within a sizeable part of American society. It is far from the only energy alternative to be associated with strong feelings, but it is qualitatively different in the frequency with which these stong feelings are implacably negative.

This situation raises questions that are both specific and general, both pragmatic and philosophical. For example, it bears on practical decisions by vendors of nuclear technologies whether to stay in that business, and it affects decisions by electric utilities about how to generate electricity. At the same time it offers us an interesting body of experimental evidence about how technology decisions are made in our society, and the implications may be important in policy fields other than energy alone.

In their general concern with spatial structure, geographers deal with several factors that may have a bearing on these questions. The most obvious is location. Clearly, locational choice affects the acceptability of particular nuclear energy facilities, at least in the sense that (1) different sites expose different people to hazards, (2) different sites involve different types and levels of hazard (e.g., dangers of earthquakes), and (3) different sites vary in their relationships to local resources (e.g., water requirements or housing for a construction work force). These issues have been treated in fair detail in the literature and need not be addressed again here.

Another topic of interest to geographers, however, is the scale of an activity. Obviously, the components of any particular spatial structure are of different sizes. These sizes, together with their spatial arrangement, define the geographic pattern involved; they shape the spatial processes: growth and decline, interaction and flow, areal subdivision, environmental relationships, and the like. In geography, scale is

9

therefore a central part of such conceptual frameworks as interaction models, location theory, and central place theory.

This paper is a rather speculative exploration of scale as it may affect the acceptability of nuclear energy. In our utilization of this energy option, how does large vs. small relate to attitudes toward it, and what can we learn from this about technology choices in the United States more generally? In order to address such a question, several stepping-stones are needed. First, scale is defined for the purposes of the paper. Second, our recent experience with nuclear energy is reviewed: trends in the scale of use, the current status of nuclear energy as an option, and the social context for its acceptance problems. Third, conventional notions about the importance of scale in electricity generation are summarized. With these preliminaries out of the way, the paper then discusses apparent relationships between scale and the accepance of nuclear energy and suggests some policy implications of these preliminary findings. Finally, some comments are offered about general relationships between scale and technology choice.

THE MEANINGS OF SCALE

For this paper, scale is defined as a measure of magnitude, such that the magnitudes of different observations can be compared and evaluated. It is a relative attribute in two ways: (1) it has no particular meaning for one observation without reference to another; for example, a 3000 Megawatt (MW) power plant is "bigger than" most existing power plants; and (2) it has little meaning without reference to the context of the observation; for example, a 3000 MWe power plant is big for an electric utility whose total baseload capacity is only 6000 MWe, or it is big for a very small town that might be proposed as a site. As with such relative terms as "emergency" or "crowding," this relative nature of scale complicates any attempt to build theory by saying, for example, that large-scale always has certain implications. All one can usually say is that larger vs. smaller generally has a certain qualitative significance.

Stressing the relativity of the concept, however, can be misleading. Figure 1.1 displays an artist's rendering of two coal-fired power plants placed beside an outline of the Empire State Building. One plant has an electricity-generating capacity of 500 MW ("moderate" scale), the other a capacity of 3000 MW ("large" scale). Although the terms "moderate" and "large" are themselves relative and thus changeable in their meaning, the difference in the magnitudes of the two plants

11

FIGURE 1.1
Artist's Rendition of Size of Power Plants Relative to the Empire
State Building (Ford, 1980).

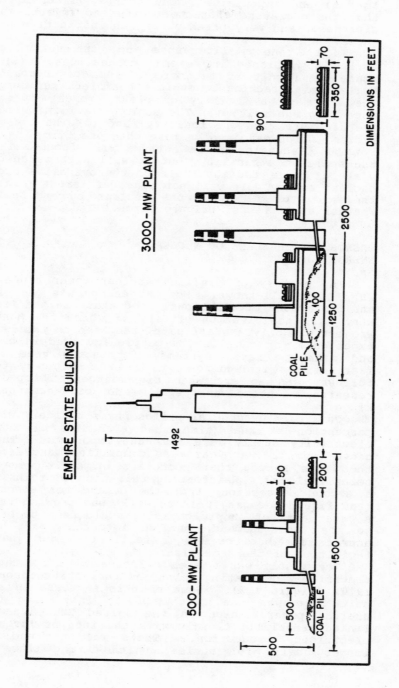

is real and impressive. One gets a visceral feeling that the operating characteristics and impacts must be different, and very often one forms instinctive preferences.

In the energy literature and the energy debate, scale is mentioned in several connections: (1) The scale of a unit of technology, such as a coal-fired boiler or a cracking tower in a petroleum refinery; (2) The scale of a facility or plant, which may include more than one unit; (3) The scale of a firm, which may operate more than one facility; and (4) The scale of a system, which may involve more than one firm.

The discussion which follows will focus mainly on the scale of a facility, but it will also touch on the scale of units and the scale of the overall system for nuclear energy. It will not consider issues related to the scale of firms in the nuclear energy business, whether utilities or private firms.[1]

TRENDS IN THE SCALE OF NUCLEAR
ENERGY SUPPLY

The history of civilian nuclear power since World War II is well-known. Nuclear energy was one of many manifestations of a national mood that emerged from the war. The mobilization effort, of which the Manhattan Project was only a small part, had been an exhilerating experience for many of the people involved in creating and managing large systems, imprinting them with a fascination with doing things in a big way. What followed was an era when big management (operations research, systems analysis), and big science were kings.

This sort of general philosophy fit nicely with the perception at that time that economies of scale in electricity supply were real and important. The evidence from pre-war years was compelling, and data from the 1950s showed that economies of scale were still being realized. Reinforcing this kind of thinking was a second perception: that the demand for electricity (and liquid and gas fuels as well) was growing rapidly. A great deal of supply capacity would need to be added to meet this demand, and an important part of the answer seemed to be larger facilities, not just more facilities of the same size.

The result was a steady increase in the scale of electricity generating units and facilities through the 1970s (Figure 1.2). Some of this increase in average unit size was the result of an increasing use of nuclear power, which in the United States has been linked especially closely with the idea of large-scale electricity production. There was a trend toward larger scales for nuclear units during this period

FIGURE 1.2
Average Unit Sizes (from Messing and others, 1979).

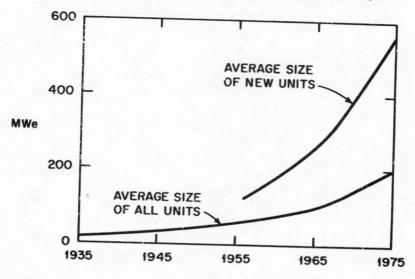

(Figure 1.3); a trend toward larger facilities was even more pronounced, as multi-unit nuclear plants came into use during the 1970s. Table 1.1 summarizes the overall growth in energy supply from nuclear power plants.

As recently as the early 1970s, these increases were expected to continue or even accelerate for a considerable period of time. In 1970, for example, the Federal Power Commission predicted that nuclear power plants would reach the size of 5000 MWe by 1980 and 10,000 MWe by 1990 (Ford, 1980). In 1974, the Atomic Energy Commission predicted that the electricity supplied by nuclear energy in the United States would contribute more than half the total supply by the year 2000, while total electricity generation capacity would triple. This implied the construction of about 600 new generating units in 25 years (assuming an average unit capacity of 1200 MWe): a nuclear energy industry eventually exceeding 700 GWe in total capacity (U.S. Nuclear Regulatory Commission, 1976).

14

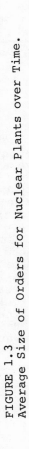

FIGURE 1.3
Average Size of Orders for Nuclear Plants over Time.

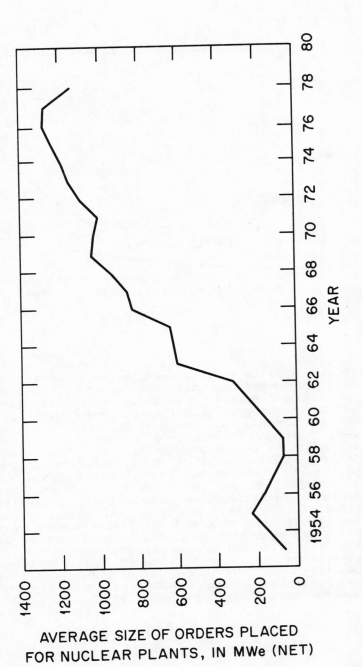

THE CURRENT STATUS OF NUCLEAR
ENERGY SUPPLY

The reality in 1983 is much different, of course.
Continuing the trends that began to appear in the mid-
1970s (Figure 1.3 and Table 1.1), nuclear power is
leveling out as a source of energy supply in the United
States. In effect, the commercial reactor orders of
the mid-1960s have been fully realized; but about half
the new plants ordered in the 1970s have since been
cancelled, in most cases before construction actually
began. For half a decade now, nuclear energy has held
steady at about 12 percent of total U.S. electricity
generation, with a total net current capacity of just
over 55 GWe (Tables 1.1 and 1.2). Power supplied in
1981 was about 272 billion kilowatt-hours, a return to
the 1978 level (the Three Mile Island Accident led to
widespread plant shutdowns for equipment modifications
in 1979 and 1980). As plants now under construction
are placed in operation, nuclear power generation is
now expected to increase to about twice its present
capacity (120-130 GWe) by the end of the decade. No
further units are on order, which means that--given
present schedules for permitting and construction--
nuclear generating capacity in the year 2000 is unlikely to be substantially above the level for 1990.
The current status of the industry is summarized in
Table 1.3.

The most important reason for this departure from
earlier expectations has been that the growth rate in
demand for electricity has eased, reflecting such realities as post-1973 energy price increases, government
advocacy of energy conservation during the 1970s, and
the increasing availability of technologies and
services designed to improve the efficiency of energy
use. Lacking the need for a great deal of additional
generating capacity, utilities have sharply modified
their construction plans, and nuclear energy-- which
had been expected to supply a major part of the
additional electricity, growing in its overall share--
has been affected more than any other single energy
alternative.

But more than this has happened to nuclear energy
in the past decade. Behind the agony and contention of
the energy policy process in the late 1970s, some
strong currents were flowing in search of consensus
(Lakoff, 1983); and by the end of the decade, considerable progress had been made in this direction (e.g.,
Kash and Rycroft, forthcoming).

TABLE 1.1
U.S. Nuclear Power, 1953-81. (Data from U.S. Department of Energy, U.S. Commercial Nuclear Power, DOE/EIA-0315, March 1982.)

	Total Orders Placed*	Net MWe Capacity	Average Capacity in Net MWe	Year-End No. of Reactors	Year-End Net MWe Capacity	Average Capacity in Net MWe
1953	1	60	60			
1954	2	465	233			
1955	0	0	0			
1956	1	175	175			
1957	0	0	0	1	60	60
1958	1	65	65	1	60	60
1959	1	72	72	1	60	60
1960	0	0	0	2	260	130
1961	0	0	0	3	435	145
1962	2	630	315	4	700	175
1963	5	3,018	604	7	910	130
1964	0	0	0	9	885	98
1965	7	4,475	639	10	901	90
1966	20	16,526	826	11	1,789	163
1967	31	26,462	854	10	1,763	176
1968	15	14,018	935	10	2,730	273
1969	7	7,203	1,029	13	4,040	311
1970	14	14,264	1,019	16	5,088	318
1971	21	20,957	998	21	8,467	403
1972	38	41,313	1,087	27	13,171	488
1973	38	43,319	1,140	36	20,051	557
1974	34	40,015	1,177	44	29,207	664
1975	4	4,148	1,037	53	36,885	696

1976	3	3,804	57	40,542	711
1977	4	5,040	65	47,013	723
1978	2	2,240	68	49,957	735
1979	0	0	70	51,082	730
1980	0	0	72	52,937	735
1981	0	0	72**	55,061**	765**

*Many of these orders, especially among those placed in the period 1971-74, were eventually cancelled.

*Preliminary data.

TABLE 1.2
Status of Nuclear Power Plants December 31, 1981. (Data from U.S. Department of Energy, U.S. Commercial Nuclear Power, DOE/EIA-0315, March 1982.)

Status	Number of Reactors				Capacity, Net Electrical (Megawatts)	
	Boiling Water Reactor	Pressurized Water Reactor	Other[a]	Total	Total	Average
Operating or licensed	25	48	3	76	58,495	770
(Fully operable)	(24)	(45)	(3)	(72)	(55,061)	
(In power ascension)	(0)	(2)	(0)	(2)	(2,328)	
(Shut down indefinitely)[b]	(1)	(1)	(0)	(2)	(1,106)	
Construction permit granted	26	49	0	75	83,288	1,110
(20 percent or better)	(21)	(38)	(0)	(59)	(64,870)	
(Less than 20 percent complete)	(5)	(9)	(0)	(14)	(15,858)	
(No construction)	(0)	(2)	(0)	(2)	(2,560)	
Construction permit pending	5	5	1	11	12,714	1,160
Order placed for plant	0	3	0	3	3,510	1,170
TOTAL	56	105	4	165	158,007	960

[a] Includes two DOE-owned reactors (Shippingport and Hanford N), and one high-temperature gas-cooled reactor (Fort Saint Vrain), all in operation. Also includes one liquid metal fast breeder reactor (Clinch River), not yet authorized for construction.

[b] Does not include Indian Point I (265MWe), shut down since 1974, for which the operating license was withdrawn in 1980; nor Humboldt Bay (65MWe), for which future operation is questionable due to the owner's recent withdrawal of the application to restart the unit. Units shutdown indefinitely include Three Mile Island 2 (906MWE) and Dresden 1 (200MWe).

Note: During the first two months of 1982, seven reactors in the above tally were cancelled. These included two units totaling 2,458MWe which had been granted construction permits, and five other reactors totaling 6,140 MWe for which construction permits were pending.

TABLE 1.3
Status of Nuclear Reactor Units in the United States (From U.S. Department of Energy, Monthly Energy Review, January 1983, Energy Information Administration.)

Year	Reactors Licensed for Commercial Operation	Construction Permits Granted	Construction Permits Pending	Reactor Units on Order	Reactor Units Announced	Total Reactor Units	Total Design Capacity in Net GWe
1976	62	72	66	16	19	235	236
1979	71	91	21	3	0	186	180
October 1982	78	64	3	2	0	147	138

A part of this emerging consensus was that our environment and health should not be put seriously at risk by our decisions about resource and technology use, that we do not have to accept tradeoffs of this sort to assure the abundance that we have come to expect as a key part of the American way of life. Technologies are only to be deployed if the risks are "acceptable," however we define risks. And it is the job of our institutions to assure this level of protection for--and through--the mixture of technologies that they select and operate. This means, of course, that technologies involving greater risks are at a substantial disadvantage compared with other options for meeting the same needs.

Because a significant part of American society considers nuclear energy especially risky, one result is that this energy option has in many respects become America's option of last resort. As long as we have other options, such as coal and natural gas, that are in the same ballpark economically, nuclear energy is not going to be our technology of choice very often. Unless our attitudes change--radiation exposure gets decoupled in people's thinking from the risk of catastrophe, civil uses of nuclear energy from the risk of nuclear war, institutions for radiation protection from perceptions of a promotional nuclear establishment--we are likely to continue the kinds of policy conditions that put nuclear energy at a disadvantage. As Komanoff suggests (Komanoff, 1981), society is apparently concerned not with limiting the risks from individual nuclear power plants--risks per facility-- but with limiting risks per unit of time, say a year, or per unit of geographic area, say a state or region. This means that the more plants built, the tighter the standards must be for each plant in order to keep the total temporal or spatial risk from getting unacceptable. Eventually, the cost of achieving the ever-tighter standards may become too great, compared with alternatives, and that approach to meeting energy needs is no longer selected.

THE SOCIAL CONTEXT OF ATTITUDES
TOWARD NUCLEAR POWER

To understand the current situation as it pertains to nuclear energy specifically, and not just electricity supply more generally, it helps to take a quick look at several aspects of the social history of the United States in the past several decades.

A Change in How We Make Decisions [2]

To some extent, dissensus has always been a part of the American scene, a corollary of our democratic ideals. But a fundamental kind of sociopolitical change in the past several decades has made it a more salient issue in energy policy. As recently as two decades ago, it was generally assumed that major decisions in the United States could be grouped into distinct categories, in each of which a limited number of groups had a right to participate--usually those with direct economic, regulatory, or technical roles. For instance, it was quite clear who made oil policy decisions and utility policy decisions and national defense decisions. As long as the participants agreed, action could be taken. By the end of the 1950s, in fact, the ability of these decision-making consortia, often made up of big business and big government, was so unbridled that President Eisenhower felt it necessary to warn the country about the power of a "military-industrial complex."

But a number of important events during the 1960s, including civil rights struggles, Vietnam, and the Santa Barbara oil spill, convinced many people that decisions being made within traditional frameworks were affecting individuals and groups outside those frameworks. As a result, the demand grew for broader participation in decision-making, ranging from pressures for consumer representation on corporate boards to student participation in promotion decisions for university professors. The "environmental movement" was the most visible indicator of this change, but it involved more than environmental interests alone.

Because of this change, energy policy decisions now involve a wide range of groups and interests as parties to the decisions, and it takes a broad consensus (or at least acquiescence) among the parties to put major actions into effect (Kash, et al., 1976; Schurr, et al, 1979). The change is probably irreversible, and it is in many respects the way a democratic process is supposed to work. But it does leave us with a serious problem. As Lewis Branscomb (1978) has suggested, our decision-making structures--designed for a different time and a different set of conditions--have broken down under the current conditions, and we do not have a new structure to replace them. Without it, our old structures turn uncertainties into disagreements, and disagreements into antagonisms, and prospects for action fade away time after time.

Social Perspectives on Nuclear Energy

No single category of energy technology has been buffeted by this turbulence more than nuclear energy, for reasons which also have historical roots (See Earl Cook, Chapter 2). At the risk of oversimplification, consider two very different perspectives toward nuclear energy that have developed in American society during the past four decades.

First, there is the perspective of the nuclear enterprise. The energy R&D community came out of World War II with a real sense of mission: developing the world's first major new energy source since the discovery of fossil hydrocarbons, producing energy "too cheap to meter," removing energy supply as a constraint on world economic development. It is hard to exaggerate the excitement of this, the sense that history would long remember the achievements of this enterprise. It attracted many of the best young scientists and engineers in the U.S. in the 1940s and 1950s, just as the space program did in the 1960s. It shaped people's lives. And in many ways our current energy R&D establishment, at least in the public sector, is still dominated by the people and personal relationships that were a part of that experience (consider, in 1981 and 1982, the links between Kenneth Davis, Deputy Secretary of the Department of Energy (DOE), and Lou Roddis, Chairman of DOE's Energy Research Advisory Board; Davis and Roddis worked together closely in the Atomic Energy Commission). Here was a powerful perspective, very positive toward nuclear energy, optimistic about the prospects of solving problems and controlling adverse impacts, and focused on most likely events and outcomes.

Second, there is the perspective of "the public" (recognizing that, strictly speaking, there is no such thing as a homogeneous "public"). Recent research on the history of "nuclear fear" in the United States suggests the following kind of picture.[3] In the years after World War II, public attitudes toward nuclear power were largely positive, influenced by the views of such media advocates as Bill Laurence of the New York times. But during the 1950s, the public was barraged with information about the dangers of atomic bombs. This was the time of growing civil defense programs, drills in schools, the book and movie On the Beach. Some of the talk about impending doom was an expression of doubts and concerns in the scientific community, in a few cases representing second thoughts about their involvement in nuclear weapons development; some of it was a part of an effort by civil defense advocates to get people to pay attention to their programs; some of it was a political strategy by people who believed we should be properly scared of the Russians. But the

general effect was to frighten large numbers of people
about things nuclear, to create the impression that
nuclear war could destroy civilization, or at least
most of it--not just from physical destruction alone
but also from such effects as genetic mutation. Weart
says: "From interviews today, we know that many of the
children who were taught about atomic warfare back in
the 1950s never quite forgot the fears that came to
them: (Weart, 1982:77).

How could people communicate these fears and convert them into policy actions? One tangible thing they
could insist on was that we control exposure to radiation, and a lot of this concern was initially converted
into opposition to nuclear testing as a cause of radioactive fallout. The real worry was not so much about
nuclear testing as such as about dangers from the entire nuclear enterprise; this was just a handy way to
demonstrate concern and to try to come to grips with
some concrete issues. But one of the consequences was
to forge a strong link in people's minds between radiation and catastrophe.

As this effort began to bear fruit in the form of
bans on surface or atmospheric testing of nuclear weapons, the concern was redirected toward a new target,
because the underlying worry had not gone away. That
target was civil nuclear activity, at least as it was
taking the form of large nuclear plants. Again, the
operational focus was on radiation exposure, especially
in connection with the possibility of major accidents.
In contrast to the nuclear enterprise perspective, this
point of view concentrated on consequences not of most
likely events and outcomes but of worst case events and
outcomes.

As these two perspectives have come together in a
social and political environment that has increasingly
stressed participation, the result has been a kind of
stalemate--not between knowledge and ignorance of nuclear energy, as some would see it, but between two
different sets of cultural attitudes. Clearly, as we
view our options in society, technological or otherwise, what we choose to worry about is a function of
our culture, not of intrinsic characteristics of the
options themselves (Douglas and Wildavsky, 1982); and
here we have a cultural division between those who
choose to worry a great deal about radiation exposure,
sort of as a surrogate for a much more general concern
about nuclear catastrophes, and those who choose to
worry about constraints on economic progress because of
a failure to realize the potential of new technologies.

In this new environment, any search for areas of
agreement has been impeded by a fundamental institutional problem. Simply stated, public confidence in
our institutions responsible for nuclear energy has
been consistently undermined by the perception that

these institutions are the nuclear enterprise: that
they are expressions of just one of the perspectives;
that they cannot be counted on to approach the issues
in a balanced way; that by being technological opti-
mists and by focusing on the "most likely," they will
permit risks of radiation exposure to grow to unacceptable levels (regardless what their leaders may say in
order to get a particular decision made); that they
continue to symbolize a commitment to both nuclear
energy and nuclear weapons. Skeptics ask us to look at
the links between energy and weapons that are implied,
for example, by discussions of nuclear proliferation
and by the responsibilities of first the Atomic Energy
Commission and now the Department of Energy.

SCALE AND EVALUATION OF
ELECTRICITY SUPPLY OPTIONS

Before trying to relate the scale of nuclear energy
supply to this social context, it is useful to review
how scale is normally treated in the evaluation of
electricity supply options. The most basic assumptions
are about economies of scale for generating units--
essentially, the elasticity of total operating costs
for a unit with respect to its level of output.
It is widely known that such economies do exist.
For example, larger boilers involve smaller materials
requirements per unit of capacity. Similarly, maintenance costs per unit of capacity should be less, as
well as administrative requirements, land use, and
support facilities (e.g., access roads, and fuel handling facilities.) A larger activity should also be
better suited to the use of efficient specialized
equipment and people, such as pollution control technologies and highly-skilled planners, and it should be
able to save money by purchasing fuel and other inputs
in larger lots.
These savings have traditionally been represented
as follows:

$$C_{(s)} = Ks^n$$

where $C_{(s)}$ is the cost of a unit of production at
size S

K is a constant, normally the cost of a
unit at some standard size

s is an expression of unit size relative
to the base of comparison

n is the effect of scale on unit costs

If economies of scale exist, n is less than one; the farther its value is from one, the greater are the economies of scale. The standard rule of thumb is n = 0.6 (commonly referred to as "the six-tenth role"). In other words, doubling the production capacity of a unit increases its cost by a little more than one-half. Obviously, this implies substantial economies of scale, and such a "rule" places no limits on the size at which savings are realized. No wonder engineers have often tended to think that bigger is better.

Such a feeling has had other bases as well. For instance, it can be shown that optimal size is likely to relate to market size, and it has generally been assumed that markets would continue to grow more or less indefinitely. In addition, both to technologists and policymakers, bigger is often easier to get excited about; perhaps this translates into a size bias in utility and government R&D funding and technology choices (Hammond, 1972), not only developing more large-scale options but helping these options to be more feasible. Moreover, as siting and other technology decisions have become enmeshed in our institutional encounters with participation, there has been a tendency to think that larger facilities mean fewer facilities, thus fewer difficult decisions.

This picture of substantial and continuing economies of scale, reflecting a wide range of technological, economic, political, and social conditions, only started to be questioned seriously in the 1970s. One source of criticism was from those who argued on philosophical grounds that "small is beautiful" and big is not; that technological bigness begets institutional bigness, and this shuts out the kind of broad participation in major decisions that society is seeking; that an orientation toward bigness unthinkingly puts us on a centralized "hard path" which is likely to change our society in undesirable ways.

A second basis for re-evaluation has come from individuals taking a hard look at evidence from the 1960s, which include more large units than periods such as the 1950s which showed such strong indications of economies of scale, and at changes in the late 1970s and early 1980s. At least for fossil-fueled electricity generating units, the 1960s data indicated that economies of scale were certainly real and important up to some size, but beyond that size they leveled out quickly. At a size of, say, 400-600 MWe for coal-fired units, these economies were essentially exhausted (Abdulkarim and Lucas, 1977, and Loose and Flaim, 1980). Studies of economies of scale in energy firms during that same period (mainly electric utilities) showed much the same thing (Christensen and Green, 1976, and Huettner and Landon, 1978).

Together, these two bodies of thought resulted in a more concerted effort to investigate possible diseconomies of scale with very large units or very large plants, and many examples of such diseconomies were found. For example, evidence from the 1960s and 1970s showed that larger plants had more "downtime" for repair; they spent more of their time out of operation (Table 1.4). When capital costs were compared with actual output, the large units were therefore not as advantageous economically as they appeared from analyses based on comparisons with designed capacity.[4] One British study, for instance, estimated that building two 650 MWe generating units rather than one 1300 MWe unit would reduce downtime enough to save 20 percent in the overall cost of electricity generation (Lucas, 1979). Part of the reason is that, with the bigger units going down more often, and with these withdrawals of energy supply coming in larger chunks, a system needs a bigger and more expensive reserve—its backup capability, its margin of safety.

TABLE 1.4
Fossil Unit Forced Outage Rate Versus Unit Size
1967-1976

Capacity Category (MWe)	Forced Outage Rate (%)
60- 89	2.5[1]
100-199	4.4
200-299	5.7
300-399	7.9
400-599	9.6
600-799	13.8
800 and above	16.1

[1] The 1977 EEI report contained no data on units less than 100 MWe. Data for the 69-89-MWe category were taken from Edison Electric Institute data by Ford (1980).

Another factor is that very large units take longer to construct than small ones, and sometimes a more complicated permitting process adds an additional lead time penalty. As long as demand forecasts could be made confidently and interest rates were low, this was not a problem. But under current conditions, the difference between, say, size and ten years in lead time

means a large difference in finance charges, and it
means a significant difference in uncertainty about
what the level of demand will be when the new plant is
ready for operation. For instance, a 1979 analysis
considered large vs. smaller-scale fossil-fuel options
for meeting a perceived need in 1992 for 1150 MW of
additional electricity supply capacity (ignoring any
other diseconomies besides finance charges associated
with lead time). Its conclusion was that the cost of
electricity from the larger option would be about 20
percent higher (Budwani, 1980). In times of high
interest rates, such effects are even greater for
technology options with higher front-end captial costs,
such as nuclear energy, even if their operating costs
are lower.

Other diseconomies of scale include higher average
transmission costs for the product, because average
transmission distances are greater (e.g., Wilbanks,
1983:Appendix); higher costs of custom fabrication for
components of very large plants, compared with the
possible mass production of components for smaller
ones; regulatory complexity (government typically looks
harder at very large actions than at smaller ones,
other things equal); and, in some cases for electricity
production, a need for larger step increases in rates,
which tend to be harder to get approved.

Because of factors such as these, it is increasingly accepted that electricity generation cost curves
with respect to scale are U-shaped (Figure 1.4).
Analyses to date have been concerned largely with coal-
fired generating units, where the optimum appears to be
in the 400-600 MWe range. And, in fact, several electric utilities have revised their construction plans in
the last few years to take this understanding into
account; as one example, Pacific Gas and Electric
Corporation switched from a plan to build two 800 MWe
units to three 500 MWe units (Iwier, 1980). Such a
U-shaped curve has not been conclusively demonstrated
for nuclear-powered generating units, and opinions
still differ about the optimal scale for nuclear units
and facilities; but at least one study has suggested
that a curve of this type for nuclear plants is
supported by theory (Lucas, 1979). That study
estimated the optimum to be in the 500-600 MWe range,
considerably smaller than most nuclear units being
planned in the mid-1970s. U-shaped curves also have
been shown for such other energy-related facilities as
ethanol production (Alam and Anos, 1982) and ethylene
production (Cantley, 1979).

Some people have argued that clusters of moderate-
sized units at one site represent a promising way to
combine many of the benefits of both bigness and smallness (e.g., Ford and Yabroff, 1980). This approach
offers, for instance, the reliability of a moderate

FIGURE 1.4
Electricity Generation Cost Curves Related to Scale.

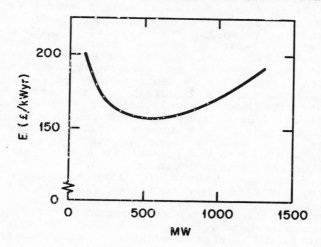

scale together with some of the management and purchasing economies of a large scale. On the other hand,
such clusters also combine some of the disadvantages of
both scales: for example, the capital costs of smaller
units and the local environmental impacts of larger
ones. So far, such issues in determining the optimal
scale of a facility have not been considered in as much
detail as the optimal scale of a production unit.

SCALE AND ATTITUDES TOWARD
THE NUCLEAR ENERGY OPTION

Such evaluations of the "optimal scale" of energy
actions revolve around economic realities; but in the
long run economic considerations are partly a function
of culture, as preferences and social concerns come to
be expressed in the form of market conditions.
Clearly, one of the implications of social concerns
about nuclear power has been to raise its (presumably)
U-shaped curve of costs with respect to scale. For
example, the Federal Energy Regulatory Commission
recently approved a utility's request for a higher rate
of return because of that utility's concentration in
nuclear power. In effect, it determined that nuclear-
oriented utilities need a higher rate of return in
order to be judged the same credit risk as non-nuclear-
oriented utilities--a social perception of risk
converted into an economic fact.

Beyond this general effect, transcending scale, it appears that our collective social concerns have effectively limited the scale of nuclear energy facilities in the United States, at least for the foreseeable future. The potential for very large, integrated nuclear energy centers (on the order of 10,000 MWe or larger, perhaps with certain processing activities located there as well) is still being discussed (e.g., Southern States Energy Board, undated); but it is difficult to imagine a utility in the 1980s choosing this path to meet its electricity generation needs, even if everyone were to agree that such a quantity of additional capacity is needed. The utility—and its sources of financing—would certainly question the prospects for such a proposal to survive the current regulatory and approval process.

This indicates the possibility of a more general connection between scale and attitudes toward nuclear energy. To lay out such a proposition, the following discussion will present the basic argument, offer some supporting evidence, and present some suggestions about the significance of the connection for nuclear energy in the United States.

Some Fundamentals of Risk Acceptance

Why are certain relatively large risks considered acceptable, while certain relatively small risks are considered unacceptable? Behavioral research has shown that risk acceptance is a function of our general attitudes toward life and the world around us, not just of scientific evaluations of the probability of harm. Some of these general attitudes (or judgmental rules or "heuristics") are rather predictable. For example:

(1) _Familiar risks are more likely to be acceptable than unfamiliar risks_. One of the ways that we evaluate the risks from an event or action is by referring to other instances of it. In effect, familiarity with that type of event increases our personal knowledge of its impacts—reducing our fear (if, in retrospect, we judge the experience to have been acceptable) that some of the impacts this time will be unacceptably calamitous. Familiarity may also have the effect of muting our sensitivities to certain risks and impacts (accepted parts of our lives, even if others would find them unacceptable: e.g., the risks of underground coal mining as an occupation).

(2) _Risks that cannot be prevented or controlled are more likely to be acceptable than risks considered unnecessary_. A key issue is "controllability". Life cannot exist without risks; we all learn to live with

risks from natural hazards (Hurricanes) and everyday activities (driving a car). Some, especially natural hazards, we accept as "acts of God". Others, we accept because effective controls would restrict our personal freedom in unacceptable ways (e.g., seat-belt buzzers or a ban on saccharin). On the other hand, we perceive that many risks can be avoided, or at least controlled. Airplanes can be safe. Manufacturing plants and city sanitation systems can be non-polluting. In particular, we perceive that technologies can and should be controlled, but we perceive a dilemma in this respect:

> Attempting to control a technology is diffcult, and not rarely impossible, because during its early stages, when it can be contolled, not enough can be known about its harmful social consequences to warrant controlling its development; but by the time these consequences are apparent, control has become costly and slow (Collingridge, 1980:19).

Perhaps the "giddy pace" of technological change (Charlesworth, 1970:41) has had a kind of cumulative impact on our instincts for survival. Perhaps our capacity for risk acceptance is limited, and the explosion of new creations of humanity has in some ways exhausted it. Perhaps we are a little disillusioned because technology has so often failed to deliver the magic claimed for it (Handler, 1979). But, like other societies through history, we still tend to view technologies—or at least some new technologies—as not natural, not normal, an intrusion, posing unnecessary risks not only to health but to society and culture as well (Douglas and Wildavsky, 1982). Unless we can be assured that these risks are controllable, we resist adoption of the technologies.

(3) _A risk of catastrophe is viewed differently from a risk that is not catastrophic_. In judging risks, people generally overestimate those associated with dramatic and sensational events and underestimate those associated with "unspectacular events which claim one victim at a time and are common in nonfatal form" (Slovic, Fischhoff, and Lichtenstein, 1979:15). But this kind of bias, resulting from differences in the "visibility" of the information we receive, should not be confused with the social judgments that we all make in weighing everyday risks vs. potential for disaster. It appears that, when lay people rank activities and technologies as to their riskiness, perceived disaster potential is a powerful factor in these evaluations (_Ibid._, 20, 36-38; Fischoff and others, 1978), and such perceptions are persistent indeed, especially when people are not convinced that the risks are necessary and unavoidable.

(4) <u>Risks are more likely to be acceptable if they are accepted voluntarily than if the exposed individual or group has no choice</u>. Several studies have shown that higher risks are more tolerable for voluntary than involuntary activities (e.g., Fischoff and others, 1978); Starr, in fact, has suggested as one of his "laws of acceptable risk" that the public is willing to accept risks from voluntary activites roughly a thousand times greater than from involuntary activities that provide the same level of benefits (Starr, 1969 and 1972). Perhaps consent is one of the most fundamental of all the dimensions of risk acceptance (Douglas and Wildavsky, 1982).

(5) <u>Risk acceptance is related to other agendas and concerns</u>. Evaluations of risk do not arise independently. They are bound in sheaves with other issues, objectives other than risk aversion alone.[5]

Our judgments may therefore be shaped by our own need for a job, our political philosophy, or our sense of being part of a social group. In addition, they are certainly affected by our belief that others have hidden agendas as they participate in decisions.

<u>Scale and Risk Acceptance</u>

The tenets of risk acceptance outlined can be related fairly directly to scale as an attribute of a contemplated new facility:

(1) <u>When they involve new technologies, very large facilities are less likely to seem familiar than small ones</u>. New technologies usually evolve through an orderly process of "scaling up"--with larger, more expensive facilities awaiting experience with demonstrations at smaller scale. This means that, as such a process develops, the preponderance of evidence is at relatively small scales; relatively large scales are less familiar. Furthermore, larger facilities are more likely to be unique in some way, involving one-of-a-kind components or arrangements. For these reasons, along with the sheer visibility of large projects (see below), government more often insists on scrutinizing projects which exceed a certain size. For instance, in Minnesota it is not necessary to obtain a certificate of need (requiring public hearings before the state energy agency) in order to seek permission to build power plants smaller than 50 MW, power lines less than 200 kV or shorter than 100 miles within the state--nor are siting hearings through the state's Environmental Quality Board necessary for plants smaller than 50 MW or lines less than 200 kV (Gerlach, 1982).

(2)   <u>Risks associated with very large facilities</u>
<u>are likely to be viewed as harder to control</u>.   From
society's point of view, controlling any technology is
difficult.   Technologies develop their own momentum;
they become associated with institutional systems that
lack flexibility and responsiveness to social concerns.
Because lack of control is associated with riskiness,
technologies which seem easier to control are usually
more acceptable.   Among the characteristics associated
with controllability is scale: "Intuitively size and
ease of control are opposed; small is controllable"
(Collingridge, 1980: 111).   In small activities, for
example, mistakes can be corrected more easily (size
can imply unstoppability: Gerlach, 1982: 89-90).   And
one might suspect that, when a suggested change
involves a small activity rather than a large one,
large organizations are less likely to have a sizable
stake in resisting it.   These concerns do not make
large facilities inherently unacceptable; but for
relatively unfamiliar technologies, they place a
greater burden on demonstrating that the risks are
really necessary.

(3)   <u>Catastrophes are more likely to result when</u>
<u>something goes wrong with a large facility than a small</u>
<u>one</u>.   Intuitively, the consequences of a "worst case"
event for a large activity are more serious than for a
smaller activity: e.g., large vs. small wars.   Social
concerns about disaster potential, therefore, usually
rise as scale increases.   One can hypothesize that this
curve of concern takes the form shown in Figure 1.5.
At very small scales, the level of concern is relatively small and relatively insensitive to, say,
doubling the size of an activity.   As scale increases,
however, social concerns increase more rapidly, with
the tail of the curve partly defined by experience and
partly by the nature of consensus or dissensus within
society.

(4)   <u>Participation is likely to be more difficult</u>
<u>to assure for very large facilities</u>.   Some technologies
are more "participatory" than others (Carroll, 1971),
because their uses are subject to fewer technical constraints.   For instance, in a society worried about
unemployment, the potential to substitute labor for
capital in deploying a technology is limited for many
technologies (e.g., a process involving very high
temperatures).   Many of these constraints are related
to scale, such as the need to work with indivisible
large or heavy objects (Forsyth, et al., 1982).   More
fundamentally, larger facilities have larger-scale
implications: more extensive connections for flows of
materials, people, and money; more complex linkages for
spreading impacts.   This means that, if participation

34

FIGURE 1.5
Risk Acceptability and Scale of Activity.

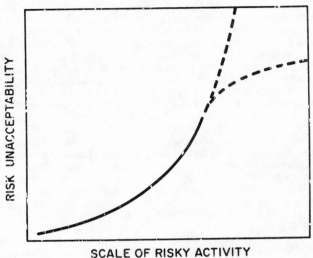

is to be offered to all those who might be affected by the activity, the scale of participation is related to the scale of the activity. And it is logical to expect that, to the extent that any sort of agreement among the participants depends on personal interaction (which is constrained by time and space), there may be limits to the social and spatial scale within which agreement can be reached, short of an unmistakable threat (Wilbanks, 1981b).

(5) Because they are more salient, very large facilities are more likely to stimulate actions on behalf of other agendas. Large facilities are more noticeable, in a sense more visibly symbolic, which increases the likelihood that they will attract attention from parties with a variety of objectives: regional development, personal wealth, political power, environmental protection, institutional growth, sociopolitical change. For instance, "the larger the scale of the project, the more likely it is to be met with effective resistance by grassroots groups which come together in...large scale networks..." (Gerlach, 1982: 89). And one result of this added complexity is that, in a participatory decision-making system, agreement is harder to find.

Scale, Risks, and Nuclear Energy

In the United States, nuclear energy has been developed as a large-scale energy supply option, even though it is the kind of option for which a large scale is most likely to cause acceptance problems:

(1) Nuclear energy is not yet a familiar technology to the lay public. We have not yet lived with it for several generations. Many of the technical details still seem mysterious. As accidents occur (even minor ones), it appears that we are still in the early part of the learning curve.

(2) Nuclear energy has raised fears about controllability. Partly because of the momentum developed by its support from big government, large firms, and large utilities (all viewed as proponents of the technology), together with a level of technological complexity which seems to require reliance on a special kind of "priesthood" (Weinberg, 1971), large-scale nuclear facilities have seemed difficult to control, difficult to manage carefully. They have stimulated an unusual amount of attention to institutional mechanisms for control (Bronfman, et al., 1979).

(3) Nuclear energy is viewed as having a significant disaster potential. Unlike certain other kinds of technologies (computers), in the case of nuclear energy a perceived connection with nuclear catastrophe has led society to assign it a nonzero probability of causing a major disaster (e.g., Fischoff and others, 1978).

(4) Partly because of the sophistication of the technology, informed participation has been a particular challenge for nuclear energy. The barriers to effective communication have been formidable.

(5) Nuclear energy has become a classic example of the impact of diverse agendas on technology choice. It has, for instance, served as the prime example of a "hard path" in technology development and use and as a rallying point for anti-establishment groups since the end of the Vietnam War.

For these reasons, if scale is associated with risk acceptance in general, it should have an especially powerful connection with risk acceptance for nuclear facilities. Perhaps nuclear energy in the United States is characterized by an acceptance domain as depicted schematically in Figure 1.6. We have reason

36

FIGURE 1.6
Perceived Risk and Acceptance Relationship.

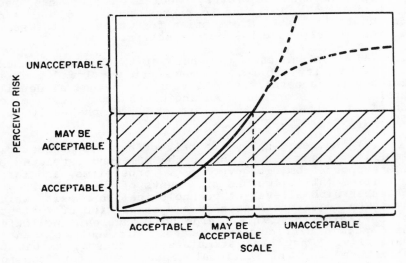

to believe, for example, that certain extremely large
scales for nuclear energy facilities are unacceptable
in the United States at this time, because they are
considered too risky. And we also have reason to
believe that facilities at very small scales are quite
acceptable. Besides the nuclear reactors in our
central-station electric power plants, over which there
is so much controversy, we operate a large number of
small research and test reactors (Table 1.5). Especially notable is the fact that forty-nine nuclear
reactors were operating (or at least operable) in
American universities, under the supervision of university faculty with students involved as a matter of
course, often within major population concentrations.
Clearly, the risks associated with nuclear power at
this scale are considered acceptable.

In between, of course, is a wide range of scales--
say, from 100 MW(t) to 5000 MWe--and we have relatively
little basis for describing the shape of the perceived
risk curve in this range (except to say that it rises
with an increase in scale) or to estimate the two kinds
of acceptance thresholds suggested in Figure 1.6. The
positioning of the thresholds is especially important
with respect to unit cost curves for nuclear facilities. In Figure 1.7, for instance, if the thresholds
are s and s , nuclear energy has a bleak future at the

TABLE 1.5
Operable Test and Research Reactors in the United
States (From U.S. Department of Energy, "Nuclear Reac-
tors Built, Being Built, or Planned in the United
States as of December 31, 1981," Technical Information
Center, June 1982).

Category of Reactor	Number	Sizes in MWt	Startup Dates
General Irradiation Test	1	250	1968
High-Power Research and Test	7	5-100	1958-67
Safety Research and Test	3	55 and smaller	1959-78
General Research (government and industry)	21	2 and smaller	1950-78
University Research	49	10 and smaller	1957-77

FIGURE 1.7
Scale and Unit Cost of Reactors

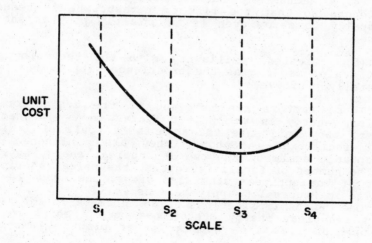

scales at which it makes most sense in terms of engineering economics. If the thresholds are s_1 and s_2, as appears to be the case, the prospects of nuclear energy at such scales are contingent on other factors, such as actions to address social concerns related to those scales. If the thresholds are s_3 and s_4, then in effect risk acceptance related to scale is not a significant issue, because it is not a constraint up to and including the scale offering the lowest engineering-economic unit cost.

Factors Bearing on the Acceptability of Risks Associated with Larger Scales

In understanding technology choice, the most interesting scale range is the one termed "maybe acceptable." Within this range, what factors are likely to be associated with a decision in our participative society that a facility at such a scale is acceptable? The literature on risk acceptance, as well as our actual experience so far, suggests that they include the following:

(1) Demonstrations that a facility of that scale is really needed.

(2) Mechanisms for broad participation in decisions, both when the facility is planned and while it is operating.

(3) Broad consensus that, in terms of a wide variety of agendas, the benefits substantially outweigh the costs and risks.

(4) Mechanisms for identifying and containing accidents so that the risk of catastrophe is reduced virtually to zero, even if an unlikely problem should arise.

(5) Greater familiarity with the technology at that scale, as long as the experience with it has been generally positive.

These factors are relevant to the evaluation of nuclear energy in the United States in two respects. First, they influence the acceptable scale of individual facilities. Second, they also influence the acceptable scale of the overall nuclear energy system. As the number of facilities grows, the perceived riskiness of the entire system also grows; the concepts embedded in Figure 1.6 apply here as well.

One of the implications of an increase in the scale of the nuclear energy system has been a shift in the balance of risk from nuclear power plants themselves toward their supporting infrastructure: a predictable trend. In a great many fields of inquiry, the effective scale of a system turns out to be limited by its dependence on linkages (Wilbanks, 1982). For example, as the scale of our analytical modeling grows, we often find that the most serious limitations is our inability to link separate components of a modeling system in ways that are scientifically valid and that we (and users) fully understand. Similarly, to paraphrase Brian Berry, we know that "technologies are systems within systems of technologies"; each is a part of a sequence of activities, a kind of "trajectory" (Kash, White, Wilbanks, and others, 1975) representing a particular combination of activities. In order for the entire system to work, each element must work, and the elements must be effectively linked.

As the scale of just about any kind of system is increased, additional linkages are required; networks for the flow of goods, services, information, and control. These linkages require resources themselves, and they create continuing logistical needs, such as supervision, monitoring, and maintenance. In addition, they are often sources of entropy-like "energy" loss in a system, contributing to waste at the same time that they work to counteract it. Many of these costs are linear, rising in proportion to the number of links required. Some of the costs, it has been argued, rise even faster, because a more complex system of linkages calls for more coordination and control (e.g., bureaucracy). Meanwhile, the benefits per additional link eventually decline, sooner or later sinking below the added costs.

Moreover, as the system grows, its health requires that the various elements be kept in balance—that providers provide more, that users use more. And this adds another major dimension to the need for coordination and control.

In nuclear energy, linkage constraints have become increasingly important in several ways. First, the issue of expanding power generation capacity has become linked with concerns about risks in the nuclear fuel cycle, most specifically with issues of radioactive waste disposal, including the movement of wastes to disposal sites. It seems likely that, even if electricity demand were to take off again, further use of nuclear energy would soon be limited unless the waste disposal issue has been resolved. Second, as the scale of nuclear plants has grown, it has been more and more necessary to seek remote sites for them, but this trend in resolving conflicts about power plant siting has made the siting of new transmission lines a major public issue—to the point that many experts believe that the remote siting of power plants is about to become a thing of the past (Friesema, 1979). Recognizing these kinds of concerns, impact assessments and decisions based upon them have come to place considerable emphasis on linkage issues. One of the consequences has been a need to face up to some real complexities in trying to keep risks within the limits of public acceptance. In discussions of participative decision-making, for example, it is increasingly recognized that decisions involving a few specific sites are relatively simple compared with decisions involving a network of movement routes.

To the extent that risks associated with linkages are limits on the acceptable scale of a nuclear energy system, the social judgment of acceptable scale is thus affected by (a) the number of linkages (reducing linkages is likely to reduce risk, but it may require reducing the scale of the system); and (b) investments in linkages to increase their reliability or acceptability, which either reduces the non-linkage resources available to the system or reduces the number of linkages it can afford, in either case placing some limits on the ultimate scale of the system.

Significance of the Relationship for Nuclear Energy Policy

To the extent it is valid, the meaning of all this for nuclear energy policy in the United States is fairly clear. Risk acceptance might be increased if this energy option were available at smaller scales, other things equal. Rather than simply assuming that nuclear energy is a large-scale central-station energy supply alternative, suitable for no other role, subject to economic laws and regulatory constraints that make any other path impractical, might it be worth looking a little more seriously at the feasibility of smaller-scale nuclear power supply?

The first issue is the technological feasibility of small-scale power production from a nuclear reactor.[6] As Table 1.4 implies, production is possible from almost any size reactor, down to below 10, or .01 MWe. Kraftwerkunion (KWU), a German firm, has offered relatively small reactors on the commercial market [100-400 MWe] as an option for small countries or regional markets; and several U.S. manufacturers have designed power plants in the 300 MWe) range, not only in the early years of the industry but quite recently. The U.S. nuclear ship Savannah was powered by a reactor in the 80 MWt range, and the 134 operable reactors in the nuclear submarines of the U.S. Navy (plus 29 being built) are in the 20-50 MWt range. The U.S. Army has built and operated much smaller pressurized water reactors (in the 1-5 MWe range). And small, lightweight nuclear-powered systems may eventually be needed for facilities in space; for example, a communication facility might need a capacity of 10-100 KW, and technologies are available to meet such a need.

Experience with small units so far has shown them to be highly reliable, even at very small scales; they can be designed so as to have low maintenance requirements, to protect against releases or accidents, and even to respond to rapid changes in power needs. Compared with large units, the major drawback of small nuclear power units [at least below 200-300 MWe] is that they are relatively inefficient thermodynamically; thus the power supplied is relatively expensive. An additional drawback is that the requirement for technical training is high, considering the modest amount of energy produced.

Suppose that, through a combination of technological progress and economies of mass production, the engineering-economic cost of nuclear power from small units were reduced to the point where it became an attractive option for commercial electricity supply. What would be the consequences? Three kinds of questions seem to arise:

(1) Would actual risks be reduced or increased? It can be argued that, although the acceptability of risk might increase, the actual risk of mistakes or accidents would also increase. Smaller facilities would presumably mean more facilities: more installations to monitor, more locations where something could go wrong. The implications of a particular mistake might be less, but the number of mistakes might be greater, and the demands on a system for oversight --to try to avoid such mistakes--might be unsupportable. In addition, a greater frequency of mistakes, even if their health impacts were relatively small, might lead to a re-evaluation of risk acceptance for the smaller facilities.

(2) Would the deployment of nuclear energy be accelerated, assuming that it is needed? If, however, we assume for the purposes of discussion that decisions about smaller facilities are more likely to be quick and to be positive, does this mean that smaller facilities would do a better job of meeting the need? Only if the acceptance penalty of a larger scale is substantial. Otherwise, in total capacity added over a given period, the more rapid rate of deployment of small facilities adds up to no more, and quite possibly less, than a slower deployment of large facilities.

(3) Would nuclear energy displace other energy options that are more desirable from a broader perspective? It can be argued that, if there is to be a long-term role for nuclear energy, it is in meeting

large, concentrated, intensive energy needs--the kinds for which diffuse renewable energy sources, such as solar energy, are least suited. Small nuclear units (say, at a scale less that 100 MWe) might stunt the development of other options that society would prefer to have available at these smaller scales. Conversely, of course, it can be argued that if small nuclear units can meet energy needs at these scales more cheaply and conveniently, society might decide it prefers them.

It may be that these issues could be resolved. For example, the experience of the U.S. Navy suggests that the risk factor might be manageable. But, because the desirability of small units for widespread commercial use must be considered an open question, even apart from a question of their economic feasibility, it is necessary to inquire further about the significance of large scale as a possible impediment to acceptability. Without knowing more than we do about the actual configuration of Figure 1.6 in the minds and actions of American society, it is difficult to be very specific, but the implications appear to include the following:

(1) The acceptance of risks associated with larger scales can probably be increased by a convincing demonstration of need, mechanisms for broad participation, a perception of significant benefits related to other agendas, and/or credible mechanisms for reducing the likelihood of catastrophe (e.g., more attention to containment, or the use of more "tolerant" technologies such as HTGRs).

(2) Even though the acceptance of large units might be improved, within the size ranges presently available for commercial use, smaller units may fare better than larger ones. It appears that such a tendency in the public acceptance of new nuclear units is consistent with the current economics of nuclear plant construction, not contradictory to it.

(3) At sites where no nuclear power has been generated before, proposals for very large new facilities are likely to face tough going.

(4) Because of the importance of "familiarity," the best prospects for adding a nuclear generating unit (if one is needed) are likely to be at existing nuclear plants with good records of risk management.

Beyond the reach of these kinds of common-sense approaches, the acceptability of nuclear energy in the United States as a major option for our energy future may depend on either a change in scale or a change in culture: e.g., a perception of greater need or a reduced level of concern about connections between nuclear energy and nuclear catastrophe.[7] Technological change might contribute to either of these (1) by making smaller facilities more attractive in terms of economics and safety, or (2) by developing technologies for larger facilities that lessen some of the present concerns.[8] But the limits of our present acceptance of nuclear energy are based in culture, not technology; advances in technology are only likely to pay off if they are sensitive to cultural imperatives.

BROADER IMPLICATIONS FOR TECHNOLOGICAL CHANGE

Obviously, nuclear energy is in some ways different from any other kind of energy option. But our experience with risk acceptance for nuclear plants may be instructive for understanding processes of technological change in the United States more generally, because it reflects broader currents in society--currents which also shape other technology decisions as well. This section will start by offering a few generalizations from recent experience, then relate these generalizations to other energy options for the United States, and finally relate them more generally to the evolving context of technology choice in this country.

Scale and Technology Choice

As mentioned before, certain truths about scale and technology choice are widely known. For example, technology choice is influenced by economies of scale. The optimal scale for a technology also depends partly on the scale of need for the technology's services; and we know that larger-scale technologies with high front-end capital costs are at a disadvantage when the cost of capital is high, even if their operating costs are extremely low. Recent experience with large energy facilities in general, and nuclear plants in particular, suggests some additional principles as well: (1) Large-scale developments are at a disadvantage in an uncertain time; (2) Large-scale developments are at a disadvantage when they are based on relatively new technologies; (3) Large-scale developments are at a disadvantage if they are perceived to risk catastrophe; (4) For large-scale developments, in spite of the inherent difficulties, participation is especially critical to acceptance in a democratic society; and (5) Large-scale developments require broad political-economic coalitions.

In addition to these points, one can suggest some further complexities that arise from our current environment for technology choice. In a diverse, and complex decision-making arena such as energy, it is difficult to establish beyond reasonable doubt the need for a particular very large facility based on a particular technology, not only because of uncertainties about future demand but because of the very wide variety of other alternatives for meeting the same need. In addition, in situations where the earnings of firms are socially limited (for example, by regulation in the case of utilities or by social pressure in the case of large energy firms), large-scale developments are at a disadvantage when the prices of facilities are rising, because a firm may be unable to assure what it considers an acceptable return on a large capital investment.

Scale and U.S. Non-Nuclear Energy Options

To a considerable degree, the principles also shape the acceptability of non-nuclear energy options in the United States at this time, not just nuclear options. For instance, large-scale new hydroelectric projects have been constrained by a lack of evidence that they are essential, and large-scale coal-burning electricity generation units have become less attractive because of demand uncertainties and the high cost of capital.

More interesting, perhaps, is scale as an issue for new non-nuclear technologies. Consider two of our major options for the future:

(1) Synfuels. In the late 1970s, many energy experts believed that the United States faced a size-able need for domestic substitutes for imported oil and that it should move quickly, almost on a wartime mobilization basis, to meet this need with an assortment of large synfuel plants (50-100,000 barrels per day of crude oil equivalent), using U.S. coal and oil shale. A number of facilites at this scale were proposed, but only one--a coal gasification plant in the Upper Great Plains--has any chance to be in operation before 1990. The first projects to be supported by the Synthetic Fuels Corporation, created to assure the development of synfuels options at a large scale (two million barrels per day by 1992), will be in a size range of 4000-6000 barrels per day. The main reasons for this switch to much smaller scales appear to be economic, related especially to uncertainties about longer-term forecasts of world oil prices, interest rates, and rates of inflation in construction costs. But in the case of proposed facilities for the direct liquefaction of coal, involving technologies which produce carcinogenic substances, the health and safety risks of large-scale production also became a public issue, especially for

the proposed SRC-II facility in Morgantown, West Virginia. As a result, technology preferences for synfuel production have swung away from the options posing risks of exposure to cancer (even though their energy efficiency potential is higher) toward options which at any scale are more likely to be perceived by society as acceptable.

(2) Solar energy. Our efforts to develop solar energy options exhibit a peculiar kind of schizophrenia with respect to scale. On the one hand, solar energy is seen as an "appropriate" energy option-- accessible, controllable, the symbol of a "soft path." And we develop technologies for space, water, and low-temperature process heating; space cooling; and relatively small-scale mechanical and electrical energy (e.g., solar thermal collectors, photovoltaic cells, windpower, and passive solar designs). On the other hand, solar energy is seen as a large-scale high-technology energy option, calling for massive efforts by aerospace and other sectors of industry. And we work to develop such technologies as solar power satellites, ocean thermal energy conversion (OTEC), and large-scale power towers and solar cell arrays. So far, at least, it is clear that the smaller technologies are doing better in the marketplace, finding occasional success at scales up to about the 10MWe level. Because none of the larger options has come very close to deployment, one can only speculate about their acceptance; but they are likely to face the same kinds of scale-related hurdles as other large energy alternatives. For instance, microwave transmissions from solar power satellites may be perceived as involving risks at least on a par with nuclear plants. And ecological impacts of OTEC facilities may prove to be as worrisome as from large coal or oil shale developments. Solar energy technology developers may find that American society's tendency to prefer solar options for energy supply, even at a slightly higher cost, is no protection from risk acceptance problems for large-scale technologies perceived as risky.

Scale and Cultural Change

All of this implies a particular kind of context for technology choices in the United States--one where caution and dissensus abound, at least for decisions about large, complex, new technologies. It appears that, in society's judgments of risk acceptance these days, perceived risks from a technology are more often a function of prior judgments about its acceptability than the converse. [9] To what kinds of forces do these prior opinions respond? Pervasive and persistent uncertainties about the future are a part of the explanation, but another part is a general uneasiness in many parts of society about the rate of technological change. Primarily, it can be argued, these elements of society--too important to ignore--reflect concerns about (1) understanding the social meaning of technology choices and (2) given that understanding, exercising appropriate control over those choices. Dubos has suggested that the widespread popularity of "small is beautiful" indicates that such a concept "must correspond to some deep human longing" (Dubos, 1980). Increasingly, without conscious attention to logic or consistency, society seems to be rejecting the Baconian idea that technology is neutral, to be viewing technologies as fundamental shapers of patterns of life, in a fragmented and disorderly way to be seeking "a fundamental re-evaluation of the place and meaning of technology in human activity" (Winner, 1978: 128; also see Winner, 1977). This is associated with a kind of instinctive caution about the implications of large, complex, sophisticated technologies. Rather than defining progress as building and perfecting such devices, it asks about alternatives to them (Ibid). It urges that smallness and simplicity be accepted as aims of science and technology (cf. Wilbanks, 1982). Such options should be easier to control, more accessible for participation, less likely to sweep us along on undesirable paths.

One possible explanation for this situation is that, in some respects, we have exhausted our reservoir of cultural preparedness for technological change. Mumford argues persuasively that major periods of such change in human history, e.g., the Industrial Revolution, were preceded by long periods of cultural change. Without the requisite cultural evolution, the technologies could not have had such a dramatic impact (Mumford, 1963). Might the rate of technological change since World War II be outstripping our rate of cultural change, so that our attitudes and institutions are simply unprepared to accept and manage what our laboratories are producing?

To the extent that this is true, the resulting concerns seem to be focused most often on large facilites and activities, not because they are more dangerous but because they are more visible. Unlike nuclear energy today, for example, where society tends to worry more about risks than does the expert community, in bioengineering (still a small-scale activity in most respects) the experts probably worry more about risks than lay people. Perhaps it is not coincidental that so many of the major new thrusts in technology development these days--bioengineering, microelectronics, and the like--have an aura of smallness, decentralization, and diversity about them. Ever-larger and more powerful central computers are vaguely alarming, but microcomputers are irresistible. Very large firms building very large plants to conduct genetic engineering at a very large scale would be vaguely alarming; but small venture-capital firms pursuing individual theories at a small scale seem innovative and exciting.

At any rate, our national context for technology choice, an expression of our culture, is as much a part of our science and technology future as our technical skills. And it appears to include a scale bias that runs counter to many of our conventions of technology development. This does not mean that we are consistent in applying such a bias; for example, a particular group may at the same time oppose large-scale nuclear plants and advocate large-scale mass transportation systems (Gerlach, 1982). But it is a very real factor in any consensual approach to making decisions about complex new technologies, and we need to consider its implications as the nation gears up for a major effort to assure our competitiveness in international high-technology markets in the 1990s and beyond.

NOTES

1. Although the scale of the firm may in fact bear some relationship to the acceptability of nuclear power, both because very large options usually require very large institutions and because the size of a firm sometimes affects social attitudes about its proper realm of decision making.
2. Adapted from Wilbanks, 1981a and 1981b.
3. Based on work by Spencer Weart, Director of the Center for History of Physics, American Institute of Physics, New York City (Weart, 1982).
4. Such comparisons were a little unfair to the technologies involved in larger units, because these tended to be newer technologies, with more bugs to be worked out. On the other hand, many of the very small units were very old, presumably requiring a disproportionate amount of maintenance.
5. Metaphor suggested by Hans Landsberg.
6. Drawn from Kasten, et al. (1982).
7. Such observers as Marchetti predict that cultural changes of these types will indeed occur as nuclear technologies become more familiar. They point out that truly new energy options nearly always move gradually into use, not quickly--and that historical experience would lead us to expect the process to be more gradual in the United States than in certain other countries (Marchetti, 1980).

48

8. The latter is the focus of a major study currently under way at the Institute for Energy Analysis, Oak Ridge, Tennessee.
9. J. Nehnevajsa, work in progress.

REFERENCES

Abdulkarim, A.J., and N.J.D. Lucas. 1977. "Economies of Scale in Electricity Generation in the United Kingdon," Energy Research, 1:223-31.

Alam, W. ul, and J.M. Amos. 1982. "The Economies of Scale of Fuel-Grade Ethanol Plants," Energy, 7:477-81.

Branscomb, L. 1978. Testimony before a joint hearing of the U.S. Senate Committee on Commerce, Science, and Transportation and the U.S. House Committee on Science and Technology, February 1978.

Bronfman, L.M., B.H. Bronfman, and J.L. Regens. 1979. "Public Information on Radioactive Waste," ORNL/OIAPA-13, October 1979.

Budwani, R.N. 1980. "Power Plant Capital Cost Analysis," Power Engineering, pp. 62-70, May.

Cantley, M.F. 1979. "Questions of Scale," Options (IIASA), 1979 (3):4-5.

Carroll, J.D. 1971. "Participatory Technology," Science, 171:647-53.

Charlesworth, J.C., ed. 1970. Harmonizing Technological Developments and Social Policy in America, Philadelphia: American Academy of Political and Social Science.

Christensen, L.R., and W.H. Green. 1976. "Economies of Scale in U.S. Electrical Power Generation," Journal of Political Economy, 84:655-76.

Collingridge, D. 1980. The Social Control of Technology, London: Frances Pinter.

Douglas, M., and A. Wildavsky. 1982. Risk and Culture, Berkeley: University of California.

Dubos, R. 1980. "The Despairing Optimist," The American Scholar, 49 (2):151-56.

Fischhoff, B., P. Slovic, S. Lichtenstein, S. Reed, and B. Combs. 1978. "How Safe is Safe Enough? A Psychometric Study of Attitudes Toward Technological Risks and Benefits," Policy Sciences, 8:127-52.

Ford, A. 1980. "A New Look at Small Power Plants," Energy Systems and Policy, 4:25-36.

Ford, A., and I.W. Yabroff. 1980. "Defending Against Uncertainty in the Electric Utility Industry," Energy Systems and Policy, 4:57-98.

Forsyth, D., N. McBain, and R. Solomon. 1982. "Technical Rigidity and Appropriate Technology in Less Developed Countries." In The Economics of New

Technology in Developing Countries, F. Steward and J. James, eds. Boulder: Westview.

Friesema, P. 1979. "Electrical Power for Southern California: The Politics of Scale and Electricity Production." Paper presented at the annual meeting of the American Society for Public Administration, San Francisco.

Gerlach, L.P. 1982. "Dueling the Devil in the Energy Wars." In _Decentralized Energy_, P. Craig and M. Levine, eds. Boulder: Westview.

Hammond, A. 1972. "Energy Options: Challenges for the Future," _Science_, 177:875-76.

Handler, P. 1979. "Science, Technology, and Social Achievements," _EPRI Journal_.

Huettner, D., and J. Landon. 1978. "Electric Utilities: Scale Economies and Diseconomies," _Southern Economic Journal_, 44:883-912.

Iwier, L. 1980. "Utilities See Future in Medium-Sized Coal Units," _Electrical World_, pp. 17-19, July 1.

Kash, D., T. Wilbanks, and others. 1976. _Our Energy Future_. Norman: University of Oklahoma Press.

Kash, D. and R. Rycroft, forthcoming. _The Evolution of U.S. Energy Policy_. Norman: University of Oklahoma.

Kash, D., I. White, T. Wilbanks, and others. 1975. _Energy Alternatives: A Comparative Analysis_. Washington: U.S. Government Printing Office.

Kasten, P.R., and others. 1982. "Small Reactor Study," working paper, Oak Ridge National Laboratory.

Komanoff, C. 1981. "The Sources of Reactor Regulatory Standards," _Nuclear Safety_, 22:435-48.

Lakoff, S. 1983. "The 'Energy Crisis' and the Liberal Tradition: Ideological Factors in the Policy Debate," _Materials and Society_, 7:453-64.

Loose, V.W., and T. Flaim. 1980. "Economies of Scale and Reliability," _Energy Systems and Policy_, 4:37-56.

Lucas, N.J.D. 1979. "Economies of Scale in Nuclear Power Plants," _Energy Research_, 3:297-300.

Marchetti, C. 1980. "Society as a Learning System: Discovery, Invention, and Innovation Cycles Revisited," _Technological Forecasting and Social Change_, 18:267-82.

Messing, M., N.P. Friesema, and D. Morell. 1979. _Centralized Power_. Washington, D.C.: Environmental Policy Institute, March.

Mumford, L. 1963. _Technics and Civilization_. New York: Harcourt, Brace and Company.

Schurr, S., and others. 1979. _Energy in America's Future_, Baltimore: Johns Hopkins.

Slovic, P., B. Fischhoff, and S. Lichtenstein. 1979. "Rating the Risks," _Environment_, 21:14-39.

50

Southern States Energy Board, undated. "Nuclear Energy Center: Site Specific Conceptual Study," Atlanta.

Starr, C. 1969. "Social Benefit versus Technological risk," Science, 165:1232-38.

Starr, C. 1972. "Benefit-cost Studies in Sociotechnical Systems," In Committee on Public Engineering Policy, Perspective on Benefit-Risk-Decision-making. Washington: National Academy of Engineering.

U.S. Nuclear Regulatory Commission. 1976. Nuclear Energy Center Site Survey--1975, Summary and Conclusions, NUREG-0001, Part I, January 1976.

Weart, S. 1982. "Nuclear Fear: A Preliminary History." In Working Papers in Science and Technology, 1:61-88.

Weinberg, A.M. 1971. "The Moral Imperatives of Nuclear Energy," Nuclear News, 14(12):33-37.

Wilbanks, T.J. 1981a "Building a Consensus About Energy Technologies," ORNL-5784, Oak Ridge National Laboratory, August 1981.

Wilbanks, T.J. 1981b. "Local Energy Initiatives and Consensus in Energy Policy," Working paper prepared for the Committee on Behavioral and Social Aspects of Energy Consumption and Production, National Academy of Sciences/National Research Council, March.

Wilbanks, T.J. 1982a. "Is Comprehensive Analysis of Critical Interactions Possible?" In Energy, Economics, and the Environment, G. Daneke, ed., Lexington, MA:D.C. Heath.

Wilbanks, T.J. 1982b. "Linkages and Energy Emergency Preparedness," Paper presented at the annual meeting of the Association of American Geographers, San Antonio, Texas, April.

Wilbanks, T.J. 1983. "Energy Self-Sufficiency as an Issue in Regional and National Development." In Energy and Regional Growth, T.R. Lakshmanan, ed., London: Gower.

Winner, L. 1977. Autonomous Technology: Technics-Out-of-Control as a Theme in Political Thought, Cambridge: MIT. Press.

Winner, L. 1978. "The Political Philosophy of Alternative Technology: Historical Roots and Present Prospects." In Essays in Humanity and Technology, D. Lovekin and D.P. Verene, eds., Dixon, IL: Sank Valley College.

2

The Role of History in the Acceptance of Nuclear Power in the U.S. and Canada[1]

Earl Cook

The political status of nuclear power differs among nations. When reviewing the history of nuclear energy and its regulation, one is led to the thought that these differences, at least among democratic countries, may be due in part to historical memory. Patterson (1976:135) emphasized the importance of history when he wrote: "Three decades after the Hiroshima and Nagasaki bombs, and two decades after the Fukuryu Maru, the Japanese distrust of nuclear energy remains as deep-seated as ever."

Lovins, Lovins, and Ross (1980) offer an opposing view, contending that a de facto moratorium on the ordering of new reactors exists in all countries save those where a repressive bureaucracy has overridden the good sense of the people, who have realized that nuclear power not only leads to nuclear proliferation but is inherently uneconomic. The two countries singled out as repressive are France and the Soviet Union. The "good sense" category includes the United States, Canada, Japan, and Sweden.

The present article reviews the status of nuclear power in the United States and Canada, outlines the nuclear histories of these continental neighbors, suggests the role that contrasts in those histories may have played in bringing the two nations to somewhat different positions on the nuclear question, and discusses two other factors that may be even more decisive: the level of confidence in government and the perception of the potential role of nuclear power in regional and national energy economics.

STATUS OF NUCLEAR POWER IN THE UNITED STATES AND CANADA

As of December 1980, there were 75 operating nuclear power plants in the United States, 82 under construction, and 15 in the planning stage (Department

51

of Energy, 1981). Almost all embody light-water reactors that use enriched uranium as fuel and ordinary light water as a moderator and coolant. The operating nuclear plants supplied 10.9 percent of the electricity generated in the nation in 1980. Since 1975, few reactors have been ordered. Many orders have been cancelled. There is a de facto moratorium on ordering nuclear reactors in the United States.

The abandonment of plans for new nuclear power plants appears due more to the virtual certainty of cost overruns related to unanticipated delays, to a decline in the forecast demand for electricity, and to unpredictable government actions than to a revised judgment about the basic economies of nuclear power. Calculated lifetime energy budgets for nuclear power plants are strongly positive (Banerjee, 1980; Price, 1974; Rotty, Perry and Reister, 1976). Operating costs of existing plants, despite the strong rise of uranium prices within the past five years, remain lower than those of coal-fired or gas-fired plants (Rossin and Rieck, 1978; Feckmann, 1981). Some recent studies of power costs (including capital costs) from future plants (Andres Otayza, 1980; Komanoff, 1981), however, project costs of nuclear power greater than for electricity from coal-fired plants. Construction costs for nuclear plants in the United States have been rising more steeply since 1970 than those for other types of power plants. There are two reasons: (1) regulatory changes in design criteria to cope with newly identified safety problems, and (2) construction delays caused by faulty performance of contractors and by intervention of concerned individuals and groups in the licensing process (Bupp et al., 1975; Montgomery and Rose, 1979).

At least in part, public intervention, changed rules and criteria, and the proliferation of uncertainty reflect public reaction to a regulatory history in which promotion overwhelmed protection, expediency overcame caution, and secrecy allied with political power was used to block legitimate inquiry. The U.S. government, for some 30 years following World War II, failed at several times and places in its responsibility to protect the public from the adverse effects of the nuclear activities it was promoting. Today, the idea seems widespread in the United States that government cannot protect the public adequately from the hazards of nuclear power. Public opposition has been expressed at hearings, through the courts, in lobbying for anti-nuclear legislation, in books and articles,[2] in demonstrations and forceful occupation of nuclear power sites, and by campaigning politicians.

In contrast to the United States, Canada has only ten operating nuclear power reactors, all in Ontario (Aspin, 1980). They produce a third of the province's electrical power, equivalent to approximately 10 percent of Canada's electricity. Others are under construction in Quebec and New Brunswick. The Canadian power reactors, of the CANDU (Canadian Deuterium-Uranium) type, use natural uranium as fuel and heavy water (deuterium oxide) as the moderator/coolant. A capable and thorough analyst of the Canadian nuclear scene (Doern, 1980a:56) concludes that "...the nuclear option for Canada is a real and viable one." In June 1981, Canada's Atomic Energy Control Board gave Ontario Hydro permission to build the Darlington nuclear power plant (3,600 MW) which, when completed, will be one of the largest in North America.

Canadian law grants broad discretionary powers to its nuclear regulatory agency (Prince, 1978b). Intervenors are not admitted to the licensing process, although the applicant is required to carry out a public information program. A bill that would have permitted public intervention in nuclear plant licensing was introduced to Parliament in 1977 (Prince, 1978a), but was not acted upon. Canadian concern about nuclear hazards, heightened by press reports of nuclear problems in the United States, has resulted in public inquiries held by regulatory agencies, by standing or select legislative committees, and by royal commissions and commissions of inquiry set up under the Public Inquiries Act (Wyatt, 1980). Since 1974, the following bodies have carried out well-publicized inquiries into uranium mining and nuclear power in general: the Ontario Royal Commission on the Health and Safety of Workers in Mines (the Ham Commission), the Ontario Royal Commission on Electric Power Planning (the Porter Commission), the Cluff Lake Board of Inquiry (Saskatchewan), the Energy Committee of the New Brunswick Legislature, and the British Columbia Royal Commission of Inquiry into Uranium Mining (the Bates Commission). There has been no national inquiry into nuclear power, although it has been called for by a national anti-nuclear organization (Jennekens, 1979).

Opposition to nuclear power in Canada has been expressed mainly through the political system. The litigious path, so striking south of the border, has not been taken. Canada's Atomic Energy Control Board has one lawyer on its staff; the U.S. Nuclear Regulatory Commission, in contrast, has approximately 100. The few anti-nuclear demonstrations have been low-key, entirely within the law, and without violence.

Canadian regulation of nuclear power is less detailed, less adversarial, and less open than in the United States.[3] The style in Canada is to specify objectives and to let the planners, managers, and regulators work out technical strategies to achieve those objectives. Licenses are issued or denied without public hearings. The role of the intervenor in licensing of individual plants appears unique to the U.S. nuclear regulatory system (Golay, 1980); it accounts for much of the adversary nature of the U.S. regulatory process, although the press, which tends to portray the "good" regulator as a policeman, severely discourages cooperation between the regulator and the regulated. The Canadian regulatory system appears to invite "coziness" between the regulated and the regulators and allows the determination of "acceptable" risks by technologists whose bias is pro-nuclear. The adversary nature of the U.S. regulatory system encourages costly delays, very detailed written regulations, and exaggerated fears of radiation hazards, but may ignore key safety factors (e.g., Kemeny, 1980). The slowness, complexity, and expense of the U.S. system are growing. In November, 1980, a federal appeals court ruled that the Nuclear Regulatory Commission (NRC) must hold a hearing on requests before changing a nuclear plant's operating license; this ruling could force the NRC to hold hundreds of public hearings each year on technical amendments to operating licenses.

The political status of nuclear power in Canada differs from that in the United States. It may differ in part because of different systems of regulation that reflect different traditions of government and different levels of trust in government. Because nuclear regulation in the United States appears to have undergone much greater change, it may be useful to review the generally contrasting experience of the two countries in regard to nuclear energy, while remembering that it would be difficult for the two not to share, in the historical sense, the grosser events, such as Hiroshima and Three Mile Island. The contrast in regulatory style between the two countries has historical roots that much precede the discovery of radioactivity; this matter, perhaps of great importance in the present political contrast, will be discussed later.

A CONTRAST: THE WEAPONS LEGACY

The U.S. nuclear power program grew out of the wartime weapons program that culminated in the appalling devastation of Hiroshima and Nagasaki. In contrast, the Canadian nuclear power development was only briefly related to a weapons program, that of the United States during World War II. A heavy-water research group came to Montreal in 1943 from Cambridge (Eggleston, 1966) and attempted to produce plutonium from a natural-uranium, heavy-water-moderated reactor. However, by the time the Montreal group had a small experimental reactor in operation at Chalk River, Ontario, the war was over and the Chalk River operation became exclusively a Canadian responsibility dedicated to the production of electricity for peaceful purposes.

A CONTRAST: THE SEPARATION OF FUNCTIONS

The U.S. Atomic Energy Commission (AEC), created in 1946, had under its authority all bomb production as well as all peaceful uses of atomic power (Cook, 1975). Not only was the protection responsibility not separated from the nuclear power promotion function, but the secrecy that enveloped the weapons aspects of the Commission's activities seemed often to become a useful cloak to prevent inquiry into its "peaceful" operations. This arrangement persisted for 29 years. In addition, the Joint Committee on Atomic Energy of the Congress in 1954 obtained an unprecedented legislative monopoly in regard to nuclear matters (revoked in 1976) and was for many years controlled by incautious enthusiasts of atomic development.

In Canada, the 1946 Parliament enacted the Atomic Energy Control Act, which established the Atomic Energy Control Board (AECB), with both promotion and protection functions. When Atomic Energy of Canada Limited (AECL) was established as a crown corporation in 1952 to promote the commercial development of nuclear power, its first president was also president of the Control Board. This arrangement, however, was not as acceptable in Canada as the AEC structure was in the United States, and in 1954, the Atomic Energy Control Act was amended to separate the AECB from AECL. A similar separation of functions was not effected in the United States until 1975.

CONTRASTS: FALLOUT, REPROCESSING, WASTE MANAGEMENT

The credibility of the AEC suffered in the 1950's because biological concentrations of radionuclides wafted northeastward from Nevada bomb tests turned out to have much more serious health consequences than the Commission had forecast in its soothing of the fears of those downwind from the tests; it now appears that some of these trusting citizens may have contracted leukemia from their fallout exposure (Lyon et al., 1979). U.S. bomb tests in the South Pacific, from 1946 into 1958, produced fatal fallout on a Japanese fishing vessel, irradiated islanders, disintegrated islands, and rendered islanders' home atolls so radioactive they cannot now, some 30 years later, be reoccupied safely (U.S. Comptroller General, 1979; Johnson, 1980). Because Canada had no weapons program, there was no fallout, radioactive or political, from bomb tests there.

In both Canada and the United States, it was the intent of early nuclear developers to reprocess the spent or irradiated fuel from power reactors in order to recover unfissioned uranium as well as plutonium formed during irradiation. In the United States, a commercial reprocessing plant operated briefly at West Valley, New York; another was built at Morris, Illinois, but never operated because of technical difficulties; still a third was constructed at Barnwell, South Carolina, but has never received an operating permit. The West Valley plant was plagued by inability to meet radioactive effluent requirements at its plant boundaries, and was finally shut down as a commercial failure. The failure of West Valley, the technical fiasco at the Illinois plant, and a growing public fear of plutonium led President Carter in 1977 to announce that civilian reprocessing of spent nuclear fuel in the United States would be deferred indefinitely.

Liquid wastes with high levels of radioactivity produced in the government weapons program have been an embarrassment to nuclear development, because no satisfactory way to dispose of them has been devised. They have been stored on site in tanks that require continuous cooling. After the war, much research and experimentation resulted in the development of ways to solidify these liquids to facilitate their disposal in salt beds or other stable geological environments. But since no acceptable disposal site has been found, these wastes continue to accumulate. Some of the older tanks have developed leaks, necessitating transfer of waste to new tanks.

The AEC wrote another chapter in this sad story by selecting a Kansas salt mine as a repository for solidified high-level radioactive liquid wastes. Such wastes should be placed in an environment that will isolate them from the biosphere for the hundreds (perhaps thousands) of years required to reduce the radioactivity of the waste to innocuous levels. The AEC staff thought they had such a site in the salt mine in which many of their waste-isolation experiments had been carried out (Cook, 1976). Unfortunately, although bedded salt deposits in geologically stable area do appear to provide an excellent disposal environment for dry radioactive wastes, this specific site had some defects. The probable presence of unmapped and unplugged holes drilled through the salt in search for oil and gas, the existence of an active solution-mining operation some 1500 feet (457 m) away in the same salt bed, and an unresolved question about the possible explosive effects of energy accumulation in salt crystals created a storm of protest that blew the AEC out of the state and added to its reputation for placing operational expediency ahead of adequate protection. After searching in vain for several years for a geologically and politically stable site, the decision was taken not to reprocess fuel, but to store spent fuel elements at the power plants while trying to find a place to put them after the storage bays at the plants are full. Meanwhile, back at the tank farms, leaks continue to develop.

Canadian nuclear power planners appear to have decided about 1955 that reprocessing spent fuel was not an immediate necessity, in view of the large reserves of uranium ore that had been discovered in their country and because natural uranium, used in Canadian reactors, would be a cheaper fuel than separated plutonium, all costs considered. Accordingly, irradiated fuel rods have—by plan—been stored at or near Canadian nuclear power stations, while a careful search is made for suitable underground sites for permanent storage.

A CONTRAST: REACTOR ACCIDENTS

Although no deaths can be attributed directly to commercial nuclear power in the United States, its development has been marred by several accidents of catastrophic potential, notably an accident at the Fermi plant in Monroe, Michigan, in 1966, which caused the plant to be scrapped; a fire at the Browns Ferry Plant in Alabama in 1975, which shut down the plant for some time; and the Three Mile Island accident in Pennsylvania in 1979, which may result in permanent closing of the plant. These accidents have raised doubts about the ability of fallible human beings to control such an unforgiving technology.

In Canada, on the other hand, there has been no serious accident related to nuclear power production since 1952, when a small demonstration reactor near Chalk River suffered a core meltdown; no one was injured in that accident and the containment shell was not breached. Although there have been several accidents and malfunctions in CANDU power stations, none has so far caused any injury, radioactive contamination outside the reactor building, or stoppage in the delivery of electric power.

A SIMILARITY: THE DAUGHTERS OF RADON

Separation of the protection function of government from that of promotion of nuclear power, although it enables adequate protection, does not guarantee it. The history of uranium miners in Colorado and Ontario illustrates this point.

In the late 1940s, when uranium mining on the Colorado Plateau got into full swing, the hazard of underground radiation was well known. The danger was revealed in the 1930's by a study of pitchblende miners of the Erzgebirge in central Europe (Peller, 1939).

Pitchblende contains uranium. About 15 years after the start of the pitchblende mining in this region, the miners began to die of lung cancer. Ultimately, about 50 percent perished from this disease, and 80 percent of the remainder died from other lung diseases. Miners in other kinds of mines showed a much lower incidence of lung cancer. Beyond statistical doubt, the high mortality rate was related to the inhalation by the miners of radon gas, a decay product of uranium. The radioactive daughters of radon are solid and tend to lodge in the lungs, where their potential for damage is great.

Despite the ominous evidence from the Erzgebirge and the documentation and interpretation of that evidence in the American scientific literature (Lorenz, 1944), when uranium mining started in the United States, the standards set for individual exposure and mine ventilation were not stringent enough to prevent a recurrence of the tragedy. In the late 1950's, tests on air samples from several hundred U.S. uranium mines showed that many far exceeded standards set by the International Commission on Radiological Protection (U.S. Public Health Service, 1961). A 1967 study of 3,414 U.S. uranium miners (Lundin et al., 1969) indicated that 46 of them had already died of lung cancer because of their mine exposure to radon and suggested that more of the studied group would have their lives shortened by their mining experience. Not until 1967, more than 20 years after the start of uranium mining in the United States, were standards set at levels that may prove adequate for protection.

What happened? Who was in charge? As it turned out, no one was. The miners and their families thought the AEC was, but the AEC disclaimed responsibility, interpreting its authority under the Atomic Energy Act to begin not in the mine, but at the mill. State and federal health agencies finally called attention to what was happening and forced changes in the exposure standards.

In Canada, a somewhat similar history came to light from the work of the Ham Commission in Ontario (Royal Commission on the Health and Safety of Workers in Mines, 1976). Until 1975, although it appears to have claimed jurisdiction over mining and milling of uranium, the AECB let exposure standards be set by "appropriate provincial government agencies" (Prince, 1976; Duncan, 1977). A study reported in 1976 (Muller and Wheeler) showed a strong excess of deaths from lung cancer of uranium miners as compared to a sample of non miners. The Ham Commission strongly criticized both federal and provincial governments for failing to protect Ontario's miners. The 1967 U.S. exposure standard had been noted in Canada, but not adopted; not until the mid-1970's did Ontario substantially reduce the allowable exposure of miners to radon and did the AECB establish a Mining Safety Advisory Committee.

Another hazard belatedly recognized in the USA and Canada is that of uranium mill tailings, which contain radium-226, thorium-230, and lead-210, all of which are radioactive (Colorado Department of Health, 1970). Radium leached from tailings has raised radium concentrations, at some places and times far above permissible levels, throughout thousands of miles of the Colorado River system. In a hearing held in Santa Fe in 1958 (U.S. Public Health Service, 1958:48), an AEC official, questioned about the practice of his agency in not penalizing mill operators who allowed their tailings to contaminate streams, gave this astoundingly candid answer: "...we will not [try] to put them out of business because that in turn would put us out of business." A clearer statement of the triumph of promotion over protection could hardly have been made. Another example of regulatory laxity came to light between 1966 and 1971, when 5,000 homes and commercial buildings in the Colorado Plateu region were found to contain anomalously high concentrations of radon because they had been built on uranium tailings taken from piles under the authority of the AEC.

In 1975, a somewhat similar landfill problem was discovered in Port Hope, Ontario. About the same time, it began to be recognized that there are large quantities of uranium tailings in Canada which pose an unsolved problem of their continued separation from the human environment.

IS HISTORY SUFFICIENT?

The history of nuclear energy and its regulation in the United States accounts at least in part for the fear of radiation hazards and the distruct of government as a protector that are strong elements (Firebaugh, 1980; Garey, 1980) in the political miring of nuclear power in this country. But the miring itself owes much to other factors. Since its separation from the promotion function, nuclear regulation in the United States has grown into a very detailed adversarial system, subject to unpredictable changes and interpretations. Legal intervention and opposition, occasionally forceful, by the public introduce further uncertainties" into utility planning. In a period when forecasts of the growth of electricity demand are being lowered and capital is costly, such uncertainties are more than sufficient to explain the lack of orders for new nuclear plants.

As for Canada, its peaceful development and more successful regulation of nuclear power still may not account for the fact that the nuclear option remains

viable there. Substantial opposition (Doern, 1980b) to
nuclear power and some distrust of government as protector (Bates, 1980) exist there also. The role of
nuclear power in the national balance of trade, in
national pride, in domestic energy problems, and in
government-business relations needs to be considered.
The CANDU reactor, which has some well-demonstrated
technical advantages over rhe light-water reactors
preferred in the United States (McIntyre, 1975),
represents an opportunity for substantial international
sales and Canadian technological prestige (Doern,
1980b). Because government agencies design and build
nuclear power plants in Canada, advocates of nuclear
expansion enjoy special access to the policy process.
Because nuclear policy is widely viewed as linked to
Canadian economic policy (it has been proposed that
Ontario build nuclear power plants dedicated to the
export of electricity to the United States), the
Canadian public may have a less polarized view of
nuclear power than does the U.S. public. Additionally,
it should be recognized that nuclear power is an
indigenous energy option of the eastern provinces, to
be considered carefully as a replacement, insofar as
technically and economically feasible, for petroleum
imported from foreign countries and from Alberta.
Finally, the fact that Canadian nuclear power is a government enterprise, the reactors being designed by a
federal agency and the plants built and operated by
provincial utility corporations, removes one of the
great sources of mistrust that vexes the U.S. nuclear
situation.

National concensus is not required to move the
nuclear program forward in Canada. Because Canadian
provinces enjoy much more autonomy than do the states
of the United States, they may follow separate nuclear
scenarios. Western Canada, rich in fossil fuels and
hydro-electric power, does not foresee a need for
nuclear-generated electricity. Although forecast
demand for electricity has been lowered in recent
analyses, nuclear power remains politically viable in
eastern Canada, not only for the reasons given above,
but because it represents insurance against the economic impact of having to pay world prices for oil,
since nuclear-generated electricity can be used to
replace fuel oil. Eastern Canada has benefitted from
federal control of the wellhead price of Canadian oil,
a control now under strong challenge from Alberta.
This challenge is part of a wider alienation of western
Canada from the resource policies of the ruling political part, only two members of whose parliamentary
majority come from the western provinces. Should
provincial power prevail, eastern Canada's oil bill
could double.

One cannot escape the impression, on reading carefully the history of nuclear controversy in the United
States and Canada, not only that there has been far
more—and more heated—controversy in the United
States, but that it reflects a much stronger distrust
of government. The seeds of distrust whose first har-
vest was the American Revolution found inhospitable
soil in Canada. South of the border, however, a low
esteem for government, reflected in and kept alive by
Jacksonian democracy, has mingled with a deep suspicion
brought to America by waves of hungry immigrants, whose
natural enemies during the Industrial Revolution were
landowners and factory bosses whom government often
seemed peculiarly designed to assist. Today, the U.S.
citizens seem much less inclined than Canadians to accept the judgment and to respect the wisdom of elected
and appointed officials.

The cases of Sweden and Japan also raise doubts
about the historical hypothesis. Sweden, which has
suffered neither from nuclear weapons nor regulatory
failures, has a potent nuclear opposition, while Japan,
which has felt nuclear power most horribly, is now accelerating the construction of nuclear plants (23 in
1980, 90 projected for 1990). Clearly, the matter
needs more thought. It may be that the stage of economic evolution plays a role: the Japanese, great
telescopers of industrial revolution, have confidence
in their ability to control technology and enjoy its
fruits; for the industrially mature Swedes, technology
and welfare may not seem so intimately related.
Perhaps even more important, the Japanese (and the
French) feel an almost desperate need to diminish their
dependence on foreign sources of energy (Abelson,
1980), whereas the Swedes (and the Americans) do not.

What, then, is the role of history in the acceptance or rejection of nuclear power?

It is probably true that only the gross events of
history are remembered longer than twenty years, unless
there is a determined effort to transmit knowledge of
them to succeeding generations. Even in Japan, opposition to nuclear power (electricity), stemming from the
use and testing of nuclear weapons, has given way to a
perceived national interest in which such technology is
required (while opposition to nuclear arms remains so
strong that it can topple a prime minister on suspicion
of laxity on this issue). Although failures of a regulatory system, amplified by an unbridled press, can
spread fear of a satanically mysterious technology and
weaken confidence in government to control that technology, different perceptions of national interest can
lead to different decisions. The countries with the
strongest nuclear opposition[4] (Sweden, West Germany,
Austria) lie between the nuclear super-powers, fear the

centralization of political control that nuclear power represents to many persons, and feel no extreme urgency to revise their energy budgets. France, on the other hand, not only has a strongly centralized political system, but one in which technology has long been respected as a tool of political strategy. Perceptions of national and regional interests may be keeping the nuclear option alive in Canada, whereas in the United States, where such perceptions (and trust in government) are weak, nuclear power is politically comatose. The role of history is important where it has influenced legislation, for example, in the nuclear and environmental regulatory system of the United States, and where its memory is kept alive through the fanning of fear by a free press in an open society. The recent enactment, by several states, of laws banning the disposal within their borders of radioactive wastes generated in other states reflects the historic failure of the national government to demonstrate an acceptably safe way of dealing with such wastes. Fear remains a great motivating force of nations as of individuals. All nations have some fear of nuclear energy. But they may fear other things more: growing poorer instead of richer, or becoming subject to the wishes of foreign energy suppliers.

NOTES

1.   This article appeared originally in the Social Science Journal, 1982. Reprinted with permission.

2.   The opinionated nuclear literature of the United States is much too voluminous to cite here. The interested reader might start with Beckmann (1976), who is pro-nuclear, Gyorgy and friends (1979), who are anti-nuclear, and the National Academy of Sciences (1979), which has attempted a critical review of the literature on risks associated with nuclear power.

3.   Such appears to be the case in European countries, also, where the regulatory systems are said to "depend, much more than those of the U.S., upon a climate of mutual trust and coooperation between the regulator and regulated and upon substantial public confidence" (Golay, 1980).

4.   Measured by restrictive legislation and referenda.

REFERENCES

Abelson, P.H. 1980. "World Energy in Transition," Science, 210 (19 December).

Andres Otayza, Jose. 1980. "Programa de Energía," Energéticos, 4 (November):1-36.

Aspin, Norman. 1980. Letter to Earl Cook, 23 September. Dr. Aspin was president of the Canadian Nuclear Association.

Banerjee, S. 1980. "The Economics of Canadian Nuclear Policy." In Canadian Nuclear Policies G.B. Doern, and R.W. Morrison, eds., Montreal; Institute for Research on Public Policy, pp. 59-81.

Bates, D.V. 1980. "Occupational and Environmental Health." In Canadian Nuclear Policies G.B. Doern and R.W. Morrison, eds., Montreal: Institute for Research on Public Policy, pp. 219-30.

Beckmann, Petr. 1976. The Health Hazards of Not Going Nuclear, Boulder, Colo.: Golem.

Beckman, Petr. 1981. Access to Energy (Newsletter), 8 (1 January):2.

Bupp, I.C., J-C Derian, M-P Donsimoni, and Robert Treitel. 1975. "The Economics of Nuclear Power," Technology Review, 77(November/December):12-25.

Colorado Department of Health. 1970. Uranium Wastes and Colorado's Environment, Denver: Colo. Dept. Health.

Cook, Earl. 1975. "Ionizing Radiation." In Environment W.W. Murdoch, ed., Sunderland, Mass. Sinauer, pp. 297-323.

Cook, Earl. 1976. Man, Energy, Society, San Francisco: W.H. Freeman.

Department of Energy. 1981. Nuclear Power Reactors in the United States December 1, 1980, Washington, D.C.:DOE.

Doern, G.B. 1980a. "The Politics of Canadian Nuclear Energy." In Canadian Nuclear Policies, Doern, G.B. and R.W. Morrison, eds., Montreal: Institute for Research on Public Policy, pp. 45-58.

Doern, G.B. 1980b. Government Intervention in the Canadian Nuclear Industry, Montreal: Institute for Research on Public Policy.

Duncan, R.M. 1977. "The Atomic Energy Control Board and the Uranium Mining Industry" AECB-1106. Mimeographed, Ottawa: Atomic Energy Control Board.

Eggleston, Wilfred. 1966. Canada's Nuclear Story, London: George G. Harrup.

Firebaugh, M.W. 1980. Public Attitudes and Information on the Nuclear Option, Oak Ridge: Institute for Energy Analysis.

Garey, R.B. 1980. Conflict in Intergovernmental Relations:The Issue of Nuclear Waste. Paper presented at the meeting of the Southwestern Political Science Association, 2 April, 1980, Houston.

Golay, M.W. 1980. "How Prometheus Came to be Bound: Nuclear Regulation in America," Technology Review, 82(June/July):29-39.

64

Gyorgy, A., and friends. 1979. *No Nukes: Everyone's Guide to Nuclear Power*, Boston: South End Press.

Jennekens, J.H.F. 1979. "Nuclear Energy in Canada: In Search of the Definitive Inquiry," AECB-1178. Mimeographed, Ottawa: Atomic Energy Control Board.

Johnson, Giff. 1980. "Nuclear Legacy," *Oceans*, 15(January/February):22-26.

Kemeny, J.G. 1980. "Saving American Democracy: The Lessons of Three Mile Island," *Technology Review*, 82(June/July):65-75.

Komanoff, Charles. 1981. *Power Plant Cost Escalation*, New York: Komanoff Energy Associates.

Lorenz, E. 1944. "Radioactivity and Lung Cancer: Critical Review of Lung Cancer in Miners of Schneeberg and Joachimsthal," *Jour. Nat. Cancer Inst.*, 5(August):1-15.

Lovins, A.B., L.H. Lovins, and L. Ross. 1980. "Nuclear Power and Nuclear Bombs," *Foreign Affairs*, 58(Summer):1137-1177.

Lundin, F.E., Jr., William Lloyd, E.M.Smith, V.E. Archer, and D.A. Holaday. 1969. "Mortality of Uranium Miners in Relation to Radiation Exposure, Hard-rock Mining and Cigarette Smoking--1950 through September 1967," *Health Physics*, 16(May):571-78.

Lyon, J.L., M.R. Klauber, J.W. Garnder, and K.S. Udall. 1979. "Childhood Leukemias Associated with Fallout from Nuclear Testing." *New England Journal of Medicine*, 300(22 Feb.):397-402.

McIntyre, H.C. 1975. "Natural-Uranium Heavy-Water Reactors," *Scientific American*, 233(October):17-27.

Montgomery, T.L., and D.J. Rose. 1979. "Some Institutional Problems of the U.S. Nuclear Industry," *Technology Review*, 81(March/April):53-62.

Muller, J., and W.C. Wheeler. 1976. *Causes of Death in Ontario Uranium Miners: Radiation Protection in Mining and Milling of Uranium and Thorium*, Geneva: Intl. Labour Office, Occupational Safety and Health Series 32.

National Academy of Sciences. 1979. *Risks Associated with Nuclear Power: A Critical Review of the Literature*, Washington, D.C.: National Academy of Sciences.

Patterson, W.C. 1976. *Nuclear Power*, Harmondsworth, England: Penguin Books.

Peller, S. 1939. "Lung Cancer among Mine Workers in Joachimsthal." *Human Biology*, 11(February):130-43.

Price, John. 1974. *Dynamic Energy Analysis and Nuclear Power*, London: Earth Resources Research.

Prince, A.T. 1976. "The Role of the Atomic Energy Control Board in Uranium Exploration and Mining," AECB-1098. Mimeographed, Ottawa: Atomic Energy Control Board.

Prince, A.T. 1978a. "Bill C-14: The Democratization of Nuclear Energy and the Regulatory Process," AECB-1123. Mimeographed, Ottawa: Atomic Energy Control Board.

Rossin, A.D. and T.A. Rieck. 1978. "Economics of Nuclear Power," _Science_ 201(18 August):582-89.

Rotty, R.M., A.M. Perry, and D.B. Reister. 1976. _Net Energy from Nuclear Power_, Washington, D.C.: Federal Energy Administration.

Royal Commission on the Health and Safety of Workers in Mines. 1976. _Report_, Toronto: Province of Ontario.

U.S. Comptroller General. 1979. _Eniwetoc Atoll-- Cleaning up Nuclear Contamination_. PSAD-79-54. 8 May, Washington, D.C.: U.S. General Accounting Office.

U.S. Public Health Service. 1958. _Transcript of Conference on Interstate Pollution of the Animas River_, Colorado-New Mexico, 29 April 1958, Santa Fe.

U.S. Public Health Service. 1961. _Summary Report of Governor's Conference on Health Hazards in Uranium Mines_, 16 December 1960, Denver.

Wyatt, Alan. 1980. "The Use and Abuse of Inquiries," _Ascent_, 2(Summer):22-25.

3
Policymaking and Safety Issues in the Development of Nuclear Power in the United Kingdom

John Fernie and Stanley Openshaw

The past history, the present situation, and the future prospects for nuclear power in the UK present several interesting contrasts to the nuclear scene in the USA, particularly in terms of the access to and contribution of the general public in decision-making. Although Britain is nominally a democratic state, none of the important nuclear power decisions have been made in a democratic manner. A large part of the nuclear industry, the power industry, and the nuclear power research organizations are not under direct public control, except in a most indirect fashion. In theory most are responsible to Parliament but in practice a combination of the Official Secrets Act, censorship, an excessive concern for commercial secrecy, an obsession about national security, and the political power of the nuclear engineering industry lobby, have successfully conspired to preclude both the prospect of meaningful public debate and public accountability. It seems that when important vested national, industrial, institutional, and political interests are at stake, the Prime Minister worries last about checking on public attitudes and acceptability. When possibly unpopular decisions are made, a slow incrementalist approach is adopted with its high degree of secrecy. This philosophy will undoubtedly continue for the foreseeable future.

The present electrical contribution is smaller than the reserve planning margin, but once this is no longer the case a much more open policy will presumably emerge because it would then be impossible to return to a non-nuclear state. For Britain this changeover point is imminent. The recent decision never again to build fossil fuel power stations requires an all nuclear future. The realized threat to mining employment resulted in the emergence of nuclear power as a minor political issue in the 1983 General Election, although there is considerable public and political apathy to the important aspects involved in the transition to a nuclear economy. It seems likely that this status quo

67

will continue until the entire subject becomes politi-
cized, perhaps as a result of a major reactor accident.
   The situation regarding nuclear power is particularly interesting at present. Currently there is no
national energy policy. National Government leaves all
responsibility for meeting projected energy demands to
the responsible nationalized industries. The Central
Electricity Generating Board (CEGB) has a statutory
responsibility to meet power demands, although how it
chooses to do it is left to board members. They are
accountable to Parliament but in practice they are
largely autonomous. The principal government control
is CEGB's need to seek Treasury approval for their
investment programs and to satisfy safety legislation.
In practice this lack of direct political accountability has allowed the CEGB to effectively ignore
public oppostion to their plans for nuclear power,
because there is no public political input into their
decision-making processes.   Influence of public
opinion is of considerably less importance in siting
power plants than economic and engineering factors.
This situation has developed over a thirty year period
and it partly explains the high degree of both
confidence and complacency usually expressed about such
matters as nuclear safety. Nuclear legislation makes
the CEGB directly responsible for the safety of their
reactors and that is largely the end of the matter as
far as national government is concerned, since the CEGB
and not government itself would be to blame should a
major accident ever occur.
   The current Sizewell PWR public inquiry should not
foster the impression of a genuine debate. During such
proceedings counsellors argue about matters on which
they are not expert and witnesses read large submitted
documents. After that is completed (for Sizewell, at
least one and possibly two years), the Government must
decide whether to accept, reject or modify the
recommendations that result from the Inquiry. Actual
deliberations are kept secret for at least thirty
years. The CEGB views the Sizewell Inquiry as a
one-time generic inquiry that will vindicate the PWR
for all time, thereby allowing its use on any site they
wish. Quite naturally, they also expect to win their
case, partly by virtue of the ten million dollars of
public money they have spent preparing their case, and
partly because their proposals are in line with current
government policy. They expect to receive ministerial
approval in due course irrespective of the outcome of
the Inquiry. At worst, a few minor design modifications may be required to preserve public credibility.
Those opposed to nuclear power have little recourse but
to hope that a conservative government intent on
reducing public expenditure may not approve a large
nuclear power program.

In this chapter we present a look at the matter of public involvement in decisions over nuclear power in the UK. The principal issues we discuss include an historical account of the events that have led to the current situation, an examination of the safety legislation as applied in the UK, a look at past and present siting policies, a review of the current safety debate, and a more general discussion of the nature of the decision making environment.

BACKGROUND

The use of nuclear power for electrical generation in Britain was originally conceived by military scientists in the late 1940s. The first Harwell power plant was tentatively proposed in 1949. It was to be a CO_2 -graphite natural uranium (no enriched uranium could be spared) reactor called PIPPA, later Calder Hall type, later still, MAGNOX. However, before it could be built, a sudden large demand for plutonium for nuclear weapons meant that the first reactors at Calder Hall (picked because there was high level waste tanks there) were optimized for plutonium production with electricity as a by-product. In 1954 the Atomic Energy Authority (AEA) was set up with exclusive responsibilities for designing and building prototype reactors. At this early date the crucial decision was made to concentrate on a family of graphite-moderated gas cooled reactors, the only exception being plans for a fast breeder reactor. The AEA's plan was to progress from the MAGNOX design to an Advanced Gas Cooled Reactor (AGR) by the early 1960s and then to a High Temperature Gas Cooled Reactor (HTR) as a temporary expedient until a commercial Fast Breeder Reactor (FBR) could be demonstrated in the late 1960s. Britain was the only country to favor gas-graphite reactors.

In 1955, the Government announced a ten year nuclear power program. This was in fact the first time the electrical power industry was informed about the possibility of nuclear power generation, and they were expected to accept up to 2000 MWe of nuclear capacity by 1965. Sites for these first nuclear power stations were selected during a hectic six month period in 1955. It was confidently predicted that by 1963 MAGNOX plants would generate cheaper electricity than fossil-fueled stations. A revised nuclear power program was announced in 1957 with 6000 MWe of capacity all based on the MAGNOX design. The AEA was now concentrating on the AGR and FBR. The CEGB was created in 1958 and inherited this situation from the former Central Supply Authority. Ever since its creation, the chairmen of the CEGB have all been pro-nuclear and usually former directors of the AEA. This probably explains the CEGB's long flirtation with nuclear power.

By 1960, a number of problems were apparent. The price of fossil-fuelled power had dramatically decreased while that predicted for the MAGNOX stations had greatly increased due to cost over-runs and delays. At the CEGB's insistence the size of the program was reduced slightly. However, this was a time when nuclear power plants were seen as a prestige symbol; there was even a petition to Parliament arguing in favor of a site located in a National Park (Trawsfynydd).

The 1960s were also the time of growing discord between the CEGB and the AEA. In 1962 the CEGB wished to base the next power program on the CANDU reactor (heavy water reactor utilizing unenriched uranium) while the AEA wanted the AGR; by 1964 the CEGB were favoring the Boiling Water Reactor (BWR). When the second nuclear power program was announced in 1964, 5000 MW of either AGR's or BWR's would be built depending on the results of a comparative evaluation of their performance by the AEA and CEGB. The result of the CEGB report was to favor the AGR even though the comparison was biased and based on incorrect assumptions (Burns, 1978). The AGR was selected without commercial scale prototype or experimental verification of the components, with little practical experience of concrete pressure vessels, and without a complete design. The 'great AGR disaster' was compounded by the subsequent decision to order four further plants based on two different designs. The resulting AGR program had a mean time over-run of ten years per plant. It seemed at the time that the CEGB had a free choice but they were clearly treated as subordinate participants in the decision made by the AEA.

In 1969 the question of reactor choice was again under review because of problems with the AGR's. The CEGB now preferred the Stream Generating Heavy Water Reactor (SGHWR) that the AEA had hurriedly commenced building as a AGR stop-gap, and the HTR. By the early 1970s the continuing AGR problems resulted in the AEA also prefering the SGHWR until the FBR could be perfected. At the time it was thought that between one and four SGHWRs would be necessary by 1980 to meet forecasted demand. The oil crises induced the CEGB to plan for eighteen PWR's and one HTR in 1973 although there were still safety problems with the PWR. In 1974 the Government announced the third nuclear power program based on 4000 MWe of SGHWRs which by now the CEGB did not want. However, there were design problems with scaling up the SGHWR, and projected costs had risen. At the same time an AEA report (by the current Chairman of the CEGB, then Director of the AEA) indicated that the PWR safety problems could be resolved. The AEA now backed a combined PWR and FBR program. The SGHWR was finally scrapped in 1978 with the announcement that two more AGR's would be ordered largely to preserve jobs in the nuclear engineering industry. The CEGB, however, only wanted PWRs.

March 1979 brought the accident at the PWR of Unit 2 at the Three Mile Island plant, soon followed by an ironic UK government announcement of a program of 15,000 MWe of PWRs over the next ten years starting in 1982, with perhaps another 10,000 MWe by 2000. Thus, there have been four basic programs (Table 3.1).

Basically, nuclear power in the UK has been a high cost energy source which so far has contributed relatively little to total energy supplies compared with the vast amounts of money spent on its research and development. It is now thought likely that only a small program of between 4000 MWe and 8000 MWe will occur in the immediate future. The only decision that is virtually certain is that sooner or later a commercial FBR will be built. The AEA's commitment to the FBR, presumably as a replacement for current obsolete plutonium reactors, is the only constant factor over the last twenty years. Currently there are twenty nuclear establishments in the UK of which eleven are in operation; five are under construction or planned in the UK (Figure 3.1) (Table 3.2). A characteristic feature is the small size of the reactors and that the declared output for most stations is based on twin reactors. It is doubtful that any of the existing plants are generating electricity at a lower cost than the best fossil-fueled stations. Depending on the

TABLE 3.1
Nuclear Power Programs

Program	Version	Year	Size (MWe)
1	1	1955–65	1500–2000
	2	1957–65	5000–6000
	3	1957–66	5000–6000
	4	1960–68	5000
2	1	1964–75	5000
	2	1965–75	8000
3	1	1974–78	4000
	2	1978	3200
4	1	1979–91	15000
	2	1980–90	10000
	3	1983–93	4000–8000

FIGURE 3.1
Location of Nuclear Generating Plants in Great Britain.

TABLE 3.2
Public Supply Nuclear Power Stations in Britain

Station	Type	Date First Reactor on Power	Declared Net Capability (MWe)	Thermal Efficiency 1979-80 (Percent)
In Service:				
Berkeley (CEGB)	Magnox	1962	276	21.83
Bradwell (CEGB)	Magnox	1962	245	23.77
Hunterston A (SSEB)	Magnox	1964	300	24.50
Hinkley Point A (CEGB)	Magnox	1965	430	24.15
Trawsfynydd (CEGB)	Magnox	1965	390	24.34
Dungeness A (CEGB)	Magnox	1965	410	27.15
Sizewell A (CEGB)	Magnox	1966	420	26.20
Oldbury (CEGB)	Magnox	1967	416	27.34
Wylfa (CEGB)	Magnox	1971	840	25.78
Hinkley Point B (CEGB)	AGR	1976	1000*	36.16
Hunterston B (SSEB)	AGR	1976	1000*	35.01

			Installed Capacity (MWe)
Under Construction:			
Dungeness B (CEGB)	AGR	1983	1200
Heysham 1 (CEGB)	AGR	1983	1320
Hartlepool (CEGB)	AGR	1983	1320
Planned:**			
Heysham 2 (CEGB)	AGR	1987	1320
Torness (SSEB)	AGR	late 1980s	1320

* Interim ratings

** Initial site preparation work has begun, although the main
construction work involving installation of the reactors has
not yet started.

Source: Central Office of Information (1981) Nuclear Energy in
Britain, HMSO, London.

accounting conventions used, almost none of the AGRs will ever operate cheaper than coal fired stations especially if the cost of replacement energy supplies during the years of delay are included in the cost calculations. Were it not for a large surplus of power generation, the failure of the nuclear plant to come on line when predicted would have produced a decade of power shortages.

SAFETY LEGISLATION

As a result of the development of nuclear weapons, the UK government was quick to recognize at an early stage the dangers of radiation. Legislation has been promulgated to protect the public from the effects of atomic research and nuclear weapon programs, or at least to define areas of responsibility and put limits on compensation. The basic idea was that no member of the public should suffer any ill-effects (1948 Radioactive Substances Act; 1954 Atomic Energy Authority Act. The latter established the principle that the licensed operator of a nuclear facility was absolutely liable for any harm or damage whether due to negligence or not; all that was required was proof of responsibility and this is very difficult twenty or thirty years after an 'event'. The Nuclear Installations Act of 1960 extended the 'safeguards' provided by the 1954 Act to the Electricity Generating Boards and private industry.

In theory these acts are comprehensive. The 1960 Act established a system of reactor licensing and inspection with the creation of the Nuclear Installations Inspectorate (NII). The NII is responsible for approving sites, assessing design safety, and performing inspections during construction, commissioning, and operation. However, the NII acknowledged from the beginning that they had insufficient expertise to check in detail all aspects of the claim made by the licensee (Charlesworth and Griffiths, 1962).

The 1960 Act also laid down the fundamental principles regarding sites. It is the site that is licensed rather than the plant. This license can be revoked at any time by the relevant cabinet minister but the licensee's responsibility is terminated only when the site has been cleared of all radioactive material. Finally, the licensee was obligated to cover third party liability up to five million British pounds. This means that the marginal costs of seeking remote sites is very large in relation to the financial liability that remote siting would reduce.

The principal safety legislation is still found in the 1965 and 1974 Acts which have provisions similar to the 1960 Act. They constitute comprehensive, powerful, and highly flexible pieces of legislation. However, it is difficult to judge how effective they have been. The work of the NII is largely secret. No site and reactor safety assessment prepared by the NII have ever been published (one is planned for the Sizewell inquiry but it will probably not be published until after the public inquiry is over). The NII are fairly rigorous in their checking except they appear to be short-staffed and they operate under the constraints of the 'as far as reasonably practicable' dictum.

Much of the NII's work involves a combination of technical knowledge and the exercise of judgement. It is relevant to note here that experts from the nuclear industry have been the chief source of recruitment for government departments concerned with licensing and regulation. Thus, 'faith' and an implicit trust in engineering practice are major ingredients in the safety business. It is hardly surprising, therefore, to discover that the nuclear industry and the regulatory bodies are always quick to deny rumors of any reactor safety problem, while the full details are seldom available because of secrecy. It was not until 1978 that the first report of the NII was published (HSE, 1978) and not until 1979 that any of their safety principles were made public (HSE, 1979).

The latter document lists a series of basic objectives, for example "...all reasonably practicable steps shall be taken to prevent accidents", "...all reasonably practical steps shall be taken to minimize the radiological consequences of any accident", and "...the exposure of persons shall be kept as low as reasonably practicable", etc. The criterion implies judgement under the constraint that the 'cure' had to be regarded as practicable by the licensee. A number of dose limits are also given as design targets rather than as standards. The setting of critical limits has always abeen resisted in the UK partly because they may not be capable of being achieved with old existing plant. Furthermore, it is also claimed that setting limits would reduce safety, whereas without fixed limits both designers and operators have a duty to reduce risks as far as practicable whilst themselves determining what is practicable or not. Obviously this safety philosophy will work best with new plants but also allows the operator considerable flexibility.

A final problem with UK safety legislation is that there are no detailed aspects as distinct from general principles which can be enforced by law. Moreover, it works least well when applied to government agencies. There is a conflation of interests when one government department is required to assess the safety characteristics of plants operated by another or a public utility, using expert advice from 'independent' government research laboratories. The latter depend on government

support for continued existence and will be reluctant
to invoke its wrath by going against government policy.
The entire business is circular, shrouded in secrecy
and conducted apart from public participation or review
regarding safety standards and siting. It is true the
whole process is subject to Parliamentary sanction but
in practice the subject is too complex for most MPs
while government ministers either leave it to their
respective departments or concentrate on the job
creation aspects. Safety policy and legislation have
not yet been the subject of any detailed open review.
There has been no real will by government for publication of information on either design, system performance, or industry practices, and no real will for any
kind of meaningful public involvement.

Site selection is based upon proprietary information. The combined legacies of military involvement
and commercial interests ensure that few details are
made public. Indeed, the CEGB operates a policy of no
substantive information. A similar strategy is
employed concerning radiation releases. The CEGB are
responsible for monitoring their own releases and only
the barest of details are published (DOE, 1978). Data
on ground level deposition and air concentration for a
spatial mesh of monitoring stations exist but are not
published.

The extent of ground level plutonium dust contamination in the region around Windscale has been
monitored by the UKAEA, but ministerial approval is
necessary for the data to be released outside of BNFL
or UKAEA and approval is unlikely. It is thought that
information of this sort may cause unwarrented public
anxiety. Whether the British public like it or not,
they must trust their government and their agencies.
The prospect of more legal claims for compensation due
to probable radiation injuries and deaths is now sufficiently real to ensure that neither the CEGB nor BNFL
nor Government are prepared to release data that may
well be used against them. Recent court cases against
BNFL have been settled out of court without any admis-
sion of BNFL liability. The 1960 Act implied that BNFL
should have been liable without proven negligence, but
the problem is that it is almost impossible to prove
that cancer deaths are due to working for BNFL vis-a-
vis merely living. Without access to critical data,
much of it geographical in nature, the spatial and
occupational pattern of risks will never be understood.
At the same time those agencies which collect the
necessary data have most to lose should any 'interesting' patterns be discovered.

SITING NUCLEAR PLANTS

The existence of a nationalized power supply industry with a monopoly position in the UK market has centralized all the important decisions. In theory this should make it easy to study the processes of site selection but the decision makers have adopted a secretive policy. The CEGB has a site planning section which identifies and monitors a set of possible sites, but this list has never been made public. The CEGB's strategy is to regionalize and thus isolate local opposition to their siting policies and the best way to achieve this implicit objective is to adopt an incremental approach. The site search process is tightly constrained by the location of the super-grid power network and the presence of cooling water. The evaluation process that is used to select a site involves a combination of engineering and system constraints, and a desire to avoid areas of the highest landscape value if possible. The process is thorough but old fashioned in the retained heavy use of 'expert judgement' and 'personal knowledge' instead of a switch toward automated search procedures or basic optimization.

The site evaluation process, including the selection of a candidate subset, occurs in secret. No public announcements are made until selection has been made. Public participation is allowed too late to have an effect on what is still regarded as a purely engineering decision. The only visible public safeguard in this process is that the selected site has to conform with siting criteria devised by the NII. However, these criteria are for guidance only and they have no statutory power except that the NII can refuse to license a site.

The remote siting criteria (effective 1955-1968) were based on maximum population limits within certain distances around a reactor irrespective of reactor size or power (see Marley and Fry, 1955; Farmer and Fletcher 1959; Farmer, 1962). The revised remote siting criteria were due to Farmer (1962) and were based on an extremely low release level of 1 curie of iodine-131 at ground level. Nonetheless, this siting criterion was soon regarded as being too severe. The NII (who should be arguing for more remote sites) were complaining that there was a shortage of suitable sites (see Gronow, 1969). There was also a growing realization amongst nuclear engineers that nuclear power was safer than they once thought likely and that, even if it were not, it made more sense to rely on built-in engineered safety than on the special population characteristics of sites (Farmer, 1979). The NII put it more subtly when two of their inspectors wrote: "The ultimate aim must be the specification of design requirements for reactors having no siting restrictions" (Charlesworth and Gronow, 1967).

The UK Government nearly achieved this goal in 1968 when it was announced that the new AGR reactors could be operated at semi-urban sites. The only restriction was the requirement that persons within 3km could be evacuated within 2 hours. The justification for such an unrestricted siting attitude was partly the superior (but untested) safety properties of the AGR reactor and partly the desire to gain large export orders (presumably to countries with a shortage of remote sites). In reality the decision to locate later AGRs close to the center of population was based on an act of faith with regard to the supposed increased levels of safety that resulted from changes in design and a desire to locate reactors close to major centers of demand. These motives notwithstanding, no shortage of acceptable sites was demonstrated (Openshaw 1980, 1982a). It is also clear that the benefits of the so-called remote siting policy were very small. For example, the Torness AGR site is very remote according to the Farmer (1962) site rating index, yet a major reactor accident there would threaten two-thirds of the population of Scotland should the wind be blowing in an unfavorable direction.

The post-1968 relaxed siting policy was eventually rationalized by the NII in a numerical form. They wanted some means of guiding prospective licensees in their search for sites (Charlesworth and Gronow, 1967; Gronow, 1969; Gronow and Gausden, 1973). The new criteria covered a 20 mile (32 km) circle and were based on a national 1000-curie release of iodine-131 at ground level under unfavorable weather conditions. A set of distance bands and 30 degree sector population limits were then derived in an arbitrary fashion. The objective was to ensure the acceptance of all existing sites and two new semi-urban sites (Hartlepool and Heysham) which had just been declared suitable by ministerial fiat.

These criteria have several deficiencies. They provide neither guidance as to the possible effects of a reactor accident nor an allowance for local age-sex deviations from national average. Moreover, the distance bands are arbitrary, no account is taken of daytime population distributions, the reference accident is unrealistic (iodine-131 has been a British preoccupation ever since the 1957 Windscale accident although it is fairly easy to handle compared with other fission products), and until recently there was no accurate population data for the 20 (32 km) mile region. In addition, it would appear that both the Hartlepool and the Heysham sites, used to 'calibrate' the population limits, actually violate the criteria.

At Hartlepool, there is too much population within the three to five mile band, and at Heysham there is too much population in a particular 30 degree sector within one mile of the reactor (Table 3.3). Yet in

TABLE 3.3
Population Distribution for the Hartlepool and Heysham AGR Sites in 1981

Distance Band (Miles)	Hartlepool Population	Heysham Population	Implicit NII Limits	Hartlepool Sector 1	Heysham Sector 1	Sector 2	NII Limits
0 - 1	844	4485	4750	0	0	3999	1944
1 - 1 1/2	3690	2629	9822	696	0	2584	4205
1 1/2 - 2	7327	4601	13382	3972	0	3309	5810
2 - 3	34304	12843	36800	22239	0	10258	16719
3 - 5	190839	63120	111789	23460	0	18895	53636
5 - 10	329266	73397	494000	26806	33659	16020	251908
10 - 20	507049	482817	1849333	193668	227284	7426	1004348
Population within 20 miles	1,073,319	643,892	2,519,876				

Notes: Sector 1 is 30 degree sector with maximum 20 mile population
Sector 2 is 30 degree sector with maximum excess population.

both cases the NII criteria can be manipulated to disguise these excess populations. By using commulative curves, an excess of population close in can be compensated by a deficiency of population further out. This completely underestimates the significance in terms of evacuation prospects of having almost twice the allowed population in the three to five mile ring at Hartlepool. This loophole could be exploited in another way. If a site was chosen such that there was zero population out to ten miles, then there could be 5.09 million people within the next ten miles; this would have obvious implications for the siting of offshore 'power islands' as currently being proposed by some manufacturers. At Heysham the loophole is different. The maximum population sector is wholly unrepresentative of the maximum excess population sectors.

The history of siting practices has been based on an extreme reluctance to regard siting as a safety measure. It is currently thought that in any case there is only an order of two magnitude variation in numbers of potential casualties between remote and urban sites. The largely unquestioning public attitude is due largely to (1) an ignorance of the magnitude of the consequences of possible mishaps, and (2) assurances of safety from government, the plant operators, and the manufacturers. It is an easy procedure for government or the CEGB to justify the selection of sites as being in the national interest and to avoid public discussion by the simple expedient of presenting no alternatives and avoiding any public explanation of why and how any particular site was selected. This may explain the surprising claim by the CEGB that there is still a shortage of sites and thus a need to accept any and all sites they identify (HCSC, 1981). This is surprising because an estimated 83 percent of the coastline satisfies the relaxed siting criteria with several feasible sites close to London. In all probability, the 'shortage of sites' argument is, therefore, merely an excuse to avoid the CEGB having to put forward alternative sites which they have never had to do before. This, of course, immediately invalidates a basic siting tenet of Keeney (1980) that there has to be at least one alternative location.

One may reasonably ask the purpose of the secrecy mentioned by Openshaw (1982b:186): "It is surely unacceptable that the CEGB should be allowed to decide on behalf of the public what is and what is not acceptable under conditions of secrecy. It is not denied that the CEGB have the expertise to evaluate sites with respect to the requirements of their power stations, but their ability to trade off these aspects against environmental and public safety considerations is in doubt." Currently at the Sizewell PWR inquiry, the CEGB refused to reveal details of how they came to pick Sizewell as a site; the relevant documents were apparently 'lost'. Likewise, the CEGB refuses to perform any environmental impact analysis on their proposals on the grounds that they have already performed equivalent studies internally and presumably found all to be well.

In terms of existing planning legislation, the CEGB is a statutory undertaker which means that it is largely exempt from planning controls unless the government decrees otherwise. However, it is time that the government exerted more control over the site selection and evaluation process. If the "national interest" is to be invoked to justify locational decisions then it is national government and not the Board members of the CEGB who should debate the pros and cons of various sites. In short there is an urgent need for a national siting policy defined in the overall context of a national energy policy; at present neither exist.

SAFETY ASSESSMENT

In sharp contrast to the wide availability of documents in the USA, currently there are no public site-specific or reactor-specific safety reports for any British nuclear power station, despite over 25 years of operation. The NII and the Health and Safety Executive consider that they do not have the power to release detailed safety reports which are the property of the CEGB. The CEGB refuses to provide public access to these studies on the grounds that they are already accountable to Parliament, that the public has no statutory right to such material, and that the material may be used against them. It is hardly surprising, therefore, that safety pronouncements by government and the CEGB is virtually all that is made public.

The Flowers Report (1976) provided a lengthy discussion of the safety of nuclear power, but the key sections were written or depended heavily on advice supplied by the UKAEA, which has links with the nuclear industry. Great emphasis was placed on the Farmer criteria (Farmer, 1962,1979) which suggest, from a design standard, that there should be a negative log linear relationship between accident frequency and the amount of radiation released. Accidents involving more than 10^6 curies should only occur with a frequency of less than 1×10^{-7} years. This is to be achieved through fault-tree analysis.

The problem, of course, is that safety measures are probabilistic and cannot guarantee that large releases will not occur. It should also be understood that the probabilities for reactor system failures are not facts of operation but statements of faith, hope, and intent.

There are currently no system statistics because there is only a small number of years of experience with the operation of identical hardware. The degree of complexity is very great, more especially because the system includes operators and is not isolated from external events. Thus the weakness of the fault-tree analyses recommended by Flowers (1976) as a principal means of ensuring nuclear safety should be taken into account when assessing the results. However, there is little evidence that this is done. The limitations appear to be ignored; particularly important are the assumptions that the tree is complete, that the probabilities are realistic rather than imagined, that management is competent, that the design does not fail, and maintenance is of a high quality. Whether safety targets, such as the Farmer criteria, can be met in practice depends on many uncertain factors. Thus there is no real way of determining whether the once every million year frequency for major accidents is meaningful as a statistic for new reactors and whether it can be achieved given human fallibility. So far, it seems that major accidents in the UK for a Magnox reactor have a frequency of less than once in 25 years (the length of current experience) and once in five years for an AGR.

A major problem in all safety studies is the current lack of realistic empirical basis on which to make casualty or accident frequency predictions. As a result it is necessary to imagine hypothetical accident scenarios so that the approximate level of consequences can be determined. A problem here is that the owners and designers of power reactors have found it very difficult to imagine possible accidents that their safety systems cannot handle and which would give rise to significant environmental impacts. It is hardly surprising, therefore, that most 'design basis accidents' or maximum credible accidents can be handled with minimal effect. MacDonald, et al., (1977) of the CEGB consider the likely consequence of the most serious accident that they imagine could happen to a Magnox reactor would release no fission products. The very worst outcome would be limited evacuation out to 1.5 miles for a steel vessel Magnox and 0.6 miles for a concrete pressure vessel design with a release of 10 to 100 curries of iodine-131. For the AGR design a loss of pressure accident would release about 10 curies and require no evacuation. The only environmental consequence would be a ban on cow milk for one mile downwind. Beattie (1981) reports that an impact similar to a steel vessel Magnox reactor might be expected for a PWR and even a commercial fast breeder design would only release one curie.

While nuclear engineers believe they can write specifications which render reactors completely safe, the hardware itself may not and does not always conform to the initial design specifications. The AGR reactors were subjected to almost continuous design modification during construction, which is one reason for their ten year delay in commencing operation, yet it still sustains a high modification rate. Consequently recent statements to the effect that a Three Mile Island-type accident could only happen to the British PWR if design limits are exceeded is meaningless. It is misleading because it is virtually impossible to ensure that design limits are never exceeded during the working life of a reactor, especially during transients; not all aspects of the design can be monitored continuously and engineering design does not cover all aspects of reactor operation. The event-tree analyses exclude the possibility of operator error (deliberate or accidental), computer error (software problems), sabotage, accidental damage, terrorist attack, and war damage. While it may be a little severe to expect designers to consider the effects of the last two, nuclear power plants are prime strategic targets and are potentially a very dangerous source of radioactive contaminants (Ramberg, 1980).

Overall, the engineering studies ignore a number of important but qualitative events that could have serious effects on the safety of their systems. For example, the Hartlepool AGR station is located near the largest concentration of petrochemical works in Western Europe. There is no protection, however, against the possibility of a gas cloud explosion from the location of important oil and gas terminals, presumably because in the 1960s when the site was selected there was no chemical industries in close proximity. Additionally, the Health and Safety Executive discounts the possibility of gas clouds travelling more than short distances. The CEGB are apparently already aware of this risk to their Hartlepool Plant and may now not build a second station there because they may be forced to retro-fit their first station with additional protection against gas cloud explosions.

For different reasons, apparently no precautions have been taken to secure the CEGB's reactors against terrorist attack. In particular, the NII warns against certain circumstances that may be beyond the control capabilities of AGRs (HSE, 1978). For example, it is thought that an AGR reactor once tripped is susceptible to the loss of certain services. A reactor trip caused by a terrorist attack on powerlines and combined with damage to the auxiliary power plant might be sufficient to induce meltdown. It is unlikely that this scenario has been considered possible but it has a probability of occurrence considerably greater than many risks that are considered plausible.

A good illustration of many aspects of current UK practice can be seen in a recent study of a hypothetical degraded core accident at the proposed Sizewell PWR (Kelly and Clarke, 1982). The NRPB has recently developed computer programs (MARC) for estimating the consequences of reactor accidents (Clarke and Kelly, 1981). Several accident scenarios involving differing radioactive releases were examined under a CEGB contract. Details of the amounts involved, their nature, and expected frequency were supplied by the reactor designer (Westinghouse Electric Corporation, 1982). The CEGB subsequently modified some of the values. The NRPB then produced casualty estimates (Table 3.4).

As might be expected, the annualized individual risks are minute. Values range from 5.1 x 10 to 5.9 x 10 per reactor year, far smaller than the widely held belief of once every million years. For an individual living 1 km from the site an annual average risk of early death is 2 x 10 (reduce by a factor of 6 for the CEGB revised releases), and 2 x 10 for fatal cancer (reduce by factor of 3 for the revised releases; at 10 km the risks are 5 x 10 (reduce by 100) and 3 x 10 (reduce by 3) (Table 3.4). Comparison of these risks with all other life-time risks suggests that nuclear power is several orders of magnitude smaller. However, it should be noted that these low risk rates result from the initial assumption of very infrequent reactor accident possibilities. Additionally

TABLE 3.4
Casualty Estimates for Two Accident Scenarios

	Accident probability 2.4 x 10^{-9} expected values		2.4 x 10^{-11} 99th percentile values	
	(a)	(b)	(a)	(b)
Early deaths	583	41	7540	982
Fatal cancers	3300	1300	31000	15000
Nonfatal cancers	10590	5150	96200	54500
Area evacuated (Km^2)	120	27	1300	320
People evacuated	24000	3600	300000	60000

Notes: (a) Westinghouse releases
(b) CEGB revised

Source: Based on Kelley and Clarke (1982) Table 41.

one should not compare probabilities for individual events (such as lightning strikes) with nuclear accidents which have an areal effect (Openshaw, 1982b). Another aspect is that the results in Table 3.4 are based on average wind directions and are not the maximum that could happen should the wind blow in the direction of the most populated 30 degree sector, or should an exceptionally severe accident occur.

Even the 99 percentile results are underestimates of a worst case accident. If you are sufficiently unfortunate to experience a once every 416 million year accident why assume favorable weather conditions? The 99 percentile results correspond to a once every 41,666 million year accident (about 8 times the age of the planet Earth). Moreover, the study ignores the prospect of a plane crash although pilots use the existing Sizewell A station as a turning point (Pearce, 1982).

The revised EEGB release figures reduce casualties and risk probabilities even further: by factors of between 20 and 40 for early deaths, fatal cancers by between 2 and 4 times, and the area to be evacuated by between 1.5 and 9 times. In short, this safety study demonstrates the meticulous but spurious precision that the nuclear industry in Britain uses to support their safety assurances.

On using these annualized risk rates, siting is largely considered unimportant. There is, of course, the alternative view that the probability studies are gross underestimates of actual (as distinct from theoretical) risks, and that should an accident ever occur there would be a strong public and political reaction. The danger is that careless or carefree siting of the current generation of major energy facilities will affect the location of future generations. Sites being selected now for AGRs and PWRs could serve the next generation of Fast Breeder Reactors and their successors. Locational inertia is intensified once the necessary transmission infrastructure is created. Thus, although it may become desirable later to abandon many sites in favor of more remote locations, it may be too late for technical reasons.

In a small urbanized country like Britain the numbers of casualties and the area to be evacuated varies tremendously according to the choice of site (Table 3.5). While the number of deaths varies by about two orders of magnitude it is worth bearing in mind that total casualties may vary by a greater amount. Furthermore, the number of persons suffering evacuation could well vary by an even larger factor. In the case of Torness, an estimated four million people would face the choice of evacuation or risk of cancer; this amounts to no less than 60 percent of the Scottish population and almost the entire productive economy (PERG, 1980). This estimate is based on Casium-137

TABLE 3.5
Predicted Deaths for Current Sites

Site	Total deaths AGR[a]	Total deaths PWR[b]
Berkeley	45248	40170
Bradwell	116699	51995
Calder Hall	29528	14439
Chapelcross	16215	8765
Dounreay	1629	1462
Dungeness A	77608	28171
Hartlepool	49776	35376
Heysham A	56046	30621
Hinkley Point	34253	28840
Hunterston A	71804	50650
Oldbury	41331	35943
Sizewell A	45477	21612
Torness	30031	13584
Trawsfynydd	29746	10682
Winfrith Heath	41996	32249
Wylfa	18685	8070

Sources: (a) PERG (1980); (b) USNRC (1975) as cited
in Fryer and Kaiser, 1978; Fryer, 1978.

contamination and would require an evacuation period of
at least twenty years. The Rasmussen Report (USNRC,
1975) suggests a reference area of about 120 square
mile (300 sq km) with a risk of 1 x 10 per reactor
year (Beattie, 1981). Evacuation areas predicted in
Table 3.5 were far smaller.

The problem presented by long term land contamination in many ways may be considered more serious than
the early deaths or cancer deaths. If the affected
area included large towns and cities the consequences
might well be devastating. Thus, if long term
radiation contamination of land is considered a
problem, then siting aspects again become an important
safety agent. It is strange to note, therefore, that
the Kemeny (1980) recommendations for remote siting
should be dismissed by the HSE as "... not relevant to
the situation in Britain and Europe more generally and
do not cast doubts on the validity of the choice of
existing British sites" (HSE, 1981:21). In fact, the
reverse is true. Many British sites are inherently
unsuitable for modern power reactors. The reaction of
a responsible safety agency should be to start computer

searches of the UK space for 'better' sites. Large numbers of possible, better locations are available (Openshaw, 1982a,b) but whether it is worthwhile to locate them depends on the perceived importance of the risks associated with current practice. It would appear at present that neither politicians nor the public properly understand the nature of the gamble that is being undertaken on their behalf.

CONCLUSIONS: TOWARDS A GEOGRAPHY OF SAFETY

The direction which civilian nuclear power has taken in the UK can be attributed to the institutional framework within which key decisions are made. Elitism in nuclear decision-making is rooted in the creation of a 'nuclear establishment' in the 1950s which successive Governments treated as sacrosanct. Thereafter, the nuclear program came under minimal scrutiny until the 1970s when the plight of the AGR program became the focus of criticism (Burn, 1978; Henderson, 1977). British nuclear nationalism (Wonder, 1977) was being questioned, and the publication of the Flowers Report (1976) acted as the catalyst to greater public debate. The report indicated that a long-term commitment to nuclear power was undesirable until the storage of high level wastes had been proven beyond all reasonable doubt, an issue which has not yet been resolved.

At the same time, US environmentalism began to express itself in Western Europe and a new Energy Minister was committed to open government. The medium of the public inquiry became the platform for debate and as a result the Windscale and Sizewell inquiries have been held. Nevertheless, nuclear nationalism prevails. The Select Committee was not allowed access to the findings of the 'Think Tank' which was instrumental in the Government's decision to order 2 AGRs in 1980. The CEGB somewhat reluctantly divulged some of their internal development reviews to the Monopolies Commission. These planning documents are internal, secret policy papers; however, if the workings of the CEGB had not been referred to the Commission, it is unlikely that the Board's duplicit attitude to policy formulation would have been revealed: on the one hand, it maintains publicly that all technological options are being considered in its future planning schemes, yet the Board's private objectives are identified as developing nuclear power, containing costs and diversifying fuel sources (Monopolies Commission 1981).

It is in this context that the PWR Sizewell inquiry is being held and the almost inevitable approval is a function of the aforementioned peculiarities of the British nuclear decision-making machinery. Indeed, public inquiry - one of the few vehicles for public participation - is seen by the Government as another obstacle delaying the nuclear program. Through leaked Cabinet documents in 1979, the incoming Conservative Government revealed that a wide ranging inquiry would be undesirable partly on the grounds that construction of the first PWR, scheduled for 1982, would be delayed. By the time the inquiry began (1983), it was clear that the debate would cover as broad a range of issues as possible. Nonetheless, objectors felt that the odds were heavily prejudiced against them. The Secretary of State for Energy refused to grant financial support for objectors to present their case; hence, the CEGB can spend an estimated ten million British pounds of public money to procure expert technical and legal assistance whilst opposition groups, with limited funds, have had to resort to a public appeal for support. This issue of aid to objectors has been raised in the US (Calzon-etti, 1981) but in most instances the developer will use private, corporate funds and thus avoid the UK question of misallocation of public resources.

It has been argued that countries with elitist political cultures and limited public accountability have been pushing ahead with nuclear expansion programs to a greater extent that more democratic regimes (The Economist, 1977); however western countries with strong centralized governments linked to public energy corpor-ations (e.g. the UK) have also advocated nuclear expan-sion. In essence, the energy decision-making process in West Germany and the USA is far more publicly accountable throughout their various stages of administration.

One could argue that in a country as densely popu-lated as the UK, it makes sense that the public bodies responsible for the nuclear program should work more harmoniously to minimize the risks of nuclear power. For example, the NII and NRPB could and should recom-mend sites which comply with their safety standards. This would promote liaison with the CEGB rather than merely responding to sites chosen by them internally.

Nuclear power is a difficult problem for decision makers. Long term risks are only partly understood and even in the short term there are large areas of uncer-tainty. The public decision-makers understand nuclear power least of all. They are confronted with one strong lobby in the form of big-business with consider-able establishment support, one weak lobby of well-meaning, but under-informed environmentalists and a fringe minority of dissidents. The advice received from the experts is often contradictory. The problem is that the subject is immensely technical and it is not easy to reach a rational scientific decision in accordance with the levels of uncertainty that are involved. This may explain the years of indecision.

The geographer's reaction should be to 'play safe' and seek to minimize the residual risks of energy generation by seeking optimally safe sites. It would be sensible to adopt current US proposals and reinstate siting as a factor in an in-depth safety philosophy and to regard siting as being independent of reactor design. If attention is focused on what are called 'Class 9' accidents (accidents beyond the design basis of a reactor which cannot be handled by existing engineering safety measures) then siting becomes a major ingredient in the safety debate. It is interesting to observe that in the US Class 9 accidents have suddenly become important in the safety debate as a result of Three Mile Island and a general lack of confidence in the reliability of engineered safety measures. In Britain, the trend is in the opposite direction. Engineers appear to believe their probability predictions and there are currently no visible attempts to incorporate Class 9 accidents in siting considerations. Government is content to delegate responsibility for siting to the CEGB and the NII.

Government has been particularly complacent regarding site selection and is seemingly prepared to believe anything the CEGB might tell them. At the very least, Government should be actively engaged in a search for, and evaluation of, feasible sites. It is simply untrue that appropriate, remote sites do not exist in Britain; instead, no one has bothered to look. There is no need to use inherently high risk sites because there are many better locations. The marginal cost benefits of close proximity to major demand centers may well be far less than cost overruns during construction or errors in forecasts of future uranium price rises. In any case it is doubtful whether proximity is relevant given a long established national grid system, and there is no evidence to indicate the extent to which locally produced electricity is consumed locally. There is also a risk that sites are being used that may have to be abandoned later should a major accident ever occur at a similar site anywhere in the world.

The foreign observer may be excused for believing that the UK is largely in the grasp of a pro-nuclear lobby. In the absence of an integrated energy policy, a national siting policy, and a Freedom of Information Act, investment strategies are fashioned according to lobbying pressures within the decision-making environment. The nuclear lobby, especially the PWR faction, has active political influence; hence, nuclear power is the strongest plank in the energy platform at the present time. The CEGB, responsible for the operation of most of British nuclear plants, has been subjected to a great deal of criticism from government advisory bodies. The Government, however, has shown a marked reluctance to heed this advice. Indeed, the Government is resistant to the introduction of a private member's bill - The Parliamentary Control of Expenditure (Reform) Bill - which would make the nationalized industries' audits accountable to Parliament, rather than Government.

The transition to a more public, accountable decision making environment could take time; meanwhile, nuclear nationalism, fostered in the 1950s and 1960s, continues to dominate the 1980s and the PWR multinationals move in to exploit a tame market.

REFERENCES

Beattie, J.R. 1981. "The Assessment of Environmental Consequences of Nuclear Reactor Accidents." In Environmental Impact of Nuclear Power, British Nuclear Energy Society, (London: Telford), pp. 214-236.

Burns, D. 1978. Nuclear Power and the Energy Crisis, London: MacMillan.

Calzonetti, F. 1981. Finding a Place for Energy: Siting Coal Conversion Facilities. Resource Papers in Geography. Washington, D.C.

Charlesworth, F.R. and T. Griffiths. 1982. "Licensing and Inspection of Nuclear Installations in the United Kingdom." In Reactor Safety and Hazard Evaluation Techniques, Vienna:IAEA, pp. 15-30.

Charlesworth, F.R. and W.S. Gronow. 1967. "A Summary of Experience in the Practical Application of Siting Policy in the United Kingdom." In Containment and Siting of Nuclear Power Plants, Vienna:IAEA, pp. 143-170.

Department of Environment. 1978. Annual Summary of Radioactive Discharges in Great Britain, 1977. HMSO, London.

The Economist. 1977. "Nuclear Man at Bay," 19 March.

Farmer, F.R. 1962. The Evaluation of Power Reactor Sites DPR/INF/266, UKAEA, Harwell.

Farmer, F.R. 1979. "A Review of the Development of Safety Philosophies," Annals of Nuclear Energy, 6, p. 261-264.

Farmer, F.R. and P.T. Fletcher. 1959. "Siting in Relation to Normal Reactor Operation and Accident Conditions." In Proceedings of the Sixth Nuclear and Electronic Engineering Conference, CNEN, Rome.

Flowers, B. 1976. Nuclear Power and the Environment, Sixth Report, Royal Commission on Environmental Pollution, HMSO, London.

Fryer, L.S. 1978. "A Guide to TIRION 4 - A Computer Code for Calculating the Consequences of Releasing Radioactive Material to the Atmosphere," Safety and Reliability Directorate SRD Report 120, HMSO, London.

Fryer, L.S. and G.D. Kaiser. 1978. A Computer Program for Use in Nuclear Safety Studies, SRD Report 134, UKAEA, Harwell.

Gronow, W.S. 1969. "Application of Safety and Siting Policy to Nuclear Plants in the United Kingdom." In Environmental Contamination by Radioactive Material, Vienna: IAEA. pp. 549-559.

Gronow, W.S. and R. Gausden. 1973. "Licensing and Regulatory Control of Thermal Reactors in the United Kingdom." In Principles and Standards of Reactor Safety, Vienna: IAEA. pp. 521-538.

Health and Safety Executive. 1978. Nuclear Establishments 1975-76. First Report of HM Nuclear Installations Inspectorate, HMSO, London.

Health and Safety Executive. 1979. Safety Aspects of the Safety of Nuclear Installations in Great Britain, HMSO, London.

Health and Safety Executive. 1981. "The Accident at Three Mile Island: Comments by the Health and Safety Executive", First Report from the Select Committee on Energy 1980-81: Minutes of Evidence, Vol. 2, pp. 19-23.

Henderson, P.D. 1977. "Two British Errors: Their Probable Size and Some Possible Lessons," Oxford Economic Papers, Vol. 29, pp. 159-205.

House of Commons Select Committee on Energy. 1981. The Government's Statement on the New Nuclear Power Program. HMSO, London.

Jeffery, J.W. 1982. "The Real Costs of Nuclear Electricity in the UK," Energy Policy, Vol. 10, No. 2, June.

Jeffery, J.W. 1983. Letter to The Guardian, 10th January.

Kasperson, R.E., G. Burke, D. Pijawka, A. Sharaf, and J. Wood. 1980. "Public Opposition to Nuclear Energy: Retrospect and Prospect," Science, Technology and Human Values, Vol. 5, No. 31, Spring.

Keeney, R.L. 1980. Siting Energy Facilities, Academic Press, New York.

Kemeny, J.G. 1980. Report of the President's Commission on the Accident at Three Mile Island, Pergammon Press, New York.

MacDonald, H.F., P.J. Ballard and I.M.G. Thompson. 1977. "Recent Developments in Emergency Monitoring Procedures at CEGB Nuclear Power Stations." In The Handling of Radiation Accidents, pp. 435-445, IAEA, Vienna.

Marley, W.G. and T.M. Fry. 1955. "The Radiological Hazards from an Escape of Fission Products and the Implications in Power Station Location." In Proceedings of International Conference on Peaceful Uses of Atomic Energy, pp. 102-105, UN, New York.

92

Monopolies and Mergers Commission. 1981. Central Electricity Generating Board: A Report on the Operation by the Board of its System for the Generation and Supply of Electricity in Bulk, HMSO, London.

Clarke, R.N. and G.N. Kelly. "MARC - the NRPB Methodology for Assessing Radiological Consequences of Accidental Releases of Activity" NRPB - R127, HMSO, London.

Kelly, G.N. and R.N. Clarke. "An Assessment of the Radiological Consequences of Releases from Degraded Core Accidents for the Sizewell PWR", NRPB - R137, HMSO, London.

Openshaw, S. 1980. "A Geographical Appraisal of Nuclear Reactor Sites" Area 12, 287-90.

Openshaw, S. 1982a. "The Geography of Reactor Siting Policies in the UK," Trans. Inst. Brt. Geogr. Vol. 7, No. 2.

Openshaw, S. 1982b. "The Siting of Nuclear Power Stations and Public Safety in the UK", Regional Studies, Vol. 6, 3, pp. 183-198.

Pearce, F. 1982. "Sizewell: Beware Low-Flying Aircraft", New Scientist, 11th November, p. 343.

Political Ecology Research Group. 1980. Safety Aspects of the Advanced Gas-cooled Reactors, PERG, Oxford.

Ramberg, B. 1980. Destruction of Nuclear Energy Facilities in War, Toronto: Lexington Books.

Shaw, J. and R. J. Palabrica. 1974. "A Critical Review and Comparison of the Nuclear Power Plant Siting Policies in the UK and USA, Ann. Nuc. Sci. Eng., 1, 241-54.

Surrey, J. and S. Thomas. 1980. "Worldwide Nuclear Plant Performance: Lessons for Technology Policy", Futures, February.

United States Nuclear Regulatory Commission. 1975. Reactor Safety Study, WASH - 1400, Washington.

UKAEA. 1976. An Assessment of the Integrity of PWR Pressure Vessels (Marshall Report), UKAEA.

Westinghouse Electric Corporation. 1982. "Sizewell B Probabilistic Safety Study", WCAP9991, Rev. 1 (Monroevilla, PA).

Williams, R. 1980. The Nuclear Power Decisions: British Policies 1953-78, London: Croom Helm.

Wonder, E.F. 1977. "Decision Making and the Re-Organization of the British Nuclear Power Industry", Research Policy 5, pp. 240-268.

Introduction

Jerome E. Dobson

As Wilbur Zelinsky stated in his excellent introduction to Section I, there are many different ways to introduce a selection of literature. Because the tone of this book leans toward comprehensive coverage, I have chosen first to discuss a broad range of topics dealing with spatial assessment and impact mitigation, and second to offer some commentary on the three chapters, primarily in terms of placing them in the broader context and evaluating how well they cover their intended territory.

SCIENTIFIC PROGRESS IN
IDENTIFYING AND MITIGATING IMPACTS

The environmental movement begun as an emotional appeal from affected residents and concerned analysts. Since then great strides have been made in bringing more rigorous and scientific approaches to the field. The first generation of environmental impact and mitigation analysts often were fondly referred to as "retread engineers"--men and women with educational backgrounds and work experience in engineering, who, by inclination or happenstance assumed responsibility for analysis of social topics. Many of these were highly competent individuals whose technical backgrounds at times brought innovative scientific approaches to the social sciences. As they became more aware of the challenges of their task, they recognized the need for disciplinary specialists in economics, sociology, demography, and political science. This new generation then recognized that the nature of the problems required even more precise specialties within their disciplines. For example, a microeconomist was not sufficient to deal with broad regional economic issues so regional economists became critical. At this stage geographers gained recognition as specialists in spatial analysis and regional integration. This third generation, comprised of well-matched specialists in a variety of disciplines, was instrumental in developing the better methods and tailoring them to the problems of technology impact assessment. In many institutions, their work attracted colleagues among the top people in their fields, and a fourth generation was born consisting of experts in new fields such as automated geography, risk analysis, and social impact assessment. It is this fourth generation that is bringing excellence to the fields of impact identification, analysis, and mitigation.

Perhaps the greatest improvement has been in the areas of spatial assessment and behavioral sciences related to siting and mitigation studies. As a result of the research that has been conducted primarily in the fields of resource and environmental analysis (with much of the funding related to nuclear issues) traditional methods have improved dramatically, and it is now possible to conduct automated geographical analyses at the local, regional and national levels. The techniques of remote sensing, computer cartography, computer graphics, quantitative spatial modeling, spatial statistics and geographical information systems have developed to the point that they can be incorporated into traditional methods of geographical analysis. This new form of automated geography can yield remarkable results providing the scientific community with the capability to address societal and technological issues of a size and scope heretofore impossible (Dobson, 1983).

A necessary part of this methodological development has been the remarkable growth of digital spatial data bases. Analysts, long accustomed to using census data in this form, can now access a variety of resource and environmental information highly detailed both topically and spatially. The best known data bases are spectral readings from satellites that can be used to infer land cover at a resolution of one acre for most of the world. In the United States the TOPOCOM data base provides elevation data at this same resolution and permits terrain analysis—i.e., calculation of slope, aspect, local relief and other terrain characteristics—for the contiguous 48 states. The transportation network is now available in digital form—i.e., railroads, airports, state and federal highways and barge routes including intracoastal waterways. Moreover, all of these transportation data bases contain useful topical information as well as spatial locators. For example, ownership and quality of track are included in the railroad data base. Population has now been mapped for the entire United States in great detail from enumeration district data. Every power plant greater than 10MW of capacity in the country is described digitally in terms of its location, ownership, capacity, fuel type, alternate fuel switching capability, cooling type and water source. Numerous measures of air quality, water quantity and quality, the distribution of Federal lands including national parks, wilderness areas, and military bases, the habitats of endangered species; and indicators of seismic activity levels -- all of these and many more data variables are available and rapidly improving to the point that analysts can conduct spatial assessments for the entire nation much as they have conducted local area analysis and planning in the past.

THE EVOLUTION OF SPATIAL ASSESSMENTS AND THE NUCLEAR INDUSTRY

The methods of spatial assessment and impact mitigation have evolved rapidly in the last decade in response to changes in the nuclear industry. For a quarter of a century there were expectations of unbridled growth of nuclear energy, and the predominant spatial assessment theme was the selection of optimal or satisfactory locations for nuclear power plants. To illustrate from my own experience at Oak Ridge National Laboratory, my first task in 1975 was to help the Nuclear Regulatory Commission evaluate the concept of Nuclear Energy Centers (Burwell, et al. 1976). The idea was to mitigate the environmental and safety impacts of nuclear power plants by concentrating ten, or perhaps as many as forty, generating units in large clusters rather than in the current dispersed pattern of one to four units. Some appealing features included mitigation of construction impacts through the establishment of a small town of laborers with permanent jobs and homes. Emergency evacuation procedures could have been standardized. We concluded, however, that the concept would not work because the concentrated impacts were too great for any small area to bear. For example, few streams in the country could supply the necessary cooling water and these tend not to be in the kind of isolated locations one would choose for minimal exposure and easy evacuation. Another problem was transmission distance and the amount of land required for corridors. The spatial context was the crucial element, and geographers, working in association with engineers, economists and other scientists, were instrumental in identifying the shortcomings that quelled interest in the concept.

The emphasis immediately shifted toward improving the methods of identifying candidate sites that would fit within the more conventional dispersed pattern. The early overlay methods were automated and supplanted by regional screening and multiobjective optimization algorithms (Dobson, 1979 and Hobbs and Voelker, 1978).

During the remainder of the 1970s, concern about the rapid growth of electricity generation led us to concentrate our attention on simulating the implications of national energy growth scenarios. This involved siting the facilities hypothetically to meet policy-oriented objectives expressed in varying levels of coal, nuclear, oil, natural gas and solar energy use. Projections of potential health and safety, water quantity and quality, air quality and other impacts were then calculated on the basis of these hypothetical siting patterns (Honea, et al., 1979). Problem areas were identified, but mitigation studies remained focused on individual sites; they were never conducted at the regional or national levels effectively.

The end of the decade witnessed a rapid reversal from an earlier expectation of high electricity growth and high nuclear contribution. Weinberg and Burwell spoke of increasing the number of units at some existing nuclear plant sites (Burwell, et al., 1979), but the interest in finding new sites had passed. This interest is not likely to revive until there is new growth in electricity demand and an unanticipated shift in public attitudes regarding the safety of nuclear reactors. In accordance with this trend, Hillsman, Alvic and Church (1983) have developed a new spatial assessment model which examines a broad spectrum of utility planning options such as whether to meet increased demand by building new generating capacity or by purchasing power from other utilities with excess capacity. In keeping with the times, their model is equally adept with facility retirements and new construction. To illustrate how rapidly the situation has changed direction in the United States, we have moved from studies of siting new 40-unit complexes to studies of excess capacity in just eight years.

THREE CONTRIBUTIONS TO THE FIELD

The literature on spatial assessment of nuclear energy is dominated by a preoccupation with the siting and public acceptance of individual power plants. To a large extent the hyperbolic cooling tower has become the visual and perceptual tip of the nuclear iceberg. This has occurred inspite of the fact that large coal-fired plants often have identical towers and some nuclear plants have no towers at all--a phenomenon determined more by the characteristics of the water source than by the characteristics of the power plant. The only instance in which the Staff of the Nuclear Regulatory Commission has recommended denial of a nuclear plant construction permit was based on the adverse aesthetic impact of the cooling tower, not on safety considerations (Petrich, 1982). Public opposition has focused on the reactors and associated cooling towers with a special concern for safety and environmental impacts, but the nuclear fuel cycle involves other equally imposing facilities for mining, enrichment, transportation and final disposition. These too have their impacts which can be diminished by proper siting or exacerbated by improper siting.

In his chapter on spatial impact dimensions of a nuclear power plant, Pasqualetti has gone a long way toward addressing our need for a broader view of the nuclear industry. His convergence/divergence schema provides a theoretical framework for understanding not only the nuclear fuel cycle but the extended impacts related to other material needs as well. He has summarized the entire nuclear system related to one power plant which he describes as a unique case. It remains to be determined whether or not general laws can be found and generalizations made about the broad distribution of nuclear systems in this country and elsewhere. Let us hope that others will follow his lead and study these systems in greater depth.

Richetto reflects the dominant themes of most spatial assessments of nuclear energy. He alludes to the broad range of spatial analysis issues for the entire nuclear system but focuses on the siting of power plants. He alludes to the world-wide distribution of nuclear power but focuses on the United States. Indeed, he believes that the high visibility of nuclear power plants per se in certain industrial areas of the United States has been a significant factor in increasing public awareness and fostering concern over the impacts of nuclear power. Methodologically his chapter describes the historical development of siting methods. He places nuclear power plant siting in the broader context of location/allocation theory and applies entropy measurement to the distribution of nuclear power plants. His contribution serves as a summary and chronicle of the current state-of-the-art and may be most appreciated in those countries where continued growth of nuclear power is still viewed as a viable option. These skills also may be highly valued if the United States elects to return to the path of nuclear expansion at some point in the future. At present their best application in the United States would be through a modification to improve the siting of nuclear waste repositories.

Metz, Daum, Pearlman and Waite present an excellent empirical examination on the subject of land use controls around nuclear power plants. Their exhaustive study explores every legal means (and some means not yet proved to be legal) for wresting control of development from private landowners. The results remind us of the awesome responsibility we face when we devise siting methods and when we or others use our products

to select sites for actual development. The lesson is
that, wherever the site is placed, landowners, residents and others may suffer severe deleterious effects
if strict attention is not paid to equity issues and
mitigation strategies. In view of the value we place
on land ownership and property rights, it is not surprising that land use controls, especially those
requiring eminent domain, are unpopular at the local
levels. It is important that cultural and behavioral
scientists provide better understanding of the
preferences and motivations of impacted populations.

TODAY'S NEEDS AND OPPORTUNITIES

Changing conditions in the nuclear power industry
have made much of the work of the last decade obsolete.
What do we have to show for our labor? First we know
that in some instances the geographers and related
social scientists affected nuclear policy enough to
prevent further pursuit of some unfruitful ideas and to
point the way toward better alternatives. Looking
ahead we can speculate that the greatest challenges
will involve managing an existing system and defining
the future role of nuclear power. The nuclear industry, like other, less controversial industries,
should run efficiently. If nuclear power is ever going
to compete unsubsidized with other fuels and technologies, the total system construction and operating
costs must be improved dramatically while public welfare is protected. This requires objective analysis
that will define optimal configurations of nuclear
facilities in the proper juxtaposition to each other,
to the resources they depend on, to labor markets, to
their customers and to transportation and transmission
facilities. The objective is to reduce risk by minimizing potential exposure and shun unacceptable risk by
avoiding critically vulnerable locations.

As we move into the twenty-first century, the
question of public welfare will require some hard-nosed
assessments to which we are as yet unaccustomed.
Today's assessments compare the costs and benefits of
nuclear power with the costs and benefits of fossil-
fuel and emerging technologies. As fossil fuels are
depleted we shall have to turn increasingly to renewable and relatively nondepletable resources. Nuclear
power will be reevaluated along with solar and other
emerging technologies, but the new assessments will
have to include scenarios of energy deprivation as
well. This will give a new complexion to public welfare issues. For example, the likelihood of deaths
from a nuclear accident or fossil-fuel induced greenhouse effect will be compared with the likelihood of
starvation that might result from a deficiency of

energy in agricultural production. The solution is likely to involve all of the energy resources and technologies today plus some innovations. The facilities likely will be deployed in an eclectic configuration that matches decentralized energy technologies with dispersed demands and central generating facilities with concentrated demands. Spatial assessment will be essential to define optimum configurations that maximize efficiency of resource use and minimize total societal risk. Simultaneously, these assessments must deal with the inevitable uncertainty of demand projections and the potential loss of one or more technologies as a result of catastrophic events (e.g., the loss of fossil fuel technologies as a result of concern over the greenhouse effect).

Our present task is to address pressing current issues while simultaneously building the knowledge and methodological base to address questions of far greater significance in the 21st century. I would suggest the following research needs as being essential to both short-term and long-term goals:

1. <u>A broader view of the nuclear fuel cycle</u>. We must examine the nuclear industry not as a set of isolated power plants but as an integrated system from extraction to processing, generation and final disposition of fuels and facilities. The entire complex must then be viewed in a holistic societal context.

2. <u>Areal expansion of behavioral studies</u>. The realm of behavioral studies--including public acceptance of nuclear power, perception of safety and other nuclear issues, response during emergencies and mitigation of various social impacts--is now, as it always has been, at the local, site-specific level. There is a pressing need to expand the scope of behavioral research to create a new capability whereby methods can be applied to behavioral issues for large regions or for the nation as a whole.

3. <u>Dealing with uncertainty</u>. It is commonly recognized that most analytical models and projections are ineffective in predicting actual events because of the large number of imponderables that cannot be resolved within the model structure or human brain. It is imperative that spatial assessments and other decision tools be improved to deal explicitly with uncertainty. This does not presuppose that the accuracy of event prediction can be greatly improved. Rather it implies that the primary sources of uncertainty can be recognized and that robust technology and siting configurations can be designed so as to facilitate maximum societal response as random events unfold.

4. _International studies_. For the most part
spatial assessments have tended to be entirely too
provincial in scope. We need to promote international
understanding of nuclear and alternative energy issues.
As one example, the power plant siting techniques which
were developed in the United States during its period
of nuclear expansion now may have much greater applicability in several foreign countries, such as France,
which continue to build new nuclear capacity. One
especially important need is to provide better advice
to the less developed countries currently considering
the nuclear path.

5. _Improved study of equity issues_. Among the
spatial impacts of nuclear development lie many equity
issues primarily related to disparities in the spatial
distribution of costs and benefits. Social scientists
should increase their attention to these issues and
improve the methods of spatial analysis to deal with
them.

6. _Continued improvement of automated methods and
data bases_. Recognizing that the spatial and topical
complexity of energy issues is too great to be handled
by traditional manual methods alone, spatial analysts
should work to improve automated methods and data bases
that will complement their work. Specific examples
would include an integrated model of the nuclear
industry and new data bases for transmission lines,
pipelines and other energy delivery systems.

Collectively the three chapters in this section
reflect both the best and the worst aspects of the
current state-of- the-art. Evaluated in terms of the
six research needs outlined above, one of the three
examines the full fuel cycle but none contains a significant discussion of behavioral issues, uncertainty,
international issues, equity, automated geographical
methods or improved spatial data bases. If these new
directions are as important as I believe they are, we
who do spatial assessments and mitigation studies have
our work cut out for us.

REFERENCES

Burwell, C.C. et al., 1976. _Nuclear Energy Center Site
    Survey - 1975 Part V. Section 4_. NUREG - 0001,
    U.S. Nuclear Regulatory Commission. January 1976.
Burwell, C.C., M.J. Ohanian, A.M. Weinberg. 1979. "A
    Siting Policy for an Acceptable Nuclear Future."
    Science Vol. 204, 4397. June 8, 1979.

Dobson, J.E. 1979. "A Regional Screening Procedure for Land Use Suitability Analysis." Geographical Review, Vol. 69(2). April, 1979, pp. 224-234.

Dobson, J.E. 1983. "Automated Geography." The Professional Geographer, Vol. 35(2). pp. 135.243.

Hillsman, E.L., D.R. Alvic, and R.R. Church. 1983. BUILD (Baseload Utility Integrated Location Decisions): A Model of the Future Spatial Distribution of Electric Power Production. Oak Ridge National Laboratory. ORNL-5969.

Hobbs, B.F. and A.H. Voelker. 1978. Analytical Multi-objective Decision-Making Techniques and Power Plant Siting: A Survey and Critique, Oak Ridge National Laboratory, ORNL-5288.

Honea, R.B., E.L. Hillsman, and R.F. Mader. 1979. Oak Ridge Siting Analysis: A Baseline Assessment Focusing on the National Energy Plan, Oak Ridge National Laboratory, ORNL/TM-6816, October 1979.

Petrich, C.H. 1982. "Assessing Aesthetic Impacts in Siting a Nuclear Power Plant: The Case of Greene County, New York, Environmental Impact Assessment Review, Vol. 3(4). pp. 311-332.

4

Locating Nuclear Electric Energy Facilities: Structural Relationships and the Environment

Jeffrey P. Richetto

Nuclear power has established a foothold in at least forty-four nations. By the middle of 1978, twenty-two countries had licensed 220 reactors totaling a generating capacity of more than one hundred million kilowatts of electricity. In addition, 320 reactors were under construction or on order. Although these figures may appear to be impressive, increasing environmental opposition has caused a serious retrenchment in the development and growth of nuclear energy and its related facility location process. Throughout the 1970s, for example, the United States faced an array of nuclear energy-related issues, including technical and environmental problems relating to spent fuel reprocessing and waste disposal and the delaying involvement of the public in the actual planning and siting of nuclear electric facilities. Although a deluge of studies attempted to respond to this growing list of concerns, the focus of this paper is limited to site-planning or the decision of where to locate nuclear electric plants. Site-planning for nuclear facilities has required and continues to require an interdisciplinary perspective; most, if not all of the research efforts undertaken may be placed in one of three categories: (1) survey/modeling (Hobbs and Voelker, 1977; Beaujean and Charpentier, 1978; Marchetti and Nakicenovic, 1979; Propoi and Zimin, 1981); (2) empirical studies (Dobson, 1977; Miller and Honea, 1978; Richetto and Semple, 1979); and (3) policy analysis (Resources for the Future 1969; Ford Foundation, 1974; Krutilla and Page, 1975; Ruedisili and Firebaugh, 1978; McConnell, et al. 1982). Although these and other studies have added to our basic knowledge of nuclear plant siting, they omit two critical areas which this chapter will address.

First, past research efforts have failed to specify the structural relationships that link the nuclear electric facility location problem to the general body of location theory. Some of these relationships include whether or not the siting of a nuclear plant is characteristically a public or private sector location problem, involves individual or centralized decision-making, is representative of a nodal or network system pattern, and is based upon a well-defined or indefinite service region.   Such structured relationships lend insight into critical site-planning issues including methodology, lead-time and level of uncertainty, conflict resolution analysis, and implementation strategies.   Thus, in order to identify the structure of the nuclear electric facility location problem it is necessary to discuss in detail the generalized location problem wherein nuclear plants may then be classified. Second, this paper investigates the present and potential role of environmental opposition as a location influence in the development and diffusion of nuclear-based electricity.   That is, the diffusion of nuclear electric facilities is examined throughout the continental United States, a pattern that is both explainable by the spatial distribution of high-grade coal deposits, sizable water supply, and population and industrial growth, as well as being explainable as a response and a catalyst of growing public environmental concern.   The problem of measuring the diffusion of nuclear electric facilities is viewed as involving the measurement of the entropy of a system and, therefore, this study employs an information statistic model.

THE GENERALIZED LOCATION PROBLEM

The traditional approach to location problems has been either in terms of public sector location (Teitz, 1968; Mumphrey and Wolpert, 1973; Dear, 1974; Massam, 1980) or private sector location (Weber, 1929; Losch, 1954; Smith, 1971; Hamilton, 1978), depending upon the type of facility under consideration.   Although both types of location problems have employed a similar methodological approach, they differ conceptually (Revelle et al., 1970) by at least four distinct criteria.

First, public facilities provide goods and services declared to be wholly or partly within the domain of government.   Private facilities are those that offer functions wherein decision-making lies primarily within the domain of private enterprise.   Second, an observable pattern of public facility locations is formed by a single public authority deemed responsible for all facilities providing a given service within its jurisdiction.   Thus, the researcher of public facility location is drawn away from the problem of locating individual facilities and toward the structure and location of the entire system of facilities within the jurisdictional area of the public authority.   In the private sector, however, locations are formed by many decentralized individuals or corporate decisions. Third, the specific objectives to be satisfied by the choice of location are not normally the same in the public and private domain. In the public sector, for example, equity in service provision is usually the stated goal, while in the private sector, profit is more important (McAllister, 1976). As a result, one is led to believe that some public facilities are less than ideally situated in terms of conventional economic analysis. The fourth and final distinction is that public sector location decisions focus on public decisions and governmental budgets in response to a set of welfare criteria in essentially a mixed market/ nonmarket setting. Without a complete set of costs and benefits that can be expressed monetarily, the practitioner must focus on the distribution of impacts resulting from a location choice. In contrast, private sector location decisions emphasize the role of choice, taste, and utility in a predominantly market context.

Although this two-fold public/private dichotomy is intuitively appealing and useful in classifying many conventional location problems, it offers only a partial yet important understanding of the location of nuclear facility systems. In order to develop the structural properties for this unique type of facility more fully, it is informative to discuss other characteristics emphasizing the geometry of facility systems. Hall (1973) states that when facility systems are represented, the most likely geometric patterns are nodal and network. Nodal or point patterns characterize many distributive services whereby the final phase of distribution is both flexible and intermittent. Police and fire protection service systems are examples of point pattern geometry. Other service systems including water, sewage, and highways require continuous connection over space, and therefore, a network characterization is appropriate. Moreover, for those facility systems representable as point patterns, a further categorization is suggested, one that describes the system in terms of its service region. That is, if a number of consumers patronize one particular facility for the purchase of a good or service, the region enclosing these customers constitutes a definite service region for that facility. If, on the other hand, individuals from any region can patronize any facility to satisfy their demand for a good or service, the service region is termed indefinite.

Related to these two formal geometric properties of facility systems are two other attributes that combine geometry with behavior. First, the probability-of-use of a facility system is a function of both the size of the service region under consideration and the probability-of-need for the good or service provided within

that region.   The second attribute, service flow,
classifies facilities as either distributors or collectors.   A distribution facility (for example, sanitation removal) is one that dispatches its good or
service to the population, while a collection facility
(for example, day care center) requires customers to
travel to the facility to obtain service.

Finally, the structure of facility systems may be
classified according to the concept of obnoxiousness.
The significance of this factor has been reflected in
the growing awareness and involvement by the general
public, especially in locating facilities perceived to
be obnoxious.   That is, although Mumphrey et al. (1971)
define an obnoxious facility as one which is likely to
have significant spatial differences in the incidence
of costs and benefits, the conception of obnoxiousness
and the extent to which it affects the location of a
facility is directly related to public perception
(Hinman, 1971).   In other words, whether or not a
facility is obnoxious depends on how it is viewed by
the public and not necessarily on self-evident characteristics.   Based upon the preceding discussion, this
paper will define which sector nuclear electric facilities falls within and identify how system attributes
characterize the spatial structure of these plants and
their related facility system.

THE NUCLEAR ELECTRIC FACILITY
LOCATION PROBLEM

The siting of nuclear electric plants and their
related facility system may be characterized as both a
public and private sector location problem.   First, the
electric utility industry provides electricity to the
general public through these facilities under the
ownership of both private investors and government.
Second, normally each utility company supplies all
electrical needs to a well-defined service region.
Consequently, the region served, along with the operations of a utility, is considered to be a regulated
monopoly.   Thus, each utility company is responsible
for the location of an integrated system of generating
facilities within its service region quite independent
of other utility companies' locational preferences in
their service regions.   The resulting national pattern
of electric generating facilities, therefore, is one
characterized by a series of individual, centralized
decisions.

Power plant siting decisions must also consider the
fact that the influence of a single utility's power
system is not necessarily confined to its geographic
region of service (Winter and Conner, 1978).   Electrical service is often provided to neighboring power

regions, particularly during peak demand periods. Moreover, each individual utility system is part of a larger collective regional and/or national utility system; the siting of new facilities in response to growing public pressure must often occur at locations far removed from their load center; and the ever-increasing cost for construction and plant size requires that several utilities enter into a coopera- tive agreement to build a single generating unit and apportion the power produced. This last consideration has been particularly characteristic of nuclear electric installations. As a result, a utility's service region may also be considered indefinite.

The final two characteristics concern the utili- ties' legal responsibilities for providing electrical service to the general public. At the same time, they are privately-owned businesses selling a service and seeking a reasonable return on their investment (Leason, 1974). Also, the information base from which nuclear power plant location decisions are made is composed of both economic and noneconomic considera- tions. On the one hand, land acquisition, construction materials cost, labor availability and wage rates, and technical requirments represent some of the major eco- nomic considerations. Noneconomic considerations, on the other hand, include aquatic effects related to thermal discharge of a nuclear plant, permissable levels of radiation and their impact upon the ecology of human life, and aesthetic effects of cooling towers, to name a few. Importantly, although the former type of considerations are amenable to quantification and rigorous economic scrutiny, often the latter type de- fies precise measurement and numerical analysis. And these kinds of considerations are especially charac- teristic of nuclear electric facilities.

In regard to the geometric structure of the nuclear electric facility system, the entire process whereby electricity is produced and consumed requires continu- ous connection of generating plants, substation, and end users, each of which is a node along a complex transmission power grid network. However, the spatial structure of nuclear electric plants is one character- ized as network design, not location on a predetermined network whereby a transmission network arises to pro- vide access to a set of identified nodes or locations (Robers, 1971). The importance of the above distinc- tion is reflected in the structure of mathematical models employed to arrive at a location solution (Revelle, et al., 1970). That is, models designed to search for locations along a predetermined network are likely to identify incorrectly where to site a gener- ating plant. However, models designed to search for locations in a plane with a finite or infinite solution space satisfy the conditions for network design and, therefore, have a greater likelihood for determining appropriate locations.

Finally, nuclear electric units provide an ordinary service much in the same way as water or gas utilities, rather than an extraordinary or emergency service as provided by police and fire departments. The difference is reflected in the probability of normal day-to-day usage of electricity by customers. Furthermore, these plants offer a dispatching service, unlike the centralized service provided by schools in which the individuals being served must travel to the facility. Also, nuclear facilities are viewed as both experimental and obnoxious. That is, the technology is not fully developed, and a significant-sized proportion of the general public perceives these facilities as having noxious characteristics. Wolpert (1970), among others, notes that the environmentally-related elements of uncertainty, conflict, and coalition-formation often envelop the location decision for those facilities classified as obnoxious, altering significantly their ultimate location. Therefore, this paper will examine the present and potential role of environmental opposition as a location factor in the continued development and diffusion of nuclear-based electricity.

THE ENVIRONMENT AS A LOCATION FACTOR

Up to 1970, the electric utility industry encountered little public opposition in the development and location of additional electrical capacity (Slater et al., 1971). Aside from early and quite specialized controversies, there was little concern expressed over the rapidly expanding electric power system. In fact, nearly all of the opposition in public hearings appeared on legal or economic grounds, not health or environmental. Typically, electric utilities projected electrical needs and then constructed new facilities to supply these needs.

However, transition in the complexity of issues affecting the location for electric power plants from that era to the present and beyond has been taking place. This pattern has been and is being influenced by the generally increased size of generating units (especially nuclear), new high voltage transmission technology, greater penetration of nuclear energy into the fuel mix, and the stronger role of the general public in the site selection and acceptance process. Although each of these concerns is shaping the present energy situation in general, the last two issues in particular have required and promise to continue to require adjustment in the electric utility's location decision-making process.

Planning, in terms of timing energy expansion, has been a major strength of electric utilities. Events of

the past fifteen years, however, have triggered concern
about the adequacy of current site-planning practices
in addressing the major utility issues of the 1980s and
beyond. In retrospect, characteristic planning of the
electric utility industry from 1940 to the early 1960s
involved only five external relationships (Figure 4.1).

FIGURE 4.1
Relationships of Electric Utility Industry.

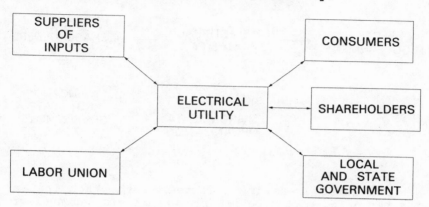

These relationships, moreover, were largely with
agencies which themselves had little, if any, interaction.
Since 1965, however, this situation has altered,
slowly at first and then at an accelerated pace.
Electric utilities developed a growing reliance upon
energy suppliers, primarily because of the notable
production economies-of-scale and lower costs of
regional power plant projects. In many instances,
these relationships created new linkages with the
federal government. In addition, utility management
found itself increasingly subject to a growing number
of external environmental influences that are becoming
ever more interrelated and complex (Figure 4.2).
As the United States continues to expand its
nuclear electric generating capability, the controversy
surrounding the development and location of nuclear
plants escalates. Citizen groups, industry, the scientific
community, utilities, and all levels of government
have become embroiled in the growing debate over
the wisdom of uncontrolled nuclear energy expansion.
Two issues which are possibly of great significance to
this debate center on the state-of-the-art of nuclear
technology and the overall spatial distribution of
nuclear electric facilities.

FIGURE 4.2
Organizational Influences of Electrical Utility
Companies.

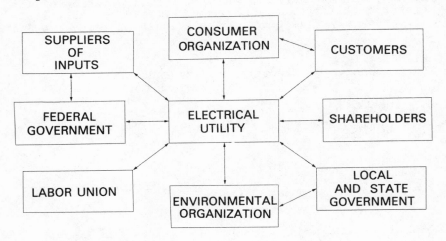

The first issue is reflected in the fact that at the beginning of the post-World War II era, no nuclear-generated electricity had been produced anywhere in the world, and the United States--then the world's only nuclear power--had no program directed for developing such a technology. Today, however, over seventy commercial nuclear power stations throughout the United States generate an estimated 300 billion kilowatt hours of electricity annually. If additional nuclear units--those under construction and in the planning stage--become operational, the United States' electric power commitment in twenty-five years could include some 150 or more nuclear installations supplying as much as thirty-five percent of the total domestic electrical need. Thus, this nation's commitment to increased reliance on nuclear-based electricity is a major one, whose impact, positive or negative, could be realized well into the future.

Unlike past opposition to power plant location that stemmed primarily from residents in the immediate vicinity of a planned site, present-day proposed sites draw broader attention from non-local citizen groups as well as government. From a technological viewpoint, for example, opponents of the nuclear industry assert that the risks involved in the production of fission energy--accidents, leakage of radioactive materials, transportation, and storage of waste, among others--are too great for society to accept and pose a major threat for the future. Proponents point out, however, that this view is inconsistent with overall experience to date, or with results of extensive research and studies bearing on involved risks. In addition, critical examination of a proposed facility frequently emphasizes its integration into regional and state development plans. Moreover, federal goals, particularly in the areas of air and water quality, have also broadened the purview of the site selection and acceptance process. Some of the consequences of such a wide-ranging, conflict-laden site approval process have been the lengthening of the time required for site approval well beyond the expected period, substantial increases in the construction cost of facility projects, and subsequent increases in consumer rates to defray additional construction expenses.

The second issue affecting the debate on nuclear energy expansion is the overall spatial arrangement of nuclear electric facilities, especially as it relates with increasing environmental opposition to develop the nuclear option further. On the one hand, the existing spatial pattern of these facilities is not only a reflection of their numerous and varied technological requirements, but is also in response to past and current expressed interests for the protection of the environment. On the other hand, the proposed spatial distribution of nuclear plants may intensify environmental concern and thereby further obscure the path of nuclear development. In an effort to investigate the above two relationships, this paper will illustrate and derive past and anticipated changes in the spatial pattern of nuclear electric facilities.

For purposes of investigating such changes between 1950 and 1992, the United States was divided into three major regions: the North, South, and West. Each of these regions was further partitioned into three census regions. The North was partitioned into East North Central, the Mid-Atlantic, and New England; the South into West South Central, East South Central, and South Atlantic; and the West into Pacific, Mountain, and West North Central (Figure 4.3).

THE INFORMATION ANALYSIS AND THE EVOLVING SPATIAL PATTERN OF NUCLEAR ELECTRIC FACILITIES

In 1957, the first commercial nuclear reactor became operational near Pittsburgh, Pennsylvania. By the early to mid 1960s, six additional reactors began operation. Interestingly, five of these pilot facilities were concentrated within the industrial north; thereafter, facility locations began to disperse (Figure 4.4). The problem of measuring the spatial dispersion of nuclear electric facilities may be viewed as involving the measurement of the entropy of a system and therefore, the study employs an information sta-

FIGURE 4.3
The United States and Its Major Regions.

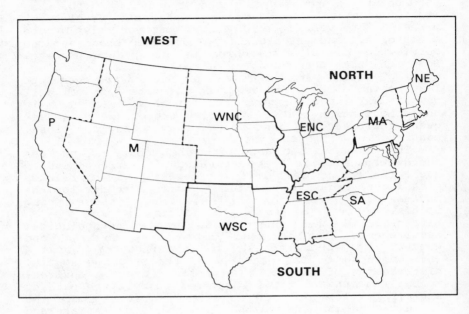

tistic model (Hexter and Snow, 1970; Horowitz, 1970).
Total dispersion occurs when system entropy is maxi-
mized, that is, when complete equality holds for a
system.   Conversely, total concentration occurs when
system entropy equals zero.   The present analysis
utilizes information statistics to measure simultaneously information concerning trends in national,
regional, and subregional levels of nuclear reactor
capacity dispersion based on existing and planned state
generation totals.   When a modification of Shannon's
original formula is utilized, the final dispersion
statistic, H(Y), is given by:

$$H(Y) = \sum_{t=1}^{T} \sum_{r=1}^{R} Y_{rt} \log_2 1/y_{rt} + \sum_{t=1}^{T} \sum_{r=1}^{R} Y_{rt} \left(\sum_{g=1}^{G} Y_{grt} \log_2 1/Y_{grt} \right)$$

$$+ \sum_{t=1}^{T} \sum_{r=1}^{R} Y_{rt} \left(\sum_{g=1}^{G} Y_{grt} \left(\sum_{k \varepsilon s_{grt}} Y_{krt}/Y_{grt} \log_2 Y_{grt}/Y_{krt} \right) \right)$$

FIGURE 4.4
The Diffusion of Nuclear Power Reactors in the United States.

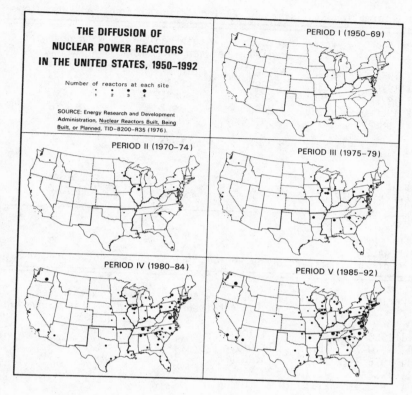

where the first term on the right measures the between region dispersion of nuclear generating capacity over all time periods, the second term measures the between subregion dispersion over all time periods, and the last expression measures the within subregion dispersion for all time periods.

Close examination of the set of maps in Figure 4.4 reveals not only an ever-increasing number of facilities scheduled for operation but a general locational pattern concentrated east of the Mississippi River, especially around the Great Lakes and along the Atlantic coast. The extent of nuclear capacity expansion throughout the five time periods, however, becomes even more evident by a thorough review of the information statistics found in Tables 4.1, 4.2, and 4.3.

TABLE 4.1
The Information Statistics by Period, Major Region, and Subregion

Period	Total	Between region	Within region Between subregion	Within region Within subregion
I 1950-69	.1982	.1950	.0027	.0004
II 1970-74	.6367	.5825	.0439	.0104
III 1975-79	.7501	.6634	.0641	.0227
IV 1980-84	1.6939	1.2186	.3472	.1281
V 1985-92	.8768	.7775	.0788	.0204
Total	4.1557	3.4370	.5367	.1820
Maximum possible dispersion	12.6617	3.9068	7.0767	1.6782

TABLE 4.2
Percent Total Dispersion by Period, Major Region, and Subregion

Period	Total	Between region	Within region Between subregion	Within region Within subregion
I 1950-69	4.77	4.70	.06	.01
II 1970-74	15.32	14.01	1.06	.25
III 1975-79	18.05	15.96	1.54	.55
IV 1980-84	40.76	29.32	8.36	3.08
V 1985-92	21.10	18.70	1.90	.50
Total	100.00	82.69	12.92	4.39

TABLE 4.3
Percent of Period Dispersion by Major Region and Subregion

Period	Total	Between region	Within region Between subregion	Within region Within subregion
I 1950-69	100.00	98.39	1.41	.20
II 1970-74	100.00	91.49	6.89	1.62
III 1975-79	100.00	88.44	8.55	3.01
IV 1980-84	100.00	71.94	20.50	7.56
V 1985-92	100.00	88.68	8.99	2.33
Total	100.00			

The smaller the value in Table 4.1, the higher
degree of concentration for a particular period or
spatial partition; the larger the value, the greater
the dispersion. The overall level of dispersion
clearly increases from .1982 in period I to 1.6939 in
period IV. Period V, however, witnesses a stage of
retrenchment in which there is an overall decline in
the number of nuclear reactors scheduled for operation.
In fact, only forty-two additional reactors are planned
for the system (Table 4.4). This decline in the number
of planned nuclear reactors is primarily reflected in
the uncertainty surrounding nuclear energy in supplying
future U.S. electrical demand. Furthermore, period III
indicates a stabilization in the number of facilities
coming on-line. Interestingly, however, the average
size (megawatt capacity) of individual units has continued to increase from 343 MWe to 1163 MWe. When the
information statistics are examined at the regional and
subregional level, a similar pattern emerges. That is,
not only has dispersion increased between major regions
(the North, South and West), it has also increased for
the subregions (for example, West: P, M, WNC), and at
the same time for the states within all subregions (for
example, Pacific: Washington, Oregon, and California).
This increase, however, has been exceedingly modest
relative to the theoretical maximum values calculated
for each spatial partition. The interpretation of such
numerical comparisons suggests strongly that the overall spatial distribution of nuclear facilities exempli-
fies a pattern of concentration as compared to dispersion. What is not immediately evident is the varying
rate at which concentration is occurring between spatial partitions. This kind of information may be
calculated by converting the values in Table 4.1 into
percentages (Tables 4.2 and 4.3).

TABLE 4.4
Total Number of Nuclear Reactors Scheduled for On-Line Operation
by Period

Period	Total number of	Average size (Megawatt capacity)
I 1950-69	12	343.1
II 1970-74	40	775.2
III 1975-79	33	968.3
IV 1980-84	79	1091.9
V 1985-92	42	1163.4

Source: Energy Research and Development Administration.
Nuclear Reactors Built, Being Built, or Planned.
TID-8200-R35 (1976), pp. 11-17.

Figures in Table 4.2, for example, are calculated by dividing each of the information statistics of Table 4.1 by the grand total and expressing them as a percentage. The percentages in Table 4.3, on the other hand, are calculated by dividing the information statistics of Table 4.1 by their corresponding row totals. Although Table 4.2 correlates closely with Table 4.1, interesting period trends emerge in Table 4.3. That is, there are significant decreases in between region percentages from period I to IV, but corresponding increases for between and within subregions. These trends indicate that although nuclear reactors are locating in a more spatially dispersed manner in each succeeding period (excluding period V), the major regional contribution to dispersion is declining relative to the dispersion between and within subregions. In other words, nuclear electric generating capacity is continuing to become somewhat less concentrated by region, but is doing so at a slightly faster rate at the subregional and state levels.

In support of these findings concerning such a spatial pattern of nuclear facilities, the literature identifies a number of underlying traditional factors. First, although the eastern regions of the United States are richly endowed with proven reserves of coal, many of these deposits contain a high sulfur content necessitating the use of expensive abatement equipment. Second, because nuclear energy facilities require a larger volume of water for the purpose of condenser cooling, the plant must be situated near a sizable water supply. And, third, current and projected growth in population as well as industrial productivity in the midwestern, southern, and eastern sections of the United States does and will continue to require substantial increases in energy consumption. In addition to the above three reasons in explaining the observed trend in siting nuclear facilities, the influence of environmental preservation should be considered. The impact such concern has had on the evolving spatial pattern of the system of nuclear plants may best be conveyed by examining the genesis of the relationship between environmental preservation and the spatial arrangement of nuclear facilities.

During the initial years in the development and location of nuclear electric facilities, the utility industry encountered little public opposition. In fact, public opposition did not begin to be voiced on such issues as thermal water pollution, radioactive emissions, safety, and waste storage until the mid-to-late 1960s. The question that arises is: why did approximately twelve years elapse prior to any organized public concern over the environmental incompatibility of nuclear electrical energy? By definition, the problems of safety, radiation, and thermal water pollution were present, if not worse, in the formative years of developing the nuclear option.

In retrospect, the time lag may have been a function of a poorly informed public with respect to the technological and operational characteristics of nuclear power. At best, however, this explanation provides only a partial understanding of the twelve-year response lag. In fact, other reasons appear more substantive in nature—for example, the number of nuclear facilities in operation, the size of the reactor measured in megawatt capacity, as well as the existing and planned spatial pattern of nuclear facilities. A reexamination of Table 4.4 shows that both the number and average size of nuclear reactors in operation have increased and will continue to increase as other reactors scheduled for operation are built (note the exception of periods III and V). The significance of such a trend lies in the fact that nuclear electric plants have become increasingly visible objects dotting the landscape. Accompanying this increased visibility has been a stepped-up awareness and concern by the public for preserving the environment. Central to understanding this present relationship and especially its future impact on the nuclear facility location decision is a knowledge of how nuclear plants have been located on the landscape and how in the future the utility industry plans to allocate them spatially.

In review, Figure 4.4, along with Table 4.4, illustrates that larger-sized and greater number of nuclear reactors are being added to the system by period, but that these new plants are being sited in already nuclear built-up areas. In other words, the landscape of nuclear plants is becoming ever more congested and increasingly more visible to the public. As a result, public concern for environmental preservation will continue to escalate; therefore, its role in the location decision will become ever more critical.

CONCLUSION

As the demand for electrical energy continues to increase; as costs for the production of electricity spiral steadily upward and nuclear energy exerts a greater penetration into the nation's fuel mix; as the availability of fossil energy resources dwindles and environmental pressure for safer, cleaner production increases; as federal, state and local government regulations become increasingly more stringent, the electric utility industry's energy development and facility location programs will encounter a number of uncertain, conflict-laden situations. These situations may manifest themselves at a time when utilities are striving to provide low-cost electrical service, even though resources are growing increasingly inaccessible and in short supply. Complicating this situation may be the fact that environmental technology could become more complex and expensive, available production sites may become increasingly more difficult to identify, and the public will be ever more prepared to question, debate, and protest. The nuclear electric industry in particular has already felt the impact of these and other concerns resulting in a significant downturn in nuclear energy development along with an ever more stringent site selection and approval process.

This paper has discussed two factors--the location structure of nuclear plants and the environment--that could allay some of the problems that beset the nuclear industry, especially in locating its facilities. However, many other factors indirectly affect facility siting and require careful attention. Some of these factors include the timing and introduction of different nuclear energy-producing technologies, the availability of uranium and spent fuel reprocessing, the development of renewable-based energy-producing technologies, and the development of a program that effectively manages and safely stores nuclear waste material. As answers to these and other issues emerge, a more comprehensive understanding about nuclear energy expansion and its facility siting process will result.

REFERENCES

Beaujean, J. and J. Charpentier. 1978. _A Review of Energy Models:_(4). IIASA: Laxenburg, Austria.

Dear, M. 1974. "A Paradigm for Public Facility Location Theory," _Antipode_, 6(1):46-50.

Dobson, J. 1977. _The Maryland Power Plant Siting Project: An Application of ORNL-land Use Screening Procedure_. ORNL/NUREG/TM-79.

Ford Foundation. 1974. _A Time to Choose_: _America's Energy Future_. Cambridge: Ballinger.

Hall, F. 1973. _Location Criteria for High Schools: Student Transportation and Racial Integration_. Chicago: University of Chicago Press.

Hamilton, F. 1978. _Contemporary Industrialization_. London: Longman.

Hexter, J. and J. Snow. 1970. "An Entropy Measure of Relative Aggregate Concentration," _Southern Economic Journal_, 36:239-243.

Hinman, J. 1971. _A Location Model for Public Facilities with Neighborhood Effects_. Discussion Paper No. 13, Department of Geography, University of Pennsylvania.

Hobbs, B. and A. Voelker. 1977. *Analytical Power Plant Siting Methodologies: A Theoretical Discussion and Survey of Current Practice.* ORNL/TM-5749.

Horowitz, I. 1970. "Employment Concentration in the Common Market: An Entropy Approach," *Journal of the Royal Statistical Society,* Series A, part 3:463-479.

Krutilla, J. and R. Page. 1975. *Towards a Responsible Energy Policy.* Resources for the Future Reprint No. 18.

Leason, J. 1974. "Capitalism's Greatest Test: The Electric Utilities," *Public Utilities Fortnightly,* 94(4):30-33.

Losch, A. 1954. *The Economics of Location.* Translated by W. Woglam, New Haven: Yale University Press.

Marchetti, C. and N. Nakicenovic. 1979. *The Dynamics of Energy Systems and the Logistic Substitution Model.* Laxenburg, Austria: IIASA.

Massam, B. 1980. *Spatial Search: Applications to Planning Problems in the Public Sector.* Oxford: Pergamon Press.

McAllister, D. 1976. "Equity and Efficiency in Public Facility Location," *Geographical Analysis,* 8:48-63.

McConnell, V., C. Harris, and J. Cumberland. 1982. "Forecasting Economic and Environmental Impacts of Energy Policy." Paper presented at the Southern Regional Science Association meeting, Knoxville, Tennessee.

Miller, C. and R. Honea. 1978. *Evaluating Air Quality/population Impacts of Fossil-fueled Power Plants: A Case Study in Frederick County, Maryland.* ORNL.

Mumphrey, A., J. Seley, and J. Wolpert. 1971. *A Decision Model for Locating Controversial Facilities.* Discussion Paper No. 11, Department of Geography. University of Pennsylvania.

Mumphrey, A. and J. Wolpert. 1973. "Equity Considerations and Concessions in the Siting of Public Facilities," *Economic Geography,* 49(2):109-121.

Propoi, A. and I. Zimin. 1981. *Dynamic Linear Programming Models of Energy, Resource, and Economic Development Systems.* Laxenburg, Austria: IIASA.

Resources for the Future. 1969. *U.S. Energy Policies.* Washington, D.C.

Revelle, C., D. Marks, and J. Liebman. 1970. "An Analysis of Private and Public Sector Location Models," *Management Science,* 16(1):692-707.

Richetto, J.P. and R.K. Semple. 1979. "The Location of Electric Generating Facilities: Conflict, Coalition and Power," *Regional Science Perspectives,* 9(1):117-138.

120

Robers, P. 1971. "Some Comments Concerning Revelle, Marks and Liebman's Article on Facility Location," Management Science, 18(1):109-111.

Ruedisili, L. and M. Firebaugh. 1978. Perspectives on Energy. New York: Oxford University Press.

Slater, H. and Niagra Mohawk Power Corporation. 1971. "Public opposition and the Nuclear Power Industry," International Atomic Energy Association, SM-146/40:847-859.

Smith, D. 1981. Industrial Location: An Economic Geographic Analysis. 2nd edition, New York: John Wiley.

Teitz, M. 1968. "Toward a Theory of Public Facility Location," Papers and Proceedings of the Regional Science Association, 21:35-51.

Thiel, H. 1967. Economics and Information Theory. New York: North-Holland Publishing Company.

Weber, A. 1929. Alfred Weber's Theory of the Location of Industries. Translated by C. Friedrick, Chicago: University of Chicago Press.

Winter, J. and D. Conner. 1978. Power Plant Siting. New York: D. Van Nostrand Reinhold Company.

Wolpert, J. 1970. "Departures from the Usual Environment in Location Analysis," Annals of the Association of American Geographers, 60:220-229.

5

A Convergence/Divergence Schema to Identify Impacts: The Case of the Palo Verde Nuclear Generating Station[1]

Martin J. Pasqualetti

A review of the environmental impact statement for a large nuclear power plant in Arizona (Arizona Public Service, 1975, 1981) reveals a focus that is almost completely local. Virtually no mention is made of more distant impacts, an omission that results from the limitations of regulatory requirements and a narrowness of the conceptual framework applied to local construction projects. While a limited context is appropriate for most such projects, it has serious shortcomings with regard to nuclear power. The reasons for such shortcomings include the enormous scale and variety of material needs originating from widely diversified and dispersed sources, the complex and separated system of fuel preparation, and the additional demand for construction workers, which is beyond the supply capacities of nearby communities.

One approach to identify more fully the topical, dimensional, and areal influences of local nuclear power plant projects is to apply the well-established geographical theme of origins and dispersals. I am adopting this basic approach in the form of "convergence and divergence" because I feel it accurately defines the impact of a nuclear power plant more completely. Geographers examining nuclear energy previously have concentrated on individual considerations. Examples include research on facility siting (Baker, 1980; Openshaw, 1982; Richetto, 1980; Semple, 1976), although the current primary concern seems to be to mitigate some potential impacts through alternative siting options. Socioeconomic impact studies, familiar to energy geographers from the early 1970s onward, continued to garner some (albeit declining) government financial support (Bergmann and Pijawka, 1981, Metz, 1981). More recently, largely since the 1979 accident at Three Mile Island, PA, a new and potentially important theme--technological risk assessment--is attracting interest vis-a-vis nuclear power. These latter efforts concentrate on risk perception and assessment

121

(Cutter, 1981; Kasperson, et al., 1980; Wolpert, 1980), emergency preparedness (Cutter, 1982), evacuation (Cutter and Barnes, 1982; Wolpert, 1977; Zeigler, et al., 1981), and especially the siting and environmental considerations of nuclear waste disposal (Brunn et al., 1980; Hare and Aikin, 1980; Metz, 1982). The latter themes are particularly geographic as they include elements of distribution and transportation. The convergence/divergence schema proposed cross-cuts many themes.

Many of the dispersed and lesser known impacts of nuclear plant construction and operation are best demonstrated by examining a single facility. My example is the Palo Verde Nuclear Generating Station near Phoenix, AZ.

THE PALO VERDE NUCLEAR GENERATING STATION

The Palo Verde Nuclear Generating Station (referred to either as PVNGS or Palo Verde) is located 34 miles (54 km) west of Phoenix's western boundary and about 50 miles (80 km) from the Central Business District (Figure 5.1). The power plant is surrounded by lightly populated desert and scattered irrigated agriculture.

FIGURE 5.1
Location of Palo Verde Nuclear Generating Station, AZ.

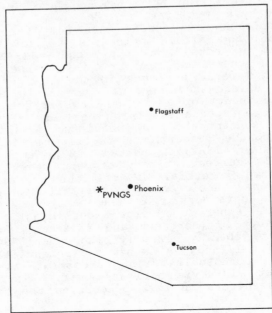

Compared to other nuclear plants, Palo Verde is somewhat unique. It is more than 35 miles (56 km) from the nearest appreciable concentration of people (many nuclear stations are much closer), it is located in a desert (unlike any other commercial station), it is cooled by effluent from a sewage treatment plant (as is no other plant), its electricity will be purchased by utilities in four large western states, and when completed it will have the largest generating capacity of any U.S. nuclear power plant. These unique characteristics have been partly responsible for the rather limited study of its physical and socioeconomic influences and impacts; federal research funds instead have gone to examine plants in the more populated sections of the U.S. where the results have more "transferability" (Chalmers, et al., 1982; Pijawka, 1983). In short, the Palo Verde power plant, like many sensitive topics of geographic interest is out of sight and largely out of mind from the federal research point of view (see Brunn, et al., 1980). These same unique conditions, however, make Palo Verde an attractive topic for investigation.

THE GEOGRAPHIC PERSPECTIVE: CONVERGENCE/DIVERGENCE

The contribution the geographic perspective can offer to the study of nuclear power plants can be illustrated by asking: once the need for Palo Verde was established, what was the process by which the power plant itself came into existence? From the nongeographic viewpoint this process consists of components (materials, equipment, people, fuel, electricity, waste) and phases (preconstruction, construction, postconstruction). But the geographer thinks in spatial terms: Where did the plant components and personnel come from? By what routes did they arrive at the construction site? How was the site selected? What was and what will likely be the impacts of the project? Where will the electricity be transmitted? These questions, framed within the context of a single isolated power plant, may be considered as questions of convergence and divergence (Table 5.1). The schema is divided into three major components: interstate convergence (involving the movement of equipment, materials, personnel, and fuel to the plant site), local divergence (the localized influence of the plant's construction and existence), interstate divergence (the outward movement of products and people when construction is

TABLE 5.1
Perspectives and Phases

Businessmen/ Economists	Contractors/ Utility Rep.	Geographers
Raw Materials	Pre-Construction/	Interstate Convergence
Assembly	Construction	Local Divergence
Distribution	Post-Construction/ Operation	Interstate Divergence

replaced by operation (Table 5.2). A fourth possible type, local convergence, is relatively insignificant in the case of PVNGS.

Examples illustrate the basic components of the schema. Interstate convergence specifically relates to the materials, people and fuel which converge on the plant site from across state lines or from abroad (Figure 5.2). The proportion of such ingredients which actually moves interstate increases directly with the isolation of the plant site and inversely with the economic and demographic complexities of the state. Thus, the appropriateness of the concept of interstate convergence per se is presumably greater in Arizona than it would be for Texas or California, although convergence at some scale always occurs.

Sometime the routes taken within the interstate convergence phase are necessarily circuitous; that is particularly true in reference to nuclear power because many of the large components are specially constructed and the nuclear fuel cycle is complex. The steam generators for the Palo Verde plant, for example, were manufactured in Chattanooga, TN. Because of their great size and weight, they were transported by barge down the Tennessee River to the Ohio and the Mississippi Rivers, across the Gulf of Mexico and through the Panama Canal, north through the Gulf of California, onshore at Puerto Penasco, Mexico, and overland on specially designed carriers across strengthened bridges to the plant site.

TABLE 5.2
Framework and Topics for Studies of Convergence
and Divergence

Framework		
Interstate Convergence	Local Divergence	Interstate Divergence
Fuel Cycle Components Employees Siting	Commuters Land & Law Risk Perception Evacuation Services Temporary Camps Land Use Change	Electricity Employees Waste

Topics

Interstate Convergence

Fuel Cycle (1)	Fuel Cycle (2)	Components (1)	Components (2)
Fuel Origins	Local Impacts	Building Materials	Local Impacts
Routes	Distribution Impacts	Description	Distribu. Impacts
Amounts	Processing Impacts	Origins	Process. Impacts
Schedules		Dispersal Routes	

Employees
  Origins
  Routes
  Numbers
  Demography
  Turnover

Siting
  Criteria
  Possible Sites
  Selected Sites

TABLE 5.2 (Continued).

Local Divergence

Commuters
  Residences
  Travel Routes
  Time Factors
  Travel Subsidies
  Distance
  Temporary Camps

Services
  Schools
  Protection
  Utilities
  Roads
  Recreation

Land and Law
  Exclusion Zones
  Low Density
  Buffer Zones

Risk Perception
  Types of Concerns
  Distribution
  Intensity

Land Use Change
  Values
  Usage
  Areal Extent

Evacuation
  Routes
  Road Conditions
  Time Require-
    ments

Interstate Divergence

Electricity
  Utilities
  Locations
  Amounts
  Lines

Employees
  Subsequent Jobs
  Location of Jobs
  Travel Routes

Waste
  Location
  Types
  Routes
  Impacts

The local divergence phase basically consists of the socioeconomic sphere of influence during construction such as worker commuting and housing; the distribution of the electricity should also be considered (Figure 5.3). The ability to filter out "noise" in the socioeconomic data is related directly to such considerations as the relative isolation of the plant site and local and regional service capacities. Because of the relatively low bulk of nuclear fuel and the absence of conventional air pollution, nuclear power plants tend to be located closer to metropolitan areas than other (particularly coal) power plants; thus a worker commuting pattern is more common than is the construction of accommodations for workers near the plant site. In the case of Palo Verde, the nearest sizable communities are an hour's drive, a time period on the margin of a maximum acceptable commute. Workers initially preferred the long commute rather than living in the isolated environs of the plant; buses were provided for

FIGURE 5.2
Interstate Convergence for PVNGS.

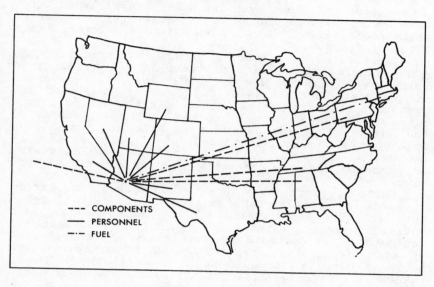

FIGURE 5.3
Local Divergence from PVNGS.   The major impact is on
the Phoenix Metropolitan Area.

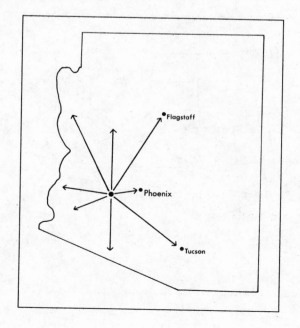

the support staff and some travel subsidies were provided for construction crews.   Later, in response to
accumulated weariness over the daily travel, dormitory-
like accommodations and a trailer park were constructed
about eight miles (11 km) northwest of the construction
site, at Tonopah, AZ.
        Interstate divergence (Figure 5.4) actually begins
prior to final completion of the plant as construction
workers move on, but the principal interstate movement
of personnel occurs when the plant is completed and
operation begins.   Not all the people who worked at the
site will leave the state, but many will.   In terms of
the electricity component the majority of power from
Palo Verde is to be transmitted out of Arizona; 47 percent will stay within Arizona, California will receive
27 percent, Texas 16 percent, and New Mexico 10 percent.
        While military waste is sent to reprocessing facil-
ities, spent fuel assemblies from all commercial reactors including Palo Verde are to be stored at each
individual reactor site (Garmon, 1981).   Palo Verde is
designed to accommodate a seventeen-year accumulation
of such material, after which it is anticipated a
national disposal program will be available.   It is
unknown what storage arrangements will be made or where
the national repositories will be located.

FIGURE 5.4
Interstate Divergence for PVNGS.

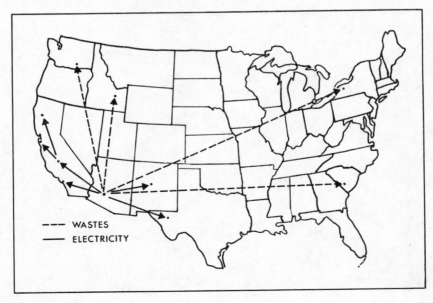

DETAILED EXAMPLES FROM PALO VERDE

Detailed examples from the Palo Verde experience illustrate the scope of convergence and divergence notions outlined above. The examples were chosen because they are representative of all nuclear power plants. The most intriguing involves fuel supply.

Fuel Supply

The fuel needed for all commercial nuclear reactors in the U.S. is uranium-235 (U-235). This fissionable isotope is found naturally combined with U-238, an unfissionable isotope. The natural proportion of U-235 to U-238 is 1:140, or about .71 percent U-235. This concentration must be increased to 3-4 percent U-235 before it can be used in conventional pressurized-water and boiling-water reactors. The complicated and energy-intensive process by which this increase is effected is part of a long series of steps which include mining the ore, milling the ore to produce "yellowcake" (U_3O_8), transportation of the U_3O_8 to the converter, conversion of the U_3O_8 to a gas (uranium hexaflouride (UF_6)), transportation of the UF_6 to a facility for isotopic enrichment, transportation of the enriched UF to a fabrication plant for conversion to uranium dioxide (UO_2) and integration into fuel assemblies, and finally transportation of the fuel assemblies to the power plant (Figure 5.5).

For the initial loading and the first reloading of the three 1270 MWe units at Palo Verde, the fuel will come from the Jackpile-Paguate mine, near Grants, NM. The mine is run by Anaconda Minerals for the parent company, ARCO. Milling takes place close to the mine and the yellowcake is trucked in 55-gallon drums by F.B. Truck line to the Nuclear Activities Division of Allied Chemical (an operating company of Allied Corporation) in Metropolis, IL in lots of 72 drums, delivering 24-25,000 pounds of yellowcake per truckload (Figure 5.6). Of the 4.0 million pounds of yellowcake needed for the initial load about 160 truckloads of yellowcake must travel from Grants to Metropolis. (Normal operations require that 53 truckloads of yellowcake be transported from the mill to the converter each year for the life of the plant.)

The gaseous uranium hexafluoride solidifies below about $150^\circ F$ ($65^\circ C$), and is transported as a solid in negative-pressure cylinders from Metropolis to one of three enrichment plants in Oak Ridge, TN; Paducah, KY; or Portsmouth, OH. Which enrichment plant the Enriching Operations Division of the U.S. Department of Energy (DOE) decides will receive the UF is based on such considerations as the marginal cost of the energy

130

FIGURE 5.5
Nuclear Fuel Cycle

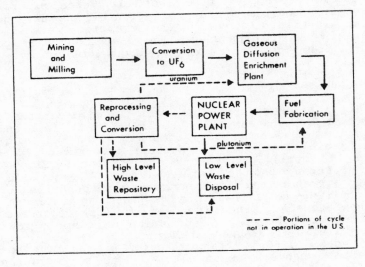

FIGURE 5.6
Nuclear Fuel Routing for PVNGS.

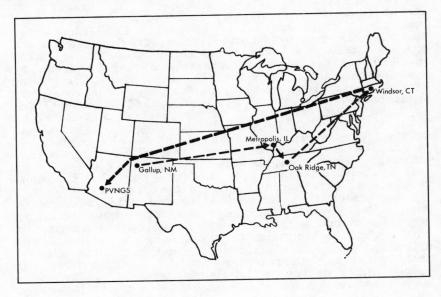

necessary to process the gas and the percentage of enrichment desired. DOE operates the enrichment facilities in a cost recovery mode and currently charges $138.65 per SWU (Separative Work Unit, a measure of the electrical work it takes to do the enrichment). All three enrichment plants are dependent upon the generation capabilities of local utilities, especially the Tennessee Valley Authority, for electricity to operate the facilities.

The fuel leaves the enrichment plants as enriched UF and is transported to the combustion-engineering fabrication plant in Windsor, CT. Once this process is completed, the original 4.5 million pounds (2045 metric tons) of yellowcake will have been reduced to about 750,000 pounds (341 metric tons) of UO_2 which arrives at the power plant as part of 723 fuel assemblies (241 per generating unit). Each year for the life of the plant 250,000 pounds (114 metric tons) of UO_2 (as pellets) must be transported to the site.

None of the environmental documents for PVNGS (e.g. APS, 1975, 1979) addresses in detail such geographical implications as steps or routes of the fuel cycle. Nor is there mention of several other geographically significant socioeconomic conditions at any of the points along the route of the fuel cycle, including housing, commuting patterns, or in-migration. Nothing is mentioned about the environmental impacts from the construction and operation of the generation facilities that provide the enormous amounts of electrical power necessary for gaseous enrichment. In the case of the Palo Verde plant, the fuel comes from a single mine source, and the cycle, though long, follows a single route. It would be more feasible to study many of the impacts of the fuel cycle now than when fuel contracts with several different mines become effective with the second reloading. Such a study would lead to a fuller understanding of the fuel supply side of the Palo Verde project and identify findings which may be applicable elsewhere.

Land Values and Land Usage

Although many significant changes in land values and usage may result from activities tied to the fuel supply of the plant (which would be included under interstate convergence), many other changes occur in the vicinity of the plant site under local divergence. The Palo Verde plant is located in a desert environment, on flat, formerly agricultural land. The site was chosen after screening the entire state for places with water availability and seismic stability. The former is a special problem for nuclear facilities owing to their relatively high waste heat rejection while the latter is a consequence of federal regulations which mandate a nuclear power plant may not be located near a "capable fault," i.e., one which has exhibited movement at or near the ground surface once within the past 35,000 years or movement of a recurring nature, within the past 500,000 years (Glasstone, 1980).

Once an appropriate site is chosen, land use considerations enter the picture in the general vicinity of the power plant itself. This emphasis results from the governing regulations of the Nuclear Regulatory Commission, specifically Title 10, Chapter 1, Code of Federal Regulations--Energy, U.S. NRC Rules and Regulations Part 100: Reactor Site Criteria (10CFR100), Section 100.11, which includes definitions for "Exclusion Areas," "Low Population Zone," and "Population Center Distance" (U.S. Nuclear Regulatory Commission, 1980). All three are related to postulated dosage rates from accidental release of radiation from the power plant. By virtue of these regulations the NRC has substantial power over the local and nearby use of the land; this control increases as one nears the plant.

In addition to 10CFR100, the following sections also are applicable here: 10CFR20, "Standards for Protection Against Radiation;" 10CFR50, "Licensing of Production and Utilization Facilities;" and 10CFR51, "Licensing and Regulatory Policy and Procedures for Environmental Protection." To aid applicants, comprehensive guides have also been prepared: "Standard Format and Content of Safety Analysis Reports for Nuclear Power Plants" (NRC Regulatory Guide 1.70) and "The Preparation of Environmental Reports for Nuclear Power Plants" (NRC Regulatory Guide 4.2). Two considerations became apparent after a review of the environmental documentation of the Palo Verde plant and the regulations governing such reports: (1) neither the documentation nor the regulations appear to deal with situations beyond the general vicinity of the plant (i.e., more than 50 miles (80 km)), except in the case of transmission lines; and (2) many local issues seem inadequately addressed. No mention was made of the commuting patterns of the 5,000 employees, land value changes near the construction project, a real expression of risk perception, or additional local services near the plant such as the complete moderization of a small local school with tax dollars resulting from PVNGS. Nor was there anything but a cursory discussion of impacts on housing and service demands. Several studies have addressed such issues elsewhere, some under funding by the NRC, but they have been subsequent to plant completion (Chalmers, et al., 1982). The majority of nuclear power plants (including PVNGS) fail to receive such attention during construction.

Power Distribution

PVNGS is one of an increasing number of power plants generating electricity largely for export to other states. In the larger western states this export inevitably requires the routing and construction of transmission lines over substantial distances. The selection of routes recently has become a very difficult and painstaking process. Two decades ago routing transmission lines from New Mexico's Four Corners coal plant to Phoenix required only two angle points on the entire 200-mile (320 km) distance. Nowadays, transmission routes from adjoining newer power plants zigzag across the landscape in efforts to reduce land use conflicts. Large, separate reports are commonly produced showing every possible route, each with its many dozens of combinations of "legs". No longer is it possible to achieve anything remotely resembling the simple routing of the Four Corners lines.

The routing of transmission lines for the Palo Verde plant is no exception to this present trend. The inability to find acceptable routes through California for the electricity produced by PVNGS Units 4 and 5 was blamed for those units being dropped from the original project. The plant required the routing and construction of primary transmission lines under three separate projects. "Project 1" passes from PVNGS to a substation northwest of Phoenix, to a fossil plant in south Tempe (a Phoenix suburb), and a fossil fuel station northwest of Tucson. "Project 3" connects to the Rio Grande substation at El Paso. Another project passes from PVNGS to the Devers substation northwest of Palm Springs, CA. The PVNGS-Devers line varied from 235 to 265 miles (376-424 km) in length depending on route selected and directly disturbed 1,742-1,915 acres (697-766 ha) (U.S. Department of Interior, 1978). The right of way was required for 50 years with the right of renewal. The line is designed to carry 585 MWe for Southern California Edison.

Of all the environmental and socioeconomic studies which are required and/or conducted in association with power plants, those involving transmission lines usually are the ones which trace impacts over the greatest distance and in some detail. Similar studies could also be carried out with regard to the fuel cycle, evacuation routes, employee movements, waste disposal transportation and repository sites, and many other topics of geographic interest.

CONCLUSIONS

Until fairly recently, geographers have contributed only slightly to research on nuclear power plants.

Most studies have dealt with "site selection," though
lately topics including socioeconomics, waste disposal,
and emergency preparedness, have been addressed. Currently NRC-mandated investigations into many of the
impacts of nuclear power plants tend to focus on local
effects, but these often are not detailed and are only
during the early planning stages. Little attention
continues to be paid to questions of socioeconomic
impacts and land use changes during or after construction. This limited approach is neither necessary nor
imaginative, cohesive nor wise, because it fails to
encompass or probe an appropriately wide geographical
range of possible contributions. This traditionally
small and parochial sphere of inquiry can be expanded
by using the conceptual framework of convergence and
divergence. Suggested examples include land use
changes in value or function in association with
various nuclear facilities, transportation and socioeconomic factors associated with the complexities of
the nuclear fuel cycle, and the exportation of fuel and
technology. The application of such a geographical
perspective in the study of PVNGS illustrates the influence (often distant and unrecognized) which can
follow a local decision to build a nuclear power plant.

NOTES

1. This chapter in slightly different form, was published originally in The Professional Geographer,
35(4):427-436. 1983. Reprinted with permission.

REFERENCES

Arizona Public Service Company. 1975. Palo Verde
    Nuclear Generating Station Units 1, 2, 3,
    Environmental Report, Construction Phase, with
    supplements. Phoenix, AZ.
Arizona Public Service Company. 1979. Palo Verde
    Nuclear Generating Station Units 1, 2, 3, Environ-
    mental Report Operating License Stage, with supple-
    ments to 1981. Phoenix, AZ.
Baker, E.L., et al. 1980. "Impact of Offshore Nuclear
    Power Plants: Forecasting Visits to Nearby
    Beaches," Environment and Behavior, (12):367-407.
Bergmann, P.A. and K.D. Pijawka. 1981. "The Socio-
    economic Impacts of Nuclear Generating Stations: An
    Analysis of the Rancho Seco and Peach Bottom Facil-
    ities," GeoJournal, Supplementary Issue on Energy,
    3:5-16.

Brunn, S.D., J.H. Johnson, Jr. and B.J. McGirr. 1980. "Locational Conflict and Attitudes Regarding the Burial of Nuclear Wastes." _East Lakes Geographer_, 15:24-40.

Chalmers, J., D. Pijawka, P. Bergmann, K. Branch and J. Flynn. 1982. _Socioeconomic Impacts of Nuclear Generating Stations_. Summary Report on the NRC Post-Licensing Studies. NUREG/CR-2750.

Cutter, S. 1982. "Coping with Nuclear Power: Emergency Preparedness and Planning for Nuclear Power Plant Accidents." Paper presented at the 78th annual meeting of the Association of American Geographers, San Antonio, Texas.

Cutter, S. and K. Barnes. 1982. "Evacuation Behavior and Three Mile Island," _Disasters_, 6:116-124.

Cutter, S. 1981. "Three Mile Island: Risk Assessment and Coping Responses of Local Residents, A Summary Report." _Discussion Paper No. 20_, Rutgers University, Department of Geography.

Garmon, L. 1981. "The Box within a Box within a Box," _Science News_, 120:396-399.

Glasstone, S. and W.H. Jordon. 1980. _Nuclear Power and its Environmental Effects_. Washington: American Nuclear Society.

Guiness, P. 1980. "The Changing Location of Power Plants in California," _Geography_, 65:217-220.

Hare, F.K. and A.M. Aikin. 1980. "Nuclear Waste Disposal: Technology and Environmental Hazards." In _Nuclear Energy and the Environment_, pp. 168-199. E.E. El-Hinnawi ed. New York: Pergamon.

Kasperson, R.E., G. Berk, D. Pijawka, A.B. Sharaf, and J. Wood. 1980. "Public Opposition to Nuclear Energy: Retrospect & Prospect," _Science, Technology, and Human Values_, 5(31), pp. 11-23.

Metz, W.C. 1981. _Construction Workforce Management: Worker Transportation and Temporary Housing Techniques_. Private publication.

Metz, W.C. 1982. "Legal Constraints to High-Level Radioactive Waste Repository Siting," _Impact Assessment Bulletin_, 1(2):55-64.

Openshaw, S. 1982. "The Siting of Nuclear Power Stations and Public Safety in the United Kingdom," _Regional Studies_, 16(3):183-198.

Pijawka, D. 1983. Personal communication.

Richetto, J.P. 1980. "The Environment as a Factor for Locating Nuclear Electrical Facilities in the United States," _Geografiska Annaler B_, 62:39-46.

Semple, R.K, and J. Richetto. 1976. "Locational Trends of an Experimental Public Facility: The Case of Nuclear Power Plants," _Professional Geographer_ 28:248-253.

U.S. Department of Energy. 1980. "Pricing of Uranium Enrichment Service."

136

U.S. Department of the Interior, Bureau of Land Management and the U.S. Nuclear Regulatory Commission. 1978. _Draft Environmental Statement Palo Verde-Devers 500 KV Transmission Line_.

U.S. Nuclear Regulatory Commission. _Rules and Regulations, Code of Federal Regulations_, Title 10. Chapter 1, revisions through 1980.

Wolpert, J. 1977. "Evacuation from the Nuclear Accident." In _Geographical Horizons_, pp. 125-129. John Odland and Robert N. Taaffe, eds. Dubuque, IA: Kendall/Hunt.

Wolpert, J. 1980. "The Dignity of Risks," _Transactions_, Institute of British Geographers, NS 5, 391-401.

Zeigler, D.J., S.D. Brunn, and J.H. Johnson, Jr. 1981. "Evacuation from a Nuclear Technological Disaster," _Geographical Review_, 71:1-16.

6
Land-Use Controls
and Nuclear Power Plants

William C. Metz, Mary L. Daum,
Kenneth T. Pearlman, and Nancy D. Waite

The Nuclear Regulatory Commission's (NRC) site evaluation procedures (10 Code of Federal Regulations (CFR)) serve as a defense-in-depth in the siting of nuclear power stations, the purpose being to foster the selection of low population density sites in case a breach occurred in the engineered safety features. Because of the potential threat of radioactive releases in the event of an accident at a nuclear power station and possible need to evacuate the local population, the NRC recommended in Part 100 of the 10 CFR that site area population density not exceed "500 persons per square mile averaged over any radial distance out to 30 miles" (NRC, 1975, 16), including a weighted transient population, at the time of initial plant operation. At the end of a plant's lifetime, area population density is not to exceed 1,000 persons per square mile (386 per square kilometer). A further requirement established a minimum distance, using a mathematical formula, between a site and a population center of over 25,000 people.

The aforementioned density and distance criteria are levels that trigger additional review in the consideration of alternative sites rather than represent upper limits of acceptability. A site exceeding the population density guidelines could nevertheless be selected and approved if no obviously superior alternative sites were identified. Part of NRC's function as a regulatory authority is to continually determine the safety significance of population density and distribution in the vicinity of nuclear stations. As yet, NRC has not assumed any active role or authority itself in ensuring that remote characteristics of a site are maintained throughout the operating life of a station.

Changes in population and land use have been occurring in the vicinity of the nation's operating nuclear power stations. Will the defense-in-depth strategy of the NRC be breached or voided through continuing rapid population growth and land use change around several of this nation's nuclear stations? As a consequence of the 1979 Three Mile Island accident, interest has been raised in the degree, extent, and causal factors of land use and population change. The NRC was required by the U.S. Congress in the 1980 NRC Appropriations Authorization (PL96-295, June 30, 1980) to "provide information and recommendations to state and local land use planning authorities having jurisdiction over the zones established under the regulations...and over areas beyond the zones which may be affected by a radiological emergency" but not to preempt state and local authorities. This concern for an assessment and establishment of NRC procedures regarding land use change, monitoring, and control had previously been discussed in an NRC Siting Policy Task Force report, which emphasized the defense-in-depth site strategy (NRC, 1979). In 1981, Brookhaven National Laboratory (BNL) undertook a study for the NRC regarding the land use and population changes that were occurring in the vicinity of operating nuclear power stations.

This chapter is concerned with the possible implementation and effectiveness of land use controls around nuclear power stations. We approach the subject from several perspectives. (1) We will examine the causes of change in land use around the 49 operational nuclear station (72 units in 1981) sites. Understanding the reasons for population growth and land use change in these site areas allows the assessment of the capability of current and future potential control mechanisms. (2) We will discuss the question of authority to regulate land use. Although the NRC has stepped up its interest in land use as it relates to population growth and hazardous activities (NRC, 1980), land use control and regulation has historically been under the authority of local and state government, and federal intrusion into this domain is usually unwelcome. (3) We will present an overview of land use controls in analogous situations and will consider examples of attempts to control population growth through land use control techniques around nuclear and non-nuclear hazardous facilities. (4) We will explore the potential of a number of local and state land use control mechanisms to limit changes in population density and land use around nuclear power facilities.

LAND USE CHANGES AROUND NUCLEAR
STATION SITES

To date, excessive population growth around nuclear plants has not become a serious problem. Although several site areas do exceed the population density siting guidelines given in NRC Regulatory Guide 4.7, they constitute fewer than ten percent of the approved sites (NRC, 1979). All sites were approved prior to the issuance of the guidelines in 1975, and only two site areas presently exceed the lifetime limit given in the guidelines. A recent survey of land use conditions around the nation's seventy-two operating units at forty-nine collective sites (BNL, 1982) revealed that low population densities continue to prevail for the time being, despite few, if any, specific local low density land use policies. According to planning officials, such changes in population and land use which have occurred so far have generally been in line with the goals and plans set forth by planning units in these areas. However, the majority of operating stations still have twenty to thirty years expected operational time remaining, well over half their estimated life expectancy. Elsewhere in this book Greenberg has shown that local areas in which nuclear plants are located have been growing more rapidly than their surrounding regions in the last 20 years (1960-1980). Growth due to local causes (as opposed to increase originating in county, regional, or national trends) has accounted for an increasing share of population increase in host local areas during that time.

What are these local causes? The BNL survey found that the presence of a major highway (cited by 40 percent of the respondents), good municipal services (sewers and drinking water, cited by 39 percent) and proximity to a major urban area 39 percent) were the main attractions to population and industrial growth in the ten-mile (16 km) Emergency Planning Zone (EPZ) around nuclear stations. Also mentioned by 25 percent of the respondents were low taxes. Although this latter tax figure is quite a bit lower than the other three, it reflects a majority of the political jurisdictions that host a nuclear facility, as opposed to those which are just in the vicinity.

In contrast, 74 percent of the respondents to the BNL questionnaire said that the local nuclear plant had had no impact at all on local area growth. Only in the southeastern United States did nuclear stations appear to play a significant role, where twenty-four percent of the respondents said the plant had a high or very high impact on local population growth. Of particular importance is the fiscal impact of taxes paid to local communities by nuclear generating stations. In the southeast, forty-two percent cited station-derived taxes as an important factor in local population growth.

In his contribution to the Brookhaven study (Burchell, 1982), Dr. Robert Burchell also found that the local fiscal impact of privately owned nuclear power plants is significantly positive because taxes remain low and public services are upgraded. (Municipally owned generating stations are tax-exempt, although many do make in-lieu payments to the host jurisdiction.)

When a municipality receives power plant revenues, its resources usually double. If a county receives revenues, its resources increase by 60 to 120 percent. School district resources increase an average of 40 percent after receipt of station revenues.

Burchell's findings parallel those of the BNL questionnaire: where the primary locational factors (for example, highway, urban proximity, public services) stimulate population growth, the effects of station tax payments are often overshadowed and play a lesser role in fostering rapid population growth. In rural areas where these locational stimuli do not exist, area growth is slow, and station revenues tend to be viewed as extremely important to any growth that may occur, especially insofar as the tax payments contribute to the enhanced quality of the school system. Instances were also found where people chose not to take advantage of the lower taxes (though possibly not significantly lower) afforded residents in host local areas, but instead preferred to locate in other communities in the EPZ where public services and other amenities were considered more attractive. Also, once construction is completed, nuclear plants have relatively small labor forces and therefore do not tend to stimulate either residential or non-residential development. BNL found that only 22 percent of respondents from the southeast said the skilled work force associated with the plant was a factor in local area growth and the positive response from other parts of the country was negligible.

In addition to the indirect effects of geographic location and the attraction of good public services at low tax rates, population, industrial, and commercial growth have also been actively encouraged in the EPZs. 63 percent of the respondents to the BNL survey reported that local government encouraged area growth, 55 percent cited county encouragement, and 45 percent said the utilities encouraged or strongly encouraged local growth.

While the number of site areas experiencing very rapid growth is limited, the potential clearly exists for increased rates of population growth around many more plants in the foreseeable future. For this reason, land use management controls are desirable to restrict excessive population growth, to protect nearby residents and to protect the plant itself from potential external hazards. They would assure that the population and land use characteristics of the original site area would be maintained during the operating life of the plant. However, the development of legally supportable land use controls to protect the public from an accident at a nuclear power plant is difficult because of the uncertainty surrounding the dangers of these plants.

The uncertainty is of two types. First, there is the uncertain probability of an accident actually occurring. Contradictory testimony[1] regarding the likelihood of an accidental release of radioactivity to the atmosphere (and the attendant risk to the public), shows the difficulty of discussing the costs and benefits of regulatory proposals.

The second type of uncertainty is that of the location of the effects of an accident. Since wind direction and other meteorological conditions at the time of a possible accident are largely unpredictable and since danger could occur over wide areas, predetermining the location of impacted areas is impossible. Thus, it may be difficult to establish evidence that a particular location has a high potential of danger that may result in severe impact on life and property and this will make legal support more difficult to find (Freilich, 1964).

AUTHORITY TO REGULATE LAND USE

Another difficulty in the implementation of regulations has been the prevailing system of land use controls in the United States. Land use controls, or their absence, are firmly entrenched at the local level of government. Although repeated calls have been made for regional regulation, there remains strong resistance to imposition of land use controls by the federal government (e.g., Manchester Environmental Coalition vs EPA, 1977) and even by the states (see for example, the case of the New Jersey moratorium on development later in this chapter). Thus, a nuclear plant may be located in one jurisdiction, but the impacts of a disaster will be felt in many others that may have their own ideas about where or how to control growth. Planning in the face of multijurisdictional interests is therefore a difficult task.

The results of the BNL questionnaire confirm this fact. More than 60 percent of the approximately 185 localities responding to the inquiry administer their own local land use plans. Only 32 percent reported county-wide master plans, and few responded that regional planning authority exists or is being contemplated.

Legally and constitutionally, the states hold land use decision-making authority, and municipalities are "creatures of the state". Over the past fifty years or so, states have delegated this authority to local governments through standard zoning enabling acts and similar state laws. A few states, however, have begun to reverse this pattern of total delegation and have passed innovative legislation giving themselves more of

a direct role (Healy, 1976).   County and municipal
governments, on the other hand, typically have shown
great reluctance to initiate any special controls over
population growth in areas surrounding nuclear power
plants.   In many cases a pro-growth orientation has
characterized the local power structure, whose own
economic interests are best served by rapid and often
dense land development.   Elsewhere, even when a few
local officials have expressed explicit concern for
radiation safety and thus perhaps for restraining population growth in the vicinity of the station, the typical response has been "if the reactor was safe enough
for the federal government to approve, who are we to
question the experts?" Though some local officials may
have been willing to support growth-control measures in
the vicinity of the nearby station, they certainly have
not been willing to take this initiative alone.   Thus,
the states have remained the primary arena for action
on population density around nuclear power stations.

A growing amount of land use control has been exer-
cised recently by state governments concerned with
population growth, environmental problems, and public
health and safety. Some control efforts are in concert
with federal legislation regarding wetlands, coastal
areas, and pollution; others are distinctly state
efforts.   Several states have adopted legislation to
regulate the use and development of critical environmental areas within the state; however, in Florida, for
example, no more than five percent of a state can be
designated as critical. States are utilizing hazardous
area zoning increasingly to protect property and population from disasters. We will examine these situations below.

PRECEDENTS

Precedents for the application of land use controls
to manage population density exist, although examples
of controls around nuclear power plants are few.
Analogous situations in areas subject to repercussions
of an accident around non-nuclear hazardous facilities
can also shed light on the applicability of the controls examined later in this chapter.

California provides us with a striking example of
explicit public policy to control post-construction
population growth around a potentially dangerous non-
nuclear facility.   It applied to areas surrounding any
new Liquified Natural Gas (LNG) terminal.   State law
(California State Senate Bill 1081, Chapter 885, 1977)
mandates that the appropriate regulatory and planning
agencies ensure that the population density guidelines
stipulated in siting regulations be maintained during
the operating life of the terminal.   Santa Barbara
County, where this law would first have applied, pro-

posed using existing zoning regulations and provisions of the Coastal Act to meet these requirements (Santa Barbara County 1980).[2]

Zoning legislation and regulations have also been applied to prohibit or regulate certain kinds of development around hazardous chemical waste facilities and airports. Federal regulations under the 1976 Resource Conservation and Recovery Act (40 CFR) require minimum buffer zones for certain hazardous waste facilities, and at least fifteen states[3] have adopted more stringent requirements. As in the case of nuclear siting regulations, however, local regulations, mandated by the state and authorized by the federal government, are applied during siting of these installations but are rarely pursued after operation begins.

Land use around airports has generally suffered from the same lack of post-siting controls. Recently, however, this laissez-faire approach began to change when the state of Missouri passed legislation restricting residential development in airport zones to single family homes on lots of at least ten acres (4 ha) (Missouri Airport Zoning Law 305.405, 1982). Such a land use control strategy, well beyond the 2 to 3 acres more common in large lot zoning, could be similarly employed around nuclear power plants.

Though not yet implemented, the Environmental Protection Agency's proposed Ground Water Protection Strategy (EPA 1980) provides a useful model of combining federal environmental goals with local land use decisions. If implemented, this strategy would induce state and local governments to reexamine and perhaps alter their land use patterns to reflect a concern for ground water protection, especially in key aquifer recharge areas, by incorporating ground water goals in the local zoning and subdivision process.

An example of land use control specifically aimed at limiting population growth around a nuclear power plant occurred in New Jersey. In contrast to the California LNG regulations, the New Jersey controls were not applied until after the facility was operating and population was increasing rapidly. Between 1965, when the Oyster Creek Nuclear Generating Station was granted its operating license, and 1976, the number of residential structures within 4 miles (6.4 km) of the plant increased 193 percent (Morell, 1977). This rapid growth can be attributed to the same factors as cited by respondents to the BNL questionnaire: good geographic location near a major highway (the Garden State Parkway), proximity to major urban centers (New York City and Philadelphia), low taxes due to the fiscal impact of the plant combined with a rural location, and encouragement of development by local officials.

In 1976, the New Jersey Department of Environmental Protection imposed a complete moratorium on any further large-scale residential development within four miles

(6.4 km) of the Oyster Creek nuclear station and within six miles (9.6 km) of the larger Salem plant located nearby. This moratorium was implemented under the authority of the state's Coastal Zone Management program. Under the state's Coastal Area Facility Review Act (CAFRA), all residential subdivisions of twenty-five or more units required a DEP permit and under the moratorium, these permits were denied.

The moratorium was a state, not a local, initiative, and local opposition, mainly from builders, was swift and loud. Adverse reaction led to strong pressure to lift the moratorium, which occurred in 1979, three years after it was imposed.[4] Although development during the time of the moratorium was not completely curtailed--subdivisions of up to twenty-four units were still permitted--it did put a damper on the extremely rapid rate of population growth the area had been experiencing.

The Warren-Alquist Act (A.B. 1575), passed in 1979, legislates land use control around nuclear plants in California. It requires that purchase of development rights or restrictive zoning be applied by the California Energy Commission to areas around nuclear stations to assure that population densities do not exceed the limits set by the U.S. Atomic Energy Commission (now the NRC). However, since the NRC has not yet developed any such detailed population standards for the area around licensed nuclear stations, California has not been able to implement this phase of the act.

No local laws seem to have been enacted specifically to regulate population density or land use because of the presence of a nuclear plant. Cases do exist, however, where land use regulations enacted for other purposes have achieved the effect. One instance has occurred in Waterford, Connecticut. There the community's desire to maintain its rural character caused it to enact large-lot zoning and restrictions on development in fragile wetlands areas. This anti-growth attitude predated the siting and construction of Millstone Nuclear Generating Station Units I and II. The town's land use controls have served to maintain a much lower population density than prevails in surrounding communities (BNL, 1982). A second example of land use controls enacted for reasons other than controlling population growth around a nuclear station is an area of industrial zoning around the Crystal River nuclear power station in Citrus County, Florida. Though not intended as a buffer for a complex which includes the nuclear plant and two other power plants, it has created a narrow ring to a distance of about four miles (6.4 km) from the complex from which residences are excluded (Morell, 1982).

In this section actual applications of land use
controls to maintain low population densities have been
surveyed. The final section presents a more theoretical consideration of the different techniques available
for controlling land use around nuclear power plants.

ZONING AND MANAGEMENT FOR LOW
POPULATION DENSITY

Available techniques fall into three categories.
First, zoning techniques are available for dealing
directly with controlling density. These techniques
are accompanied by the possibility of withholding
services and thereby discouraging development. Such
options involve direct governmental intervention.

The second category involves possible options that
affect the tenure of property by requiring the utility
to purchase either the fee or easements sufficient to
keep the population low. A major advantage of this
approach is that while the government is free to change
the zoning, contractual agreements to keep land in open
space are binding and enforceable in court.

The third option deals with incentives. It includes transferable development rights and possible tax
proposals designed to compensate industry and property
owners for restrictions on development.

The three possibilities are not, however, mutually
exclusive. They can be used individually or in some
combination, and depending on the specific circumstances, the useful combinations can be many and
varied.

Zoning and Public Facilities Control

Several types of zoning are particularly well
suited to the low-density development desired around
nuclear power plants -- agricultural, large-lot,
recreational, and industrial. These types of zoning
create low population areas because they either prohibit residential development or control the density of
development through property size and use requirements.
A favorable response to implementing low-density zoning
controls around nuclear power plants is suggested by
responses to the questionnaire. Over 50 percent of the
jurisdictions report the existence of large-lot zoning
regulations, and 25 percent report agricultural
preservation programs.

Agricultural Zoning. Agricultural zoning is zoning
which prohibits uses other than those that are compatible with an agricultural economy. In addition to nor-

mal farming activities, agricultural zoning districts generally include single-family homes on large lots, and sometimes allow industrialized agriculture such as stock feeding and canneries. Conventional subdivisions and other urban uses are, however, prohibited.

The purpose of most agricultural zoning is to preserve prime agricultural land. However, in rapidly developing areas it has been used to create holding zones in a timed-development plan so that development can be delayed until services can be extended, or until it is desirable to extend the urban area into the previously restricted zone (Freilich, 1964). Another frequent use of this zoning is to restrict development on floodplains and around airports to protect the public from hazards associated with these areas. In 1981, 271 county and municipal governments had passed agricultural zoning ordinances, a great majority in California, Oregon, Wisconsin, Minnesota, South Dakota, Iowa, Illinois, Pennsylvania, and Maryland (Coughlin, 1981).

Most nuclear sites are rural and particularly well suited to agricultural zoning, especially where agriculture is the current land use. In fact, retention of the rural/agricultural nature of land surrounding plant sites is undoubtedly the best strategy for maintaining low population density. In light of the fact that area jurisdictions reported a significant decline in land devoted to agriculture in the EPZ, the existence of land preservation schemes is encouraging. Twenty-five percent have agricultural preservation plans in place, and another 15 percent are considering them (BNL, 1982).

Legally, the designation of an area within an EPZ as an agricultural district will probably pose few problems if farming is a reasonable use of the property. A California court has even upheld an agricultural zoning ordinance when the land was not well suited for farming (Paramount Rock Co. v. County of San Diego, 1960). However, particular care must be taken when unsuitable land is zoned. Unless the compatible uses are clearly suitable and will give the landowner enough return on his land to do more than pay his taxes, the courts may declare the ordinance invalid (One Tier Beyond Ramapo, 1978).

Large-lot Zoning. Where agricultural zoning may not be practical, such as in those areas where the reactor site is less than ten miles (16 km) from a major metropolitan center of more than 25,000 population, or in jurisdictions where agricultural zoning is frowned upon if not used for agriculture, a combination of controls to provide low density development may be possible.

One of these controls, large-lot zoning, generally requires that lots on which residences are to be constructed must be equal to or greater than one acre (0.4

ha). Since homesites are dispersed by this requirement, low residential density is a result. The example
of the use of large-lot zoning in Waterford, Connecticut was cited earlier in this chapter.

Large-lot zoning has received mixed reviews in
court. The courts have examined these ordinances
carefully to determine if they artificially thwart
development and, as a consequence, discriminate against
and exclude racial minorities who cannot afford single-
family homes on large lots. Purpose, then, is an
important factor in determining the constitutionality
of large-lot zoning. If the minimum lot size is kept
small enough so that the lots are marketable,[5] and no
exclusionary effect is identified, the courts generally
uphold the validity of the zoning (Environmental
Considerations, 1974).

As a practical matter, both demographic data on the
active reactor sites and the existence of large-lot
zoning ordinances in more than one-half of the affected
jurisdictions lend support to this technique for
controlling land use around nuclear plants. According
to 1970 census figures, ten of the eleven active
reactor sites located within ten miles (16 km) of a
metropolitan center had an average population density
of less than one inhabitant per acre (2.5 per ha)
averaged over the EPZ. Projected figures to the year
2000 indicate that this ratio is expected to increase
to 1.5 per acre (3.8 per ha) for only three of the ten
(NRC 1979). These figures strongly suggest the existence of some areas of very low density development
despite the presence of one or more metropolitan
centers. If guidelines were to be established setting
a maximum density allowable in specified zones around
the reactor site, minimum lot standards based on these
guidelines might be acceptable and compatible with
these regions.

Alternatively, lot sizes could be based on an
average maximum density over the entire EPZ, thus giving developers an opportunity to cluster residences on
small lots while maintaining the density standards
inherent in a large-lot zoning district. While this
strategy would produce a greater population density at
the site of development, it would maintain overall low
density requirements through the use of open space
around the development. A concentration of population
in one compact area could thus facilitate evacuation in
case of an emergency. This concept has also found
favor with almost 60 percent of the jurisdictions
around the power plants. Although challenged in the
courts on many grounds, the cluster technique has been
widely upheld.

Industrial and Recreational Zoning. In addition to
the residential zoning possibilities mentioned in the
preceding discussion, some areas of the low population

zone might be suitable for industrial development. If the nuclear power plant is located in an area zoned for industrial use, the inclusion of compatible industries in this district should not be objectionable (cf. Crystal River, Florida, cited above). Some restrictions on the number of employees who work in the industrial zone might be necessary, however, to assure safe and prompt evacuation if required.

Recreational zoning could allow for parks, golf courses, playing fields, riding trails, tennis clubs, camping grounds, and other facilities. These uses are frequently permitted in floodplains and around airports to allow use of the land, while protecting the public from health and safety problems associated with other types of permanent development.

Since construction of the nuclear power plants began, the amount of land area actually zoned or authorized for commercial, industrial, and recreational uses increased in all regions. Fifty-three percent of the jurisdictions reported an increase in commercial, 48 percent in industrial, and 39 percent in recreational zoning, an indication of a widespread use of these zoning classifications in the study areas.

Most of the types of low density zoning discussed thus far are often included under the broad heading of "open space." Agricultural land, houses located on large lots, public parks, and private recreational facilities such as golf courses or riding trails are examples of open space. In all but the public parks example, this open space has been created by private landowners, often under the requirements of zoning regulations that have been upheld in the courts despite charges of severe reductions in property values. However, a point exists beyond which the police power cannot be used to accomplish development goals. If property is "taken" through regulation so that no permitted use or value remains, the governing body must pursue other avenues. Thus, if the goal of low-density development around nuclear power plants cannot be accomplished by utilizing zoning techniques alone, a combination of regulations, public facilities control, and incentives may be required to keep the land substantially undeveloped for the extended periods of time demanded by the operating lifetime of the nuclear power plant.

Public Facilities Extension Control

A growth management device that could complement and help enforce zoning regulations around nuclear power plants is the controlled location of extensions of public facilities, particularly sewers and highways. Governments can direct development to areas where higher density development is more desirable by providing public facilities and services to areas outside the designated "low population zone" around the reactor site. Governments can also restrict development in this area to non-intensive uses such as low-density residential or agriculture by withholding public facilities and services from the control zone (EPZ). Several regional and one statewide growth management system utilize this technique as a critical part of their programs.[6] These programs have tried to neutralize the legal problems inherent in any attempts to withhold public services by carefully defining urban growth boundaries. The existence of strong state statutes and the consistent policies of commissions that administer the programs have further strengthened their legal position (Coughlin, 1981).

Two utility extension control strategies have been proposed to manage and direct growth. First, the providers of public facilities could refuse to extend services to areas within the service area where development is not wanted. Second, connection fees to users could be based on a graduated scale according to the "desirability of the development from a land use planning viewpoint" (Deutsch, 1978). "Desirability" would be defined in terms of location, density, need, and other factors.

In the absence of utility-related reasons, such as lack of capacity or financial crisis, the first strategy is likely to suffer in the courts at the hands of opponents who argue that public service providers "must as a matter of equal protection and common-law duty extend facilities" (Freilich, 1964). The common law duty of public service providers, to provide and extend service to everyone in the service area has been well established regardless of whether it is to public utility corporations, service districts, or municipalities. The only exceptions to this rule apply in cases where there is lack of capacity or a financial crisis, or when the expense of providing the service would far exceed the revenues to be gained.

Municipalities that provide public services have argued unsuccessfully that the rules which apply to municipal corporations should also prevail in considering service extensions. They argue that cities should have the right to predicate service extensions on sound land use planning considerations. But the courts continue to uphold the prevailing view that in their capacity as public service providers, cities can only withhold services for utility-related reasons.

The second strategy, differential connection fees calculated on the desirability of the project and not on the present cost-based system, has not received wide attention in the literature, and attempts to impose "desirability-based" fees are not well documented. Differential connection fees based on service-related factors have received wide acceptance throughout the United States and have been sustained in the courts. However, when these fees are unrelated to service costs, the courts have denied authority to impose them. How the courts will view "desirability-based" connection fees may turn on whether they are "just and equitable." But court acceptance will not be obtained easily.

The body of case law relating to these two strategies does not reveal any softening of the hard line taken against manipulating utility extensions to accomplish land use planning objectives. Nonetheless, the courts might conclude that the potential danger from accident at a nuclear plant is sufficiently high that an exception to the prevailing rules could be made. If so, this method would be a powerful growth management tool in urbanizing areas around reactor sites.

LAND TENURE: EASEMENTS, FULL FEE ACQUISITION, AND EMINENT DOMAIN

Other control methods are available for that property around nuclear power plants where zoning alone may prove inadequate to control development. Two land tenure devices that would provide varying degrees of control over land development around reactor sites are the purchase of easements and the acquisition of full-fee interests in property. Although these techniques offer more permanent control than zoning, their effectiveness may possibly be realized only if these property interests can be obtained through exercise of the power of eminent domain. Thus, an analysis of the eminent domain issue for each technique is included.

Easements

The purchase by the utility company of a less-than-fee interest in property around the reactor site could take the form of development rights easements. These easements would place a restriction on the land that would prohibit development of the property beyond that deemed appropriate for maintenance of low population density. California, the only state to adopt land use controls around nuclear power plants, requires purchase of development rights when the State Energy Commission determines that existing land use controls are insufficient to guarantee maintenance of population levels and land use during the lifetime of the plant. A court recently declined to rule on the legality of this requirement by concluding that it was not "ripe" for discussion.

The restriction of development on private property through easement agreements has been utilized by state and federal governments and upheld in the courts to preserve scenic vistas, to provide protected buffers near national monuments and parks, to conserve natural conditions around seashore and lakeshore areas, and to protect scenic and wild rivers and other natural resources. Taking a cue from the success of these programs, many states have recently instituted easement programs to preserve farmland and open space (Roe 1976).

An easement agreement may be attractive to the utility company for several reasons: (1) it may ease the financial burden of acquiring the fee interest; (2) title and possession, as well as the expense of maintaining the land, remain with the owner; and (3) land so restricted remains usable within the constraints of the low density development requirements. For the landowner, an easement may be preferable to outright purchase because he keeps title and possession of his property and at the same time may realize a significant reduction in either income or property taxes due to the decreased development value of his land. Legally, the type of development rights easement that the utility would purchase would differ by common law definition from the conservation easements mentioned previously. Thus, the potential legal problems envisioned by the proponents of conservation easements should not be encountered.[7]

This plan is not without costs, however. Purchasing a development rights easement in an area subject to strong development pressure may be as expensive as the full fee. Reducing property or income taxes for the landowner reduces the tax base of local communities and places a heavier burden on other taxpayers.

Another potential problem with the easement method is that some owners will invariably be reluctant to surrender development rights even when the value of their rights is low. If the utility company must rely on the discretionary acquisition of easements, gaining control of development in designated sectors may be haphazard. Therefore, the power to acquire easements by eminent domain may be essential. In most states, gas and electric light and power companies have been granted the power of eminent domain by statute. Although these statutes generally include the right to condemn an easement or right-of-way over private land for erection of transmission lines, the right to condemn property for the plant site, and the right of entry, only California has specially addressed the

condemnation of development rights. However, a survey
of eminent domain legislation granting utilities this
power reveals that many of the state statutes could be
construed to permit the acquisition of development
rights if acquisition could be proven to be necessary
to the operation of the plant. In those states with
more specific language, amendments to the existing
statute may be required to extend the power of eminent
domain to the acquisition of development rights easements.

An alternative that could be considered is the
acquisition of these easements by a government entity.
Maryland has given the Secretary of the Department of
Natural Resources the power to condemn and purchase
potential power plant sites, which are then available
for purchase or lease to public utilities. Funds for
this acquisition are collected by a surcharge on kilowatt hours generated. The adaptation of this type of
scheme to the acquisition of development rights easements could be investigated. The legal issues raised
by private/public cooperation have been aired recently
in the courts with mixed results. However, in other
contexts in the past, this power has been upheld.

Whichever path is taken--utility or government
acquisition--the purchase of development rights easements offers an equitable means of controlling population density around nuclear power plants, particularly
in those areas with high development pressure. Easements offer a more reliable and permanent control than
zoning and would provide tighter controls for those
sectors requiring it, such as areas nearest the reactor
site. The permanency of easements, plus the power to
obtain them by eminent domain, would assure complete
control of designated sectors over the lifetime of the
plant.

The principal disadvantage of this method is also
one of the principal advantages--the issue of eminent
domain. Condemnation is expensive, unpopular, and time
consuming. Other disadvantages include the substantial
impact on utility rates which might occur as a result
of acquisition costs, and the necessity of acquiring
easements far in advance of the construction and
operation of the plant so as to assure maintenance of
the site's low density characteristics.

Utilization of this method of controlling density
around nuclear power plants is not contingent on local
legislation, but it is interesting to note that conservation easement programs in jurisdictions around these
plants are growing in all regions. According to
respondents to the questionnaire, more than 30 percent
have easement programs, with entities in the northeast
approaching forty percent involvement. This indicates
that institution of a development rights easement
program might receive support.

Full Fee Acquisition

Customarily, utility companies have acquired the full fee to property within the required exclusion area around their reactor sites. The extension of acquisition beyond this perimeter could be one of the mix of land use control devices around the site. One of the advantages of this technique is readily apparent; the utility company would acquire complete control over activities on the surrounding property. Other advantages which might accrue to property owners are: (1) The owners of land in the impacted zones would not be reluctant or captive property owners in a potentially hazardous area. They could choose to remain on the land by leasing it from the power company with an option to buy it back when the site is abandoned, or they could choose to leave. (2) Property owners who fear that their property would be reduced in value due to its proximity to the plant site may be able to prevent a loss by selling to the utility company directly.

The arguments against this technique, however, are numerous: (1) Control over development can be achieved through less costly means such as those mentioned elsewhere in this chapter. (2) Unlike other techniques, title would transfer to the utility company and along with it, the necessity of upkeep, the responsibility for paying real estate taxes, and the potential for becoming landlord over thousands of acres. (3) The exercise of eminent domain would almost certainly be required, and acquisition of the full fee through condemnation may present legal obstacles not present in less-than-fee techniques. (4) As with easement acquisition, costs may have a substantial impact on utility rates.

If this technique were to obtain the support of the NRC and the utility company, successful implementation undoubtedly would hinge on the acquisition of the fee through condemnation. As mentioned in the previous paragraph, however, legal obstacles appear to exist. The principal guideline used by the courts in determining whether the full fee or an easement can be acquired is whether an easement will satisfy the needs of the condemner; the courts prefer the taking of an easement to the taking of a fee simple interest. Since other techniques already exist that can accomplish the goal of low density development without full fee acquisition, proving that a full fee interest is essential may be difficult. In order to legally sustain the use of the power of eminent domain to purchase the fee to property outside the exclusion zone, the NRC may have to promulgate regulations requiring the extension of exclusion zone controls to a wider area. However, due to the substantial amount of land that may fall under the impacted zone, full fee acquisition is probably not feasible nor warranted beyond the exclusion zone, unless unusual circumstances exist.

FINANCIAL INCENTIVES: TRANSFERABLE DEVELOPMENT RIGHTS AND PREFERENTIAL TAX ASSESSMENT

Two types of financial incentives that have been used in land conservation and preservation schemes are also potentially applicable to land controls around nuclear power plant sites. In exchange for restricting development on their property, land owners could receive transferable development rights and/or preferential tax assessment. The former would allow owners to realize some of the development potential of their property either by transferring the development rights to another piece of their property or by selling them to another property owner outside the impact zone. The latter would permit the property to be assessed at its use value rather than development value and result in either lower real estate taxes or income tax adjustments.

Transferable Development Rights

Landmark agricultural and open space preservation laws occasionally use transferable development rights (TDRs) as a means of compensating property owners for leaving their property undeveloped. By severing the development rights from the property and allowing them to be sold and transferred to developers in other designated zoning districts, TDRs compensate the property owner for the restrictions placed on his property.

Several cities and counties in the U.S. have enacted preservation ordinances with a provision for TDRs. Between 1972 and 1980, ten municipalities and two counties reportedly incorporated the TDR scheme in programs to preserve farmland and other open space (Coughlin, 1981). However, as of mid-1981, only four TDR transactions had occurred. New York City was a leader in implementing the scheme in the early 1970s by making TDR a provision of its Landmarks Preservation Law and Special Park District ordinance. Both of these ordinances were challenged in the courts, but the resulting opinions left the constitutionality of the TDR scheme in doubt. Supporters of TDRs, however, believe they will eventually pass the constitutionality test in the courts. In a 1983 Florida Court of Appeals case, the court validated the TDRs provision of a growth management plan designed to reduce density along a city's coastal property, lending support for this belief (City of Hollywood v. Hollywood So., 1983).

However, whether TDRs reach their potential will be determined as much by their ability to achieve marketability as by an affirmative ruling for constitutionality.

If the difficulties plaguing TDR plans can be resolved, this technique appears to have potential application for controlling land use around nuclear power plants. Property owners within the low population zone could be allowed to transfer their development rights to land outside the zone. If the prevailing winds were such that a particular sector within the control zone required stricter density controls than other sectors, the development rights might be transferred to another sector within the control zone as well as outside the zone.

Although the TDR concept has been gaining popularity, only fifteen percent of the 163 jurisdictions responding to this question in the BNL survey reported TDR programs in existence or development. Although this finding does not preclude implementation of such a program to maintain low-density development around the plants, the lack of implementation elsewhere indicates that various difficulties may exist with the TDR scheme. Recommending this scheme for controlling land use around these plants before the difficulties are identified and resolved would therefore be premature.

Preferential Tax Assessment

Whether development rights are surrendered temporarily through easement agreements or zoning regulations, or permanently through transfer, most conservation and preservation programs offer tax reductions as both incentives and compensation for relinquishing these rights.

To property owners whose lands are subject to severe developmental pressure, the reduction in property taxes can be a compelling incentive to participate in a development rights control program. As the market value of surrounding properties escalates, it forces the value of all property to rise with a corresponding escalation in real estate taxes. This substantial property tax burden often forces farmers to sell their land to developers rather than continue to farm. With development rights severed from the property, real estate tax assessments reflect the decrease in market value as well as the elimination of potential market appreciation.

Most property tax reductions are conditioned on a pledge that the land will remain undeveloped for a stated period of time. If the property owner sells his land for development before the agreed upon termination date, he may be penalized by having to pay all or some of the deferred taxes, or a real estate transfer tax may be imposed. Unfortunately, these penalties have not been found to be a particularly effective deterrent, as the profit to be made from development generally far exceeds the penalty.

Regardless of the questionable effectiveness of differential taxation as a land preservation technique, its popularity has prompted forty states to pass legislation authorizing differential taxation to accomplish land use goals, including several states which have had to pass constitutional amendments in order to overcome legal barriers.[8] Implementation of this technique has filtered down to the areas around nuclear power plants. Nearly 30 percent of the jurisdictions in the BNL survey reported the existence or development of a preferential tax assessment program.

Since state courts have been willing to recognize the authority of state and local legislatures to classify, state legislatures are allowed latitude in differentiating among different types of land use for taxing purposes. By employing this latitude the legislature could possibly create a classification such as "Low Density Development Due to Hazardous Conditions" District, and all landowners in such a district would be given preferential assessment. This approach might be considered for areas under severe developmental pressure, but because most active sites are located in rural or low density residential areas, existing programs will probably suffice for controlling density around reactor sites.

Without question, differential tax treatment should be considered as full or partial compensation to landowners whose property use is severely restricted as a result of their proximity to a reactor site, especially since no legal obstacles apparently exist to implementing such a plan.

CONCLUSIONS

Changes in population density and land use are occurring in the vicinity of nuclear generating stations. Some of these changes have led to an erosion of the AEC/NRC's original intent to use population distribution and distance as a defense-in-depth to supplement engineered safety features. The majority of station areas are experiencing change that is in line with the predictions made during siting, but a few station areas are experiencing unanticipated changes. Recent concern over the safety of nuclear energy has surfaced with incidents at Three Mile Island (see Cutter, Chapter 10), Salem, Browns Ferry, Ginna, and Indian Point, as well as construction problems at South Texas, Zimmer, and Diablo Canyon (see Pijawka, Chapter 9). Rapid

population growth in recent years, coupled with concern
over insufficient highway capacity has sparked heated
debate over the licensing of Shoreham (See Johnson and
Zeigler, Chapter 12). Therefore, interest is being
expressed in what factors cause land use change in the
vicinity of nuclear stations and what options are
available to utilities and state and local governments
for maintaining a low population density or ensuring
slow population growth.

Nuclear plants themselves do not appear to be major
instigators of local population growth in their immediate areas, although station tax payments may enable
communities to provide amenities which serve to attract
growth. In recent years, however, local area growth
around nuclear facilities has increased at an accelerated rate, and most currently operating plants have the
majority of their operational years remaining. Careful
assessment of available land use control mechanisms is
therefore both necessary and timely.

Land use controls to limit population density
around several types of hazardous facilities have been
used or are implemented throughout the United States.
Zoning regulations have been proposed and enacted for
use around potentially dangerous waste facilities, liquified natural gas terminals in California,
and airports in Missouri, as well as for the protection
of aquifers and environmentally sensitive areas. A
temporary moratorium on large-scale residential development existed for several years around two nuclear
sites in New Jersey. And inadvertent controls, enacted
for reasons not linked to the presence of a nuclear
power plant, have limited population densities around
one station in Connecticut and another in Florida.

We have examined a variety of land use control
options and have discussed their potential for use by
utilities and state and local governments around
nuclear facilities. These controls include zoning and
public facilities control, land tenure mechanisms, and
financial incentives. A mixed strategy utilizing
different types of controls in different situations
appears to be the most useful method of achieving the
desired goal of maintaining or limiting population (see
Table 6.1).

Techniques alone, however, are insufficient. As
seen in the New Jersey example, belated application of
growth control mechanisms in already rapidly-growing
areas can create political havoc and ultimately lead to
the abandonment of the controls. An overall strategy,
initiated during the siting stage of nuclear power
station development and pursued through the operating
life of the plant, seems essential. If population
growth is to be controlled successfully in concert with
NRC safety goals, a carefully orchestrated mixture of
site selection criteria, benefit distribution, plan-

TABLE 6.1
Land Use Controls Available for Growth Control

Technique	Rationale	Advantages	Disadvantages
	1. Land Use Controls That Create Low Density Development		
Open Space			
Agricultural Zoning	Often used to contain urban areas. Has been applied around airports. Large lot zoning often permitted under this designation.	Allows land to be income-producing.	Land may not be suitable for agriculture. Some restrictions on type of agriculture may be needed.
Large-lot Zoning	Promotes low density development.	Allows development but restricts density.	Widely scattered populus might complicate evacuation procedures.
Industrial Zoning	Prevents residential development.	Allows use of land.	Land may not be suitable for industrial use. May not be a demand for land in this zoning classification.
Recreational Zoning	Prevents residential development.	Allows use of land.	Some recreational uses such as golf, riding, or hiking would scatter users over terrain and complicate emergency notification.

Cluster Development	Allows developer to cluster residences more densely in exchange for open space. May be suitable in outer rim of low population zone.	Any high density development regardless of location, may not be desirable.
Withholding Public Utility and Municipal Services	Without public services, only low density development could proceed.	Land would remain in unintensive use.

2. Land Tenure Devices

Easements	Allows control of development through purchase of development rights to property.	Purchase of property not required, leaving title and possession with private owner. Often less expensive than purchase of full fee. Landowner may qualify for reduction in taxes. Easements can be held in perpetuity.	Easement purchase is sometimes as costly as full fee. Some easements may have to be acquired through eminent domain.
Eminent Domain	Allows development control of all property in designated zone, even when owners are reluctant to relinquish rights.	Permits effective control of all development in designated zone.	Costly, unpopular, and time consuming.

(Continued)

TABLE 6.1
Land Use Controls Available for Growth Control (Cont.)

Technique	Rationale	Advantages	Disadvantages
Full Fee Acquisition	Allows utility company to determine all activities on the property.	Would provide the ultimate control because the utility company would determine all activities on the property, including the right to exclude or evacuate persons	May present legal obstacles if condemnation of full fee is required. Most expensive alternative. Would transfer title of thousands of acres from private owners to utility.
	3. Techniques Used to Compensate Property Owners for Severe Regulation		
TDRs	Allows transfer of development potential from one zoning district to another. Very good possibility with Open Space zoning.	Allows property owners to retain some value of development potential.	No sites may be available for transfer. May be problem with marketability of TDRs.
Preferential Tax Assessment	Land zoned for agriculture and open space are sometimes given special assessments.	Resulting lower taxes are often considered adequate compensation for prohibiting development.	Taxing districts have decrease in revenues resulting in heavier burden on other taxpayers.

4. Land Use Controls That Protect the Public

Airport Zoning	Used to protect public from dangers associated with aircraft operations as well as excessive noise levels.	Often allows large-lot development, as well as agricultural or industrial use restricted zone.	Property owner may challenge on grounds of reduced development value of property, or argue that permitted uses are not possible.
Floodplain Zoning	Used to protect public health and safety, persons and property.	Some uses can generally be made of the land--agriculture, recreation.	Severe impact on property owner may result in "taking" challenge to restrictions. Courts require high forseeability of danger before upholding ordinance.
Development Control Around Hazardous Facilities	Achieve protection for the public from dangers associated with hazardous facilities, such as hazardous waste sites and liquified natural gas terminals.	Prevents development before the fact by prohibiting the location of a site in a densely populated area.	Due to the nature of the hazard, no viable use of land may be permitted resulting in need to purchase title to land outright.

ning, and legal guidelines by all concerned utility companies and state, local, and federal governments is necessary.

NOTES

1. An NRC reactor Safety Study (NRC 1974) predicted that melting of the reactor core is expected once per 20,000 years of reactor operation. Yet recently, two public interest scientists estimated that the study could be wrong by a factor of 100 or more (Wald, 1982) and that a major release of radioactivity in an area could render an area of 5300 square miles (13,727 sq km) uninhabitable for decades.
2. The long-term population control provisions of the LNG Terminal Siting Act remain untested because, for various reasons, the terminal proposed in 1978 has not been built.
3. Arkansas, California, Georgia, Illinois, Kansas, Kentucky, Louisiana, Maine, Michigan, Mississippi, Missouri, New Jersey, Oklahoma, South Carolina, and Wisconsin.
4. Safety modifications undertaken at the station also played a role in lifting the moratorium.
5. A comment in Urban Law Annals (1978, p. 377) suggests a five acre (2 ha) maximum lot size for nonagricultural land to ensure marketability. A California court (Eldridge v. City of Palo Alto) invalidated a ten acre (4 ha) minimum lot size, principally due to its concern about the marketability of the lots.
6. State of Oregon; Metropolitan-Dade County, Florida; Lexington-Fayette Urban County, Kentucky; Twin Cities, Minnesota.
7. Conservation easements are easements "in gross" under the common law because they do not benefit another parcel of land. Problems of enforceability and assignability of these types of easements have led to the passage of statutory authority for conservation easements in more than forty states to avoid these problems. However, the development rights easements the utility would purchase would be defined as "appurtenant," benefitting another parcel of land; under common law these easements run with the land and are assigned with relative ease.
8. States that passed constitutional amendments are Kansas, Maryland, and California.

REFERENCES

Brookhaven National Laboratory. 1982. Under a grant from the NRC in 1982, Brookhaven National Laboratory conducted a survey of 204 entities within a ten mile (16 km) radius (the Emergency Planning Zone) of the nation's forty-nine operting nuclear power plant sites.

Burchell, R.W. with J. Nemeth. 1982. Fiscal Conditions of Nuclear Plant Sites -- A Survey of Fiscal Conditions (I); Case Studies of Revenue and Expenditure Disposition (II). Report submitted to Brookhaven National Laboratory.

City of Hollywood v. Hollywood, Inc. 432 S. 2d 1332 (1983)

10 Code of Federal Regulations. Atomic Energy Act of 1954. Reactor Site Criteria, parts 100.10, 100.11 (rev. 1980).

40 Code of Federal Regulations. Resource Conservation and Recovery Act of 1976. Parts 265.176, 265.198B (rev. 1980).

Coughlin, R.E. and J.C. Keene. 1981. "The Protection oF farmland: An Analysis of Various State and Local Approaches," Land Use Law and Zoning Digest, 33(6):8.

Deutsch, S.L. 1978. "Capital Improvement Controls as Land Use Control Devices." Environmental Law 9:70.

Eldridge v. City of Palo Alto. 1976. California Reporter 129:575-584.

Environmental Considerations: New Arguments for Large Lot Zoning (comment), Urban Law Annual, 8:377.

Freilich, R.H. and J.W. Ragsdale. 1964. "Time and Sequential Controls -- the Essential Basis for Effective Regional Planning: an Analysis of the New Directions for Land Use Control in the Minneapolis/ St. Paul Metropolitan Region, Minnesota Law Review, 58:1009, 1079-1080, quote on p. 1077.

Healy, R.G. 1976. Land Use and the States. Washington, D.C.: The Johns Hopkins University Press for Resources for the Future, Inc.

Manchester Environmental Coalition v. EPA. 1977. Environment Reporter Cases 14:1004-1008.

Morell, D. 1977. Housing Growth Around the Oyster Creek Nuclear Power Plant. Unpublished manuscript.

Morell, D. and G. Singer. 1982. Land Use Controls in Communities Near Nuclear Power Plants -- a Policy Analysis. Report submitted to Brookhaven National Laboratory.

One Tier Beyond Ramapo: Open Space and the Urban Reserve (comment). 1978. San Diego Law Review, 15:1211.

Paramount Rock Co. v. County of San Diego. 1960. California Reporter, 4:317.

Santa Barbara County. 1980. _Coastal Plan_. Santa
    Barbara County Planning Department.
United States Environmental Protection Agency (EPA).
    1980. _Proposed Ground Water Protection Strategy_.
United States Nuclear Regulatory Commission (NRC).
    1979. _Demographic Statistics Pertaining to Nuclear
    Power Reactor Sites_. Office of Nuclear Reactor
    Regulation, NUREG-0348.
United States Nuclear Regulatory Commission (NRC).
    1974. _Reactor Safety Study: An Assessment of
    Accident Risks in U.S. Commercial Nuclear Power
    Plants_. WASH-1400.
United States Nuclear Regulatory Commission (NRC).
    1975. _General Site Suitability Criteria for Nuclear
    Power Stations_. Office of Standards Development,
    Regulatory Guide 4.7.
United States Nuclear Regulatory Commission (NRC).
    1979. _Report of the Siting Policy Task Force_.
    Office of Nuclear Reactor Regulation, NUREG-0625.
United States Nuclear Regulatory Commission (NRC).
    1980. NRC Appropriation Authorization, Section
    108(e). PL96-295.
Wald, M.L. 1982. "2 Scientists Cite Nuclear Plant
    Perils," _New York Times_, July 9, 1982, p. B3.

Introduction

Clark Prichard

It is not usual to find discussions of nuclear power in the context of geography. One would have a much greater expectation of finding the subject of nuclear power treated in the context of engineering and the physical sciences, or possibly, technology assessment. However, upon reflection, there are good reasons why the geographic aspects of nuclear power plants should be worthy of interest. The location of a nuclear electric generating station is a key consideration for a variety of reasons. The rationale for the construction of a nuclear power plant is to produce electric power to run homes, factories, businesses, public services, and farms. The closer the generating station is to these load centers, as concentrations of electric energy users are called, the more efficient is the delivery of electric power from producer to consumer. Typically though, densely populated areas correspond to load centers. For safety reasons, nuclear power plants are restricted from locations where population density is too great. Thus, nuclear power plants tend to be located near, but not too near, major metropolitan areas.

Numerous safety and environmental factors constrain nuclear power plant siting. For example, nuclear power plants are large users of water for cooling. They need to be located on or near a large source of water unless special provisions are made in construction. Most nuclear power plant sites are on the coasts of oceans or lakes or by major rivers. Also, geologic features, such as major faults or earthquake-prone areas, rule out many locations.

All of the factors which help determine the location of a nuclear power plant site ultimately influence to some degree the social, demographic, and economic impacts of the plants, the general topic of the papers in this section. Focusing on these types of impacts for facilities such as nuclear power plants became commonplace during the 1970s. The "energy crisis" of

165

the early part of that decade spawned many energy-related construction projects. An upsurge in the number of planned nuclear power plants was only one aspect of this; other major trends included an increase in oil and gas exploration projects, the opening of large coal mines in the Great Plains and Rocky Mountain regions, planned expansion of port and pipeline facilities, planned synthetic fuels projects, as well as an increase in plans for adding capacity in coal-fired electrical generation, either from new units or conversion of oil-fired units. This activity coincided with an increased awareness of environmental concerns among the public, as well as the introduction of legislatively mandated formal environmental review of major actions likely to affect environmental quality. As a result, all environmental effects of energy projects attained an importance not previously given them. Among these, the socioeconomic effects of large-scale projects became a controversial subject, in large part due to seemingly disruptive effects on local communities from western coal development. Other types of energy-related projects also came under scrutiny for potential adverse socioeconomic effects--housing shortages, fiscal strains on local governments, disruptions to the local social structure and others--such as were observed in western so-called "boom towns."

A nuclear power plant is a major energy facility, employing several thousand construction workers during peak construction, taking up to a dozen years to build, and requiring several billion dollars to finance. In addition, large portions of the public view nuclear energy as a controversial issue, regarding its safety. Expectedly, particularly that the building and operation of nuclear power plants, have a major effects on the surrounding area. In addition to the major impacts on the local economy and population from any large energy project, the special concerns about nuclear power expand the range of possible socioeconomic effects by a considerable margin. A nuclear power plant should, therefore, offer a researcher in socioeconomic impact analysis an extremely rich subject.

David Pijawka makes the most of this opportunity in the first of the three chapters included in this section. His work focuses on the socioeconomic impacts at a dozen nuclear power plant sites in the United States. The methodology used is rigorous, and allows comparisons between sites, and worthwhile overall conclusions to be drawn. While socioeconomic impacts of different types--demographic, economic, local government finances, local institutions, and community services--are summarized, the focus is on public sector effects and social impacts. The larger Post-Licensing Study from which this paper is derived is a major step forward in socioeconomic impact assessment. Careful attention is given to the questions of attribution of impacts to causal factors, baseline projections, perceptions of impacts vs. reality, and social assessment aspects which are often neglected. This chapter represents a thorough examination of the impacts on the local area of the construction and operation of nuclear power plants and provides insights on approaches to technology assessment in general.

The Greenberg, Krueckeberg, and Kaltman chapter explores trends in population change in the immediate vicinity of nuclear power station sites during the 1960 -1980 period. Nuclear plant sites have always been selected with consideration given to avoiding densely populated areas. Since a nuclear reactor receives an operating license for forty years, projections of population growth in the vicinity of the reactor are an important part of the licensing process. The principal questions regarding population growth in the vicinity of nuclear power stations are: (1) What has been the magnitude of population change at operating nuclear station sites to date? (2) To what extent does the construction and operation of a nuclear power station have an influence on population change? and (3) Have population projections made at the time of site selection and licensing turned out to be relatively accurate?

The first of these questions is one which basically asks if population change to date is so pronounced as to indicate that a safety concern may exist or is likely to exist in the near future. If what were thought to be sparsely populated areas are turning into densely populated areas, then this would seem to be the case.

Population density criteria used by the Nuclear Regulatory Commission in licensing nuclear reactors consist of an "exclusion zone" which is the area immediately around the reactor and inside which the licensee controls access. Residence within the exclusion zone is normally prohibited. Just outside the exclusion zone there is a "low population zone," which can contain a residential population. However, this population must be small enough that adequate protective measures can be taken on their behalf in the event of a serious accident. There is no specified population limit for the low population zone; the zone is specified in terms of maximum exposure to individuals at the outer boundary of the zone resulting from a release of radioactive material. There is an added criteria specifying the permitted distance of a reactor site to a population center (one containing more than about 25,000 residents). This distance must be at least one and one-third times the distance from the reactor to the outer boundary of the low population zone.

These criteria involve areas which are very close to the reactor site, usually within several miles. Population density in areas somewhat further out from the reactor site is given consideration in the alternative site review analysis done as part of the implementation of the National Environmental Policy Act as it applies to nuclear facility licensing. If the population density exceeds 500 per square mile (192 per sq km) averaged over any radial distance to 30 miles (48 km) from the reactor site, additional consideration is given to alternative sites. The 500 per square mile density threshold at the time of licensing is supplemented by a 1,000 per square mile (385 per sq km) threshold for projected populations over the lifetime of the facility.

Greenberg, Krueckeberg, and Kaltman present actual population growth data over the past two decades for forty-nine nuclear station sites. Ideally, data would be used which corresponded exactly to the population density criteria referred to, but population data are not collected in this way. Nevertheless, the authors have done a meticulous job of matching minor civil subdivision population data to nuclear station site geographic areas. Their results allow some conclusions to be drawn regarding the issue of what the actual magnitude of population growth has been.

The authors use a method which enables disaggregation of total population growth into a national, regional, county, and a purely local component. This breakdown permits the analyst to identify those sites where local factors seem to be influencing population growth to a greater degree than those associated with larger geographic areas.

A great many operating nuclear power plants have provided county and local governments with huge tax revenues. These tax payments by the utility operating the nuclear plant have enabled county and local governments to keep tax rates comparatively low, while at the same time providing an exceptionally high level of services. This situation might very well have a significant impact on population change in the local area. Low taxes combined with good schools and other local services could be a magnet for local population growth. Thus, the presence of the nuclear plant itself could be a major factor leading to above average rates of population growth in nearby communities. Pijawka's cross-site study notes that not all nuclear plants produce large local tax payments. Many states have legislation which "spreads out" utility tax payments to the remainder of the state. Also, some plants are owned by public entities and do not, therefore, pay local property taxes.

One important manifistation of socioeconomic impacts of nuclear power plants is found in attitudinal changes in the public sector. The final chapter in this section, "Public Response to the Diablo Canyon Nuclear Generating Station" is a longitudinal study of such changes, from planning stages to the present. Attitudes of local residents toward the Diablo Canyon plant underwent drastic changes over a twenty year period. Pijawka explains these changes by linking them to various causal factors. The study differentiates attitudinal changes which are due to national trends in attitudes about the environment and nuclear technology from those which seem to be due to events specific to the Diablo Canyon plant, such as the controversy over the seismic safety issue.

7

Public Sector Effects and Social Impact Assessment of Nuclear Generating Facilities: Information for Community Mitigation Management[1]

K. David Pijawka

The socioeconomic effects of nuclear generating stations emerged as a major regulatory concern in the late 1970s. This was a reflection of three interrelated factors. First, while public concerns and fears were expressed over potential environmental and safety impacts of plant operations, there was no comprehensive information base available on how the public, local economies and social institutions were affected by nearby operating nuclear facilities. In addition, there was a growing awareness that the few available studies (1) failed to address the complete array of impacts (particularly social and evaluative consequences), (2) transferred findings from impacts of major energy development in the Western U.S. to nuclear power plants, which had the effect of overestimating the enconomic advantages of nuclear plants, and (3) were inconsistent in methodology for assessing projective impacts (Chalmers et al., 1982; Bergmann and Pijawka, 1981; Pijawka and Chalmers, 1983).

Second, by the late 1970s, socioeconomic assessments began to place emphasis on those elements that were important to impact mitigation planning and growth management. For regulatory agencies and local communities faced with plant sitings, it became critical to understand the community effects of nuclear plants throughout their construction and operation and during accident events. This understanding, however, was lacking. The trend toward socioeconomic impact monitoring presented a need for identifying the salient adverse effects for mitigation on a continuing basis. Third, the view that the 1979 accident at the Three Mile Island nuclear station would alter many of the existing socioeconomic patterns around nuclear facilities (because of expanded plant retrofitting efforts, new regulations, emergency planning and land use controls and changing public perceptions of risks and benefits) catapulted socioeconomics both as a worthwhile area of study in "technology assessment" and as a critical element in policy formation.

This chapter reports the findings of a major socio-economic impact assessment of nuclear power plants for the U.S. Nuclear Regulatory Commission. There were three prime objectives of the study: (1) to investigate the nature and magnitude of the effects of construction and operation of twelve commercial nuclear facilities through retrospective analyses; (2) to develop an operational approach that would link measurements of traditional economic-demographic elements (employment, income, population housing, and public sector effects) to aspects of social structure, public evaluation and political response; and (3) to test the validity of the methodology and to evaluate other approaches to be used in projective assessments of impacts of hazardous technology.

The assessment of nuclear plant impacts as described in this chapter focuses on three areas that are particularly germane to nuclear plants both as an industrial facility and as a hazardous technology and are relevant to technology assessment issues in general. The first describes the effects of the plants on the provision of community public services in terms of the revenue-expenditure balance. The socioeconomic impact literature is seriously concerned about community coping and the management of growth from energy projects and particular attention is being paid to industry-community strategies for scheduling and determining the magnitude of revenues needed for mitigation planning. Closely related to public sector impacts are the questions of how these plants have affected social structure and political processes and the public perception of nuclear technology and facilities in the context of risks and benefits. The social impact component of the study included the effects of the nuclear plants on community social structure, an evaluative/perceptual dimension, and public response.

PUBLIC SECTOR IMPACTS

One of the major issues in community impact management is the gap between revenues generated by energy projects and expenditures for public facilities and services because of project-induced growth. Of issue is the experience of communities experiencing rapid growth where project revenues are not generated until operations commence and yet, considerable investments are needed to accommodate growth during the construction phase. Such revenue imbalances have resulted in communities demanding "up-front" capital investments or revenue prior to and during construction. However, with the construction and operation of nuclear facilities, the few available studies have found substantial revenue gains allocated to local jurisdictions and little adverse expenditure effects. The analyses of twelve nuclear stations found that the demands for new and expanded public facilities and the social services attributable to the plants were generally small, that adverse impacts were controllable and mitigatable, and that utility revenue payments varied substantially among the host areas.

Lack of Significant Project-Related Expenditures

The findings show that the nearest city of at least 50,000 persons was usually located less than 50 miles (80 km) from the nuclear sites, and that even at nuclear sites located in rural areas a number of communities for residential accommodation were available for workers and their families who inmigrated. Because residential alternatives were available within commuting distance of the sites, a dispersed settlement pattern of inmigrants was observed. In addition, most of the plants were found to be located within commuting range of large laborsheds. Such locational characteristics had the effect of reducing the level of worker in-migration and concentration, thus diminishing potential adverse effects on provision of public services and social patterns of the host communities. However, a large commuter work force and a dispersed residential pattern of inmigrants also resulted in a dispersion of economic benefits.

Research on socioeconomic impacts of energy projects has shown that the degree of geographic isolation of a host community is an important factor in explaining the magnitude of impacts. The major impacts occurring in energy "boom towns," for example, have been attributed to the concentration of workers and their dependents into a single, relatively isolated community. Subsequent impacts were caused by the lack of absorptive capacity of the community's infrastructure to accommodate the population influx and attendant demands. Pressure on existing public services was less when inmigrants were geographically dispersed.

Early studies of the socioeconomic effects of nuclear generating stations erroneously applied the findings of major energy resources development to areas in which nuclear plants were to be sited, and the impacts were consequently estimated to be substantial. Expansion of our understanding of the effects of nuclear projects showed that the impacts of the development of nonnuclear energy resources, particularly in the western states, was irrelevant for the study of nuclear power plants. Nuclear power plants, in contrast to fossil plants were located relatively close to large metropolitan areas, and in areas with well-developed infrastructures that have accommodated nuclear facilities without excessive strain on existing community services. The location of the facilities tended to result in substantial worker commutation rather than worker in-migration, and this, in turn, reduced the level of project demands on local areas.

In rural areas, the number of local skilled workers was generally small and the hiring of local workers was lower than had been originally expected. Union halls were often located in the large metropolitan areas and worker commutation to the site was significant. The commuting work force averaged 68 percent of the total work force at the twelve sites. In addition, communities examined near the twelve sites were absorbing inmigrating workers because of high occupancy rates or adjustments to the housing stock. This had the effect of reducing potential adverse effects to any one community.

At peak construction, the average of total project-related inmigration (basic and nonbasic employment effects) and diminished out-migration for the twelve study areas was 1,200 persons. Relative to the size of the existing host populations, the percent increase in population directly attributable to the nuclear plants ranged from 1.3 percent to 19 percent. For the twelve sites, the average population change as a percent of the study area was only 3.7 percent. Population change was assessed to be a minor impact factor in terms of relative size and because of its lack of disrupting existing social patterns. Because demographic change was relatively small and not permanent, the increase was readily absorbed in all cases.

The assessment of nuclear plant impacts found that project-induced inmigration was small and controllable because of high commutation levels and the dispersed settlement pattern of the work force among a number of communities. As a result, pressures on existing infrastructure was low and manageable and no adverse effects were experienced by local social institutions particularly education and social services sectors.

Variation in Revenue Effects

Substantial variation exists in revenue effects from nuclear power plants because tax-generating arrangements affect the extent to which local areas benefit from utility tax payments. These include: public owned utilities that pay no taxes on nuclear plants; utilities that pay taxes for nuclear power plants to their respective states, which, in turn realocate revenues to communities based on size of population (rural areas with low populations that host nuclear facilities have the burden of risk but may not benefit from tax revenues); utilities that pay taxes directly to local taxing jurisdictions; and municipalities that have imposed local wage or income taxes to take advantage of the sizeable construction work forces at nuclear sites (Table 7.1).

The Rancho Seco plant (which is not included in Table 7.1) is a municipal utility which does not contribute to local property taxes. The Peach Bottom plant in Pennsylvania pays taxes to the state which then distributes the revenue on a per capita basis, resulting in only minor revenue increases for the local host area. However, revenues earned for the local area was from a one percent wage-income tax levied on workers at the site.

During the benchmark operations year, the nuclear plants contributed up to 50 percent of the total budget of the local jurisdictions but substantial variation in payments occurred. While the taxes paid to Surry County were the smallest for a county-level area, they accounted for almost 35 percent of the total budget. Diablo Canyon, in contrast, paid over $12 million annually to the local taxing jurisdictions, accounting for about 20 percent of the budget. The utilities at six of the twelve plants contributed over 20 percent of the total budgets to local areas in which the nuclear facilities were located.

At sites where there were few project-related inmigrants, a lack of locally purchased construction materials, and a prevalence of large income leakage due to worker commutation, the revenues generated by nuclear plants became the most significant benefit. In addition, these revenues became particularly important because large proportions were often allocated to upgrade, improve, and expand educational facilities and services. In no case was the stress on the provision of public services so large as to offset the benefits gained through tax revenues or the reduction of property tax rates. In five of the twelve case studies, overcrowding in the local school systems was evident. In these cases, however, the in-migration of workers and school-aged children was only a small proportion of the excess students and in all cases the overcrowding problem was resolved without detrimental effects on educational quality.

SOCIAL IMPACT ASSESSMENT

Background and Approach

Social impacts of large industrial projects such as nuclear facilities have been assessed in three ways. Project effects have been measured by attempts to define changes in a number of basic social activities such as changes in crime and divorce rates. The

TABLE 7.1
Fiscal Effects of Nuclear Plants on Local Areas ($000).

Nuclear Plants	1974			1978		
	Local Budget	Project Revenue	Percent of Local Budget	Local Budget	Project Revenue	Percent of Local Budget
Arkansas	7,435	3,830	51.5	13,413	6,772	50.5
Calvert Cliffs	6,483	N/A	N/A	23,614	11,267	47.7
Crystal River	26,633	1,158	4.4	42,543	3,019	7.1
Diablo Canyon	40,893	6,378	15.6	59,469	12,413	20.9
FitzPatrick	17,034	3,585	21.0	28,191	8,118	28.8
Oconee	10,249	2,678	26.1	19,514	3,722	19.1
Peach Bottom	303	186	61.4	184	30	16.7
St. Lucie	27,261	106	0.4	40,943	4,148	10.1
Surry	3,300	1,130	34.2	4,257	1,486	34.9

experience in attributing these changes to a nuclear plant or a rural industry has not been successful. In areas where major social changes have occurred (e.g., boom towns), the social indicator approach has been of some use. However, in areas where large demographic change has not occurred due to a project, this approach has not adequately addressed the problem of project attribution.

Further, while the use of various indices to measure social well-being has been used to guage the degree to which social change is positive, a set of indices or standards that is universally accepted does not exist. The attribution of such changes to an individual event, such as building a nuclear facility, poses methodological problems. Measuring social change at aggregate levels--income, education, crime rates--does not focus on the causal mechanisms that shape the change. Moreover, statistical measurements of change by means of social indicators usually have not addressed the issue of group variation in the social attribute that is being investigated.

Social impacts have also been defined on the basis of individuals' judgement of the changes and its effects on the individual or community. Thus, individuals have been asked to identify the present or probable impact and their evaluation of the change. Although such assessments are worthwhile and important, especially in light of the increased importance placed on public involvement, they are evaluative. As such, they are most valid in measuring the perceptions and evaluations of the changes at the time of the interview. This study included an evaluation component of the effects, but objective indicators of social impact were important to ascertain and measure, particularly for projective purposes.

The approach adopted for assessing the social impacts of constructing and operating nuclear generating plants was based on integrating two social impact assessment methods. These were (1) measuring changes in community social structure and organization that were attributable to the plants, and (2) evaluating the impacts or changes as expressed by community residents. The assessment of social structure was based on identifying "objective" social changes and their causal mechanisms. In order to project the magnitude and nature of social impacts from energy developments it is important to integrate economic, demographic, fiscal and other project-related changes to social impacts. Project effects are not experienced equally by all; some may gain, others lose, and the consequences may affect social organization, behavior and quality of life. Variation in project-induced effects may result in changes to community stratification, political processes, level of modernization, and group political behavior. Communities also have developed coping strategies to deal with adverse change. It is important to understand these coping processes for they can reveal much on how the political structure may respond or be incapable of responding to the effects of large industrial projects.

In the retrospective analyses, social structure is described by identifying and characterizing the major functional social groups at the beginning of the study period and the interrelationships among the groups.

The aggregation of study area residents into groups has three principal objectives: (1) to define groups which accurately reflect the functional organization of people within the study area; (2) to identify groups to which differential effects (economic, demographic, housing, or governmental) of the nuclear stations were distributed or for which the evaluation of those effects is unique; and (3) to identify groups which are discernible to study area residents and upon which they can focus in discussing the composition of the community, the economic, political, and social relationships within the community, and the distribution of project-related effects community wide.[2]

The analysis of social structure change was based on the distribution of the projects' effects among the groups over the life of the plants. These changes may occur in two ways. First, the number of identifiable groups may be altered. This occurs when inmigrants enter the community and form a distinct group along cultural, socioeconomic status, or value dimensions. Research on social impacts of energy resource development in the Western States has found evidence of traditional/newcomer splits in some communities that may have resulted in social tension. Second, social structure changes occur through alterations to the characteristics of existing groups or to the interrelationships among groups. New employment opportunities and training programs at nuclear project sites for minority groups have, in some cases, resulted in significant shifts of these groups in socioeconomic status and political control. Changes in the profiles of the groups and in their interrelationships over the study period were identified and the role of the projects in those changes were determined.

Effects on Social Structure

Despite the relatively large influx of workers during the construction period and a modest influx during plant operations, in no case did the inmigrants form a separate, identifiable group on a permanent basis. In all communities, plant construction workers could be identified, but these workers were (1) largely commuters and therefore did not have an effect on community demographic profiles; (2) located in a number of communities around the site thereby diffusing any potential demographic impact; and (3) assimilated into existing group structures. There was no evidence of inter-group tension or social conflict, nor were there sharp distinctions between traditional-newcomer groups in all case studies. In only one case (Calvert Cliffs) did the operations workers form a distinct subgroup of suburbanites due to their geographical concentration in a few residential developments. At other sites experiencing rapid and substantial suburban growth, operations workers and their families were not discernable as a subgroup.

Changes in group characteristics occurred at all sites over the study period but most were not large enough to significantly affect economic, political and social interaction patterns. There were some exceptions, however. At Calvert Cliffs, a significant proportion of blacks gained employment and training at the plant, which resulted in a significant shift in the group's economic and political mobility. At the Peach Bottom site, the elderly received important rental income but this was not permanent. Concerns expressed over plant safety at that site, however, resulted in increased political sophistication and altered participation patterns. Moreover, there was some evidence of friction between newer residents and "old-timers" over the allocation of tax revenues, but no major change in political interactions developed among groups. At the Arkansas site, the substantial revenues paid to the school district resulted in the expansion of local educational services and quality, which in turn, attracted new industry into the area and expanded employment opportunities. The local area became less provincial and the expansion of the employment base resulted in a major reversal of family out-migration (particularly for young persons) and a significant level of returnees. While the area expanded and became more diversified, the sense of community heightened and social interaction patterns among groups were enhanced.

Most of the changes due to the plants were not large enough to significantly affect the prevailing economic, political or social interaction patterns among groups. However, changes in interrelationships at some sites were noteworthy. At Diablo Canyon, the construction of the power plant had the affect of stabilizing the tourist industry during a number of slow seasons because of worker demands for housing accommodation. At the Peach Bottom site, businesses that were marginal prior to plant construction remained viable much longer than expected originally, but were ultimately replaced by larger, more centralized, and nonlocal firms. Income generated by plant construction was partially responsible for this change, and although there was some loss in social familiarity and credit availability, these changes were evaluated as positive. At Surry, stimulated by substantial utility tax revenues, a complete political reorganization occurred, such that Blacks gained political control and the traditional power base, white farmers, lost control of the county government. Finally, at several sites where tax revenues from the power plants were substantial, the level and availability of social services expanded to benefit various community groups, especially the rural elderly. While such changes were discernable, overall the construction and operation of the nuclear plants had only modest effects on social structure and played only a small role in shaping other social processes.

Few community economic, demographic, housing, and social changes were found to be attributable to the plants. While prices for housing escalated in some cases, the effect was not long-term and no controversy over housing surfaced in the communities. Significant social tensions between newcomers and the indigenous population did not develop. There were, however, a few jurisdictional disputes over the distribution of tax revenues, but these were isolated disputes that did not result in the emergence of community conflict.

In a number of cases, the nuclear plant sensitized local residents and local government officials to growth management issues, even though the nuclear projects were not necessarily responsible for the growth. In five of the case studies, greater attention was placed on zoning and land use planning subsequent to nuclear plant construction. Examination of this trend found that the public perception of the plants as growth-inducing was not an insignificant factor. Concern over community growth, concomitant with revenues from the plants, resulted in increased professional specialization and expansion of planning functions in local government.

Implications for Projective Assessments and Mitigation Planning

An important contribution of the social impact assessment was to demonstrate operational procedures by which changes in social structure could be defined, measured and their significance studied. The host area population was broken into a set of functional groups for this purpose and social structure was defined in terms of the groups, their characteristics, and the ways in which they interacted. Two principal criteria motivate the group identification process. First, there is reason to differentiate between groups that

will be affected by the construction and operation of a
nuclear generating station unequally. Second, there is
reason to differentiate between groups that would
evaluate given consequences of a station differently.
That is, groups with important value differences with
respect to the kinds of effects that might result from
the nuclear station ought to be identified and studied
individually.

The analyses demonstrated that it was not unusually
difficult to distribute the socioeconomic consequences
of a nuclear plant among groups in a study area and to
examine whether group characteristics or group interaction patterns were in any way importantly affected.
This same process could be followed in a projective
assessment. It would also not be difficult to anticipate the way in which group characteristics would be
changed knowing the magnitude of the causal elements.
Further, given the conclusion that interaction patterns
were well established and were generally unaffected by
the nuclear stations, it will probably be safe to
assume that interaction patterns will be unaffected
except under unusual and obvious circumstances (e.g.
boom town effects not evidenced with nuclear stations).
In general, social structure is extremely resilient and
will not experience structural alterations unless the
dislocations to local areas are extreme. With nuclear
power plants there were only infrequent cases of social
structure change and, in these cases, the effects were
small.

PUBLIC RESPONSE

The objective of this technology assessment was
also to describe the public response to each of the
projects. The nature, saliency, and prevalence of
plant-related issues and their articulation in the
political process was considered an important aspect of
social impacts. Thus, public response, with its underlying motivations in terms of public concerns and attitudes, constituted a social phenomenon to be addressed
and explained. Further, public involvement in the
licensing hearing process and in political activities
outside the public hearings was hypothesized to result
in changes to social and political organization or to
group interaction. Change, particularly in the areas
of political interaction and in processes was viewed as
being related to the nature (magnitude, saliency, and
prevalence) of the public response.

Public response to the construction and operation
of the nuclear plants took place over a lengthy period
of time, and the nature of the local response must be
viewed in a temporal perspective. For most of the
nuclear plants in our sample, construction began in the

mid-to-late 1960s and continued into the 1970s. Plant operations commenced during the mid-to-late 1970s, with construction and operations work overlapping in the case of sites with two or more units. Consequently, the response of local residents to the nuclear stations can be characterized in an environment of evolving environmental and safety regulations, changing foci of regional and national concerns over nuclear energy, and site-specific events resulting from construction and operations activities. The Three Mile Island accident was found to heighten existing issues and expand public concerns over nuclear safety at other nuclear sites, and to precipitate political activity, particularly over emergency planning.

Variation in Response: Contrasting Cases

The assessment found substantial variation in timing, saliency and extent of public political response. These differences are illustrated by public response at two sites - Arkansas Nuclear One and Peach Bottom. This is followed by a section which explicates the variation across sites.

Arkansas Nuclear One (ANO). Overwhelming community support for the nuclear power plant has prevailed during the preconstruction, construction, and operation phases. This support can be exemplified by events occurring during early planning phases. When several landowners were reluctant to sell their land, except at an inflated price, the business community purchased the property for the utility. A prevailing value was the importance the community, and especially the business group, placed on economic development for the area. Historically, the business community had taken initiatives to attract industry and to aid the agricultural sector. Consequently, the business community had inherited a political leadership role. Thus, the business community's active endorsement of the nuclear facility was an important catalyst for community-wide acceptance of the plant. Notwithstanding the importance of this group role, the community consistently favored the plant: the public hearings were uncontested, and only three individuals (nonlocals) presented concerns through limited appearances. Despite ANO's history of shutdowns, equipment failures, and a leak of radioactive material to the environment, no opposition emerged in the region to raise safety or environmental questions. The nuclear facility was a "nonissue" for the community.

Positive local attitudes toward the nuclear facility was found to be related to the following factors: (1) The historic role of the utility within the study

region.   Arkansas Power and Light was one of the
earliest companies to locate in Arkansas and invest in
industrial development. In addition, the company was a
major contributor to the recreational development of
the state and host area. Moreover, in a state that had
lost significant population between 1940 and 1960, the
company was viewed as a stabilizing and important
economic asset; (2) The lack of any social problems or
community stress on public infrastructure during the
construction phase concomitant with large revenue
effects that expanded social services; (3) The visibly
positive effects on the local school system which
resulted in attracting industrial development to the
area; (4) The value placed on industrial growth and
economic expansion of which the nuclear plant was
symbolic; and (5) the general acceptance of risks
inherent in nuclear technology.

Peach Bottom. Strong public support for the construction of two additional units at the Peach Bottom
site was evidenced during the 1966-1968 preconstruction
period. There were two major factors to explain this
support. First, the area was experiencing a serious
downward economic trend, and the construction of two
nuclear units was seen as the stimulus needed to
reverse the trend. Second, Peach Bottom Unit 1, a
small reactor, had already been established and
accepted by the community, and the two additional units
were viewed as an extension of the first. No residents
expressed concern at the construction permit hearings
but, prior to the hearings, area agriculturalists
voiced concern that the cooling structures could have
an adverse effect on the recreational uses of the
Sesquehanna River. There was also some resentment by a
number of farmers that the transmission lines would
cross their properties.
In the early 1970s, during the operating license
hearings, local concerns had shifted, reflecting growing opposition to nuclear power development in the
state of Pennsylvania. The regional concerns focused
on the cumulative impacts on the quality of the Susque-
hanna River since the river was already well used for
industrial cooling purposes. The four interveners at
the operating license hearings represented larger
regional interests--they were not from the host area,
and received only minimal support from those residents.
Representatives of the agricultural community made
"limited appearances" at the operating permit hearings.
Their concern focused on the protection of milk products, and they demanded safeguards for the local dairy
industry. The farmers also indicated that the host
township was the recipient of a great burden of risks
and that the relatively high wages paid at the site
contributed to difficulties in procuring seasonal farm

laborers. While opposition to the plant was centered in the farming community, townspeople generally supported the plant. The dominant value expressed by area farmers was the preservation of the "agricultural way of life" and, as a group, they felt that the plant would result in a rapid urban encroachment of the rural area. Furthermore, they were concerned over possible depreciation of land values because of the proximity of the plant and its possible effects on agricultural products. The fact that the utility proposed to build another plant across the river from Peach Bottom township and to expropriate considerable farm land heightened the farmers' opposition to the Peach Bottom plant. The concerns by area residents were over potential effects of the plant that were site specific; generic safety concerns were expressed by groups from the larger region. There was a strong relationship between traditional values held by the farmers and their predispositions toward growth and the public response to the plant.

After TMI the effect of public response was evidenced in the involvement of the township in political matters outside of its traditional parochial concerns with agriculture, particularly over emergency planning issues. This resulted in political affiliations with entities outside the local host area and increased professional sophistication over planning issues and state political processes.

Determinants of Variation Across Sites

Local Values. Existing research on community conflict and public response to nuclear facilities have noted that a major explanatory factor for the prevalence and duration of response was related to local values. Areas that hold "traditional" values have been hypothesized to be more likely to support nuclear generating facilities than "modern" areas where concerns over environmental impacts and the safety of the technology may prevail, and where public participation in the decision-making process are relatively open. The findings of this assessment support the view that local values were major determinants of attitude formation and public response to nuclear generating stations.

At those sites where local pro-growth/pro-industry values prevail, it is likely that the construction of a nuclear station will be supported strongly. Support will be heightened in those areas that have historically experienced economic decline and out-migration. In those areas, nuclear facilities were perceived as valuable community elements--they tend to induce growth and stabilize the economy. Thus, the Arkansas Nuclear

One plant became strong symbols of community renewal and economic prosperity. In these cases, visible economic gains, such as expanded employment opportunities for local residents and major fiscal benefits heightened the perceived economic importance of the plant while the risks were underestimated.

A community's value system may be described by the values held by the community's social groups. The case studies found that, although variations existed in the response to nuclear plants within the local areas, community group conflict did not emerge because of the facilities. While disagreements may have occurred in the political sector, it was rare that the controversy over the nuclear plants resulted in significant changes in political structure and organization. Rather, controversy over nuclear plants reinforced, heightened, and polarized values and group political positions. Where active nuclear opposition existed, there was also a tendency for involvement in other environmental issues.

While support for a plant may be extensive in an area due to prevailing pro-growth community norms, opposition is seldom universal in the host areas that were examined. In all twelve areas, active political opposition was undertaken by one or a few small organizations. The dominant concerns were motivated by environmental issues, parochial/self-interest issues (adverse economic effects), issues over technology control and decentralization, and questions over safety. In areas characterized by active environmentalism, there was a strong likelihood that oppositon to nuclear plants would emerge. In many of the host areas, conservation and environmental organizations were the nucleus for the formation of antinuclear organizations.

Where social groups have a strong political base, their values with respect to nuclear plants will tend to be expressed in the political arena. Thus, in Pope County, Arkansas, and Calvert County, Maryland, the business community supported the nuclear plants strongly enough to influence community attitudes significantly. The farmers of Peach Bottom, Pennsylvania, who perceived potential detrimental harm to agriculture and their way of life, actively opposed the plant at the licensing hearings and subsequently passed resolutions to improve emergency planning. Opposition to nuclear plants may be less likely in areas that have the following combination of characteristics: reliance on established government procedures; political values which belittle protest; a prevailing pro-growth attitude; an important community leadership role played by the business community; and, perceived benefits from plant construction and operation.

Site-Specific Factors and Public Response.
Analyses of public response at the twelve sites found
that underlying causal factors in the variation of
response can be explained by project-specific and
site-specific factors. Projects that are characterized
by a high level of risk uncertainty with respect to
either project-specific events (leaks at Peach Bottom)
or general risk phenomena (earthquake risk at Diablo
Canyon) will have a high likelihood of generating
public concern and possible political response. Such
concerns were found to be heightened in situations
where other hazardous events occurred at about the same
time, such as the closing of a plant, an accident at a
plant, or the disclosure of improper procedures at a
plant. Major visible economic and public revenue
benefits were found to temper concerns over risks. The
fact that Arkansas Nuclear One provided substantial tax
revenues to the local school district was viewed by
residents as the major positive effect of the plant,
while the effects of a serious leak of radiated water
was viewed as more remote and controllable, and therefore of lower importance. Where the utility and plant
were major employers in a local area and generated
large revenues, local public response was consistently
in favor of the nuclear station, as in the case of the
Crystal River station.

In most of the case studies, local opponents who
challenged the nuclear stations did so through the
legally instituted channels of the hearings process.
Once a decision had been made about the plant, particularly regarding an operations permit, public
activity and opposition dissipated to a considerable
degree. In only two of the case studies, Diablo Canyon
and Peach Bottom, did local political activity have
lasting effects on local decision-making processes. At
the Diablo Canyon site, the controversy over the nuclear plant (see Chapter 9) resulted in strengthening
political differences between environmentalists and
pro-growth advocates. Moreover, the controversy
heightened the general level of environmental awareness, and consequently, concerns over nonnuclear
environmental problems increased.

Although the projection of anticipated public
response to a technology is fraught with difficulties,
assessments can be undertaken to provide indicators of
potential response. Information on how communities
have historically responded to stress and crises, and
actions or inactions toward industrial development,
environmental quality or growth issues, will provide
clues as to possible response to siting nuclear plants.
Public participation styles, the role of leadership in
the community, and the number of public interest groups
and organizations in the area and their concerns, are
important data elements for projective assessments of

public response. Attitude and opinion surveys are sometimes used as indicators of potential public behavior. Although questions with respect to saliency of concern and possible political action can be included, these may not result in reliable information on actual behavior.

NOTES

1. This study was supported by the U.S. Nuclear Regulatory Commission. The following persons were principal investigators on the larger study: K. Branch, J. Chalmers, C. Flynn, J. Flynn, and D. Pijawka.
2. Based on a review of the literature on community organizations and social effects of large-scale energy projects the attributes specified included the following: size of group, livelihood elements, demographic characteristics, pattern of interaction among group members (cohesion), and values toward growth, environment and community political participation. The patterns of interaction among group members were examined for three spheres of activity--economic (employment and income), political (political control, participation) and social (participation, formal and informal, contact).

REFERENCES

Bergmann, P. and D. Pijawka. 1981. "The Socioeconomic Impacts of Nuclear Generating Stations: An Analysis of the Rancho Seco and Peach Bottom Facilities," GeoJournal, Supplementary Issue 3.

Chalmers, J., D. Pijawka, K, Branch, P. Bergmann, J. Flynn, and C. Flynn. 1982. Socioeconomic Impacts of Nuclear Generating Stations. Summary Report. NUREG/CR-2750. U.S. Nuclear Regulatory Commission.

Flynn C., J. Flynn, J. Chalmers, D. Pijawka. Forthcoming. "An Integrated Methodology for Socioeconomic Impact Assessment and Planning." In Methodology of Social Impact Assessment, K. Finsterbusch and C.P. Wolf, eds. Community Development Series. Hutchinson and Ross, Inc.

Pijawka, D. and J. Chalmers. 1983. Impacts of Nuclear Generating Plants on Local Areas," Economic Geography, Volume 59(1):66-80.

8
Population Trends Around Nuclear Power Plants[1]

Michael Greenberg, Donald A. Krueckeberg,
and Michael Kaltman

REGULATORY BACKGROUND

Population considerations play a major role in the licensing of nuclear power stations. Yet, existing regulatory research does not provide systematically collected follow-up data as a basis for policy evaluation. This research indicates that local areas hosting nuclear power stations experienced population growth during the period 1960-1980 substantially in excess of the national rate. During the 1960s, national and especially regional growth forces can account for the growth in the host local areas. During the 1970s, however, the decade when most of the existing nuclear facilities were opened, an even stronger local growth component arose, unexplained either by national, regional or county growth trends. This development is contrary to well developed regulatory objectives of selecting low population sites. Problematical sites and explanatory factors are isolated for more detailed investigation.

Projections of population and population change in the vicinity of nuclear power stations are elements in the regulation of these facilities by the U.S. Nuclear Regulatory Commission (NRC). Present population criteria, the product of an evolutionary process involving the NRC and its predecessor agency, the Atomic Energy Commission, are based on the recognition that the siting of nuclear reactors presents some degree of residual accident risk. Therefore, such siting has been encouraged to locate in relatively low density areas to keep this risk low and to facilitate implementation of emergency protective measures in the event of a low probability emergency. In addition, the Commission has looked favorably upon reactor sites that are at some distance from large urban concentrations where the implementation of either evacuation or sheltering would not be difficult (Price, 1968; Buchanan, 1975).

Current regulations involving population are described in Part 100 of the <u>Code of Federal Regulations</u>, Title 10, which defines an exclusion area around the plants, a larger, low population zone beyond the exclusion area, and a population center distance. In addition, the staff of the NRC considers the density of resident and transient population. This latter consideration, while not a regulation, calls upon the staff and applicants to consider alternative sites with lower population densities if the proposed site exceeds the established population density threshold. At present, the established density threshold is 500 persons per square mile (192 per sq km) averaged over any radial distance out to thirty miles at start up of operations, and 1000 persons per square mile (385 per sq km) over the lifetime of the facility.

Applicants for licenses to operate nuclear reactors are required to estimate population growth and distribution in the areas surrounding their proposed sites and then to project these populations for a forty-year period, the time frame in which a nuclear reactor will be constructed and operated. Applicants typically employ the latest available sources of population information, including projections. For one site where rapid future population growth is expected, NRC has required periodic population updates from the utility. NRC has also routinely conducted a site inspection every three years to determine whether unexpected population or land use changes have occurred.

EXISTING RESEARCH

Few studies have examined the impact of nuclear power plant siting on population growth in adjacent communities. One such study attempted to determine the extent to which 12 operating nuclear power stations reduced out-migration and increased in-migration in small study areas; however, this study did not look at these trends in a regional context (Chalmers, et al., 1982). Another study utilized the general industrial development literature to extrapolate possible socio-economic effects of nuclear power station development projects. This study also tentatively concluded that the nuclear facility "will operate to stem the population decline generally associated with rural areas" (Policy Research Associates, 1977).

But if nuclear power plants are similar to other industrial firms, they differ in at least two respects that may have an impact on population growth. The first factor relates to the size of local tax revenue benefits. Although all industrial firms, including nuclear power plants, provide jobs, increases in per capita income, and income multiplication, nuclear power plants also significantly augment local tax bases because of their large capital investments.[2] A recent study indicated that taxes paid by seven nuclear plants produced local tax revenues which ranged from 7 to 95 percent of total revenues in the local jurisdiction (Chalmers, et al. 1982).

Increased tax revenues permit communities to expand services, reduce local tax rates while maintaining services, or adopt a combined strategy (Purdy, et al., 1977; Shields, et al., 1977; Chalmers, et al., 1982). When high levels of local public services, especially education, are combined with relatively low taxes, community attractiveness should be higher than in surrounding communities and should, _ceteris paribus_, lead to in-migration and population growth. However, households desiring to consume higher levels of public services would tend to bid up property values in communities with low tax but high quality services, thereby limiting demand. In short, predicting _a priori_ the growth patterns of a community based on the analysis of its local tax characteristics is difficult (Oates, 1969; Edel and Sclar, 1974).

The second factor which could affect population growth and development and which differentiates nuclear power plants from industry is risk perception. Nuclear power is perceived to be at the extreme of all characteristics associated with risk, particularly the attributes of dreadedness and the severity of consequences (Slovic, et al., 1979). In a survey of selected social groups and 1500 members of the general public, at least three-fourths of each group agreed that no guarantee exists against a catastrophic accident (Marsh and McCennan, 1980). When asked to specify a minimum distance between their homes and five installations--a ten story office building, a large factory, a coal fired power plant, a nuclear plant, and a hazardous waste disposal site--before they would "want to move to another place or to actively protest..." people objected most to nuclear power plants[3] (Council on Environmental Quality, 1980).

The ready conclusion is that popular fears of nuclear power would act as a deterrent to growth in communities hosting these facilities. However, people encounter risk in their everyday lives and deal with it through acceptance, denial, or discounting. The disaster literature suggests that people return to their former communities after a disaster despite the exposure to subsequent risk (Haas, et al., 1977; Friesema, et al., 1979; Houts and Goldhaber, 1981). Finally, a recent study of home buyers in California showed little measurable buyer response--avoidance or the adoption of mitigation measures--to disclosure of proximity to earthquake faults (Palm, 1981).

Not only are the risk and economic factors complex, but each by itself probably does not offer an explanation for population growth around nuclear power plants. Julian Wolpert's migration model (1964, 1965, 1966), which emphasizes cognitive dimensions of human adjustment, may offer insight into population growth around nuclear plants. Using a satisficer theory of decision-making, Wolpert suggests that individuals and households will accept environmental conditions in residential areas until a threshold is reached. In this model a resident or inmigrant will evaluate the alternatives within a market and choose a location where the balance between dissatisfaction (risk) and satisfaction (low taxes, good educational services) is weighted in favor of the latter. One is forced to conclude, however, from a review of existing theory and studies of behavior that current research has not provided the factual data either for predicting population growth around nuclear reactors or for policy formulation.

OVERVIEW OF THE RESEARCH DESIGN

In the absence of systematic research, an understanding of what population changes have occurred in the vicinity of nuclear power stations is difficult. Likewise, the role played by nuclear stations in generating population growth or decline is unknown. Therefore, the NRC funded the research upon which this paper is based. The initial phase of the research, presented here, was designed to determine if local governments containing operational nuclear power stations experienced unusual demographic changes during the 1960-1980 period.

The following four tasks describe the major efforts in the research:

1. Compare population changes from 1960 to 1980 in minor civil divisions hosting the stations with demographic change in three larger units (host counties, adjacent non-host counties, and the nation).

2. Apportion local change to national, regional (adjacent non-host counties), county (host county minus host community), and local (host minor civil division) components.

3. Classify host local areas into categories based upon the extent to which local change may be explained by national, regional, host county, and local area factors.

4. Explain the variations in population change using other available data.

Method

The method of population analysis utilized in this study is a four-equation model that apportions local population change to components that can be accounted for by national, regional, county, and local trends. Briefly, the model proceeds by first extracting national, then regional, then county growth trends from the gross rate of growth of the local area. Local factors are thus defined as the residual of population change unaccounted for by national, regional, and county trends. The model is a conservative one. By subtracting national, regional, and county factors to control for their effects, it may actually understate unique local factors contributing to change in some local areas.

The key assumption underlying the method is that in an increasingly homogeneous nation like the United States, some of the causes of population change are similar across the nation. The first equation attributes the initial component of change to national factors. Assuming that some regional factors are distinct from national factors, the second equation isolates a regional component from the counties adjacent to the host county. The third equation accounts for the host county component (host county minus host community). The local community component is the residual effect calculated in the fourth equation.

1. $NATIONCH_{ijk} = (INTLPOP_{ijk}) \times (CHANGE_{ink})$

2. $REGIONCH_{ijk} = (INTLPOP_{ijk}) \times (CHANGE_{irk} - CHANGE_{ink})$

3. $COUNTYCH_{ijk} = (INTLPOP_{ijk}) \times (a)$ or (b) where

   (a) $(CHANGE_{ick} - CHANGE_{irk})$ if $CHANGE_{irk} > CHANGE_{ink}$ and
   (b) $(CHANGE_{ick} - CHANGE_{ink})$ if $CHANGE_{ink} > CHANGE_{irk}$

4. $LOCALCH_{ijk} = (TOTALCH_{ijk}) - (NATIONCH_{ijk}) - (REGIONCH_{ijk}) - (COUNTYCH_{ijk})$ where

NATIONCH$_{ijk}$ is the national component of change of population group i in local area j during time period k;

INTLPOP$_{ijk}$ is the initial population of group i in local area j when period k began;

CHANGE$_{ink}$ is the proportional change in the population of group i in the nation during period k;

REGIONCH$_{ijk}$ is the regional (non-host counties) component of population change of group i in local government j during period k;

CHANGE$_{irk}$ is the proportional change in the population of group i in the region during period k;

COUNTYCH$_{ijk}$ is the host county component of change of population group i in local area j during period k (host county minus host community);

CHANGE$_{ick}$ is the proportional change in the population of group i in the host county during period k (host county minus host community);

LOCALCH$_{ijk}$ is the local component of change of population group i in local area j during study period k;

TOTALCH$_{ijk}$ is the change in population group i in local area j during study period k.

THE DATA AND THEIR LIMITATIONS

Population counts for 1960, 1970, and 1980 for host communities, host counties and adjacent counties were compiled from U.S. Census data by Brookhaven National Laboratory (BNL). Two of the forty-nine sites posed special problems: Fort Calhoun 1 and St. Lucie 1.

In both of these cases, the precise boundaries of the local area were difficult to determine due to census boundary adjustments over the 20-year period, a difficulty requiring BNL to define a large local host area that included more than 80 percent of the host county's population. Computing a host county's component of change under these conditions would be inappropriate because such a small percentage of the county population resides in the non-host communities of the host county. Therefore, local areas and host counties were considered interchangeable and host county components of change were not calculated for these two sites.

The analysis of St. Lucie was finally set aside from the other host sites because of an additional problem--the impact it had in the initial results. The St. Lucie host community, as defined by the data, was extraordinarily large. It contained a population equal to nearly 15 percent of the sum of the population of all forty-nine host communities. Therefore, the following results are for forty-eight sites, and a special presentation of the St. Lucie analysis follows the results for the other forty-eight sites.

Four other data issues were also relevant. One is the accuracy of population counts. While this issue is germaine, we have no reason to believe that the study regions have any unusual problems.

The second involves the definition of region. The ideal region would consist of a small number of counties with economic and social profiles similar to the host community prior to the construction of the nuclear power facilities. Alternatives considered included states, BEA economic regions, SMSAs and neighboring counties. The state is a poor choice because it is too large a unit and is not necessarily a functional economic unit. The BEA has two sets of regions, both of which are defined on the basis of similar economic and social profiles. One set of regions is quite large and thus, like the state, was rejected. The second set of BEA regions is smaller, but unfortunately, too small in some cases. Some of the BEA regions are only a single county, a fact which precludes the consistent use of these regions across the United States. For example, Ocean County New Jersey, the site of the Oyster Creek facilities, is a BEA state economic area. No obvious method exists for deciding which of the neighboring BEA areas or counties is the appropriate region. The SMSA

was also rejected because not all counties are part of
an SMSA, and some one-county SMSAs exist. The last
method, and the one selected for the study, was to use
all the counties adjacent to the host county as the
region. The virtue of defining regions in terms of
adjacent counties is that the definition can be applied
consistently across the nation. However, some instances may exist where counties which are closely
economically integrated with the host county have been
included with counties that are not as closely related
to the host county.

The third data issue is the definition of local
areas. The first analysis was made on the basis of
political jurisdiction. Later analyses may be made for
school districts, which may be appropriate for testing
some explanations of population change around the
facilities.

The final data issue is the study period. Since
the plants were constructed during the 1960s, 1970s and
1980s, the period 1960-1980 has been used.

AGGREGATE POPULATION CHANGE
AROUND THE SITES, 1960-1980

Application of the model to the population data
show a large component of unique local growth around
the forty-eight sites during the 1970s. In 1960, about
274,000 people resided in the forty-eight local areas
that currently host nuclear power stations in the U.S.
That population grew by 60,000 to 334,000, during the
decade of the 1960s. An even greater increment of
population growth was added during the 1970s. Overall,
176,000 people were added to the population of the host
local area during the two decades, an increase of 64
percent (Table 8.1). In comparison, the nation's population increased only 26 percent over the same period.

Population Change by Site, 1960-1980

A good deal of variation exists among the host
local areas in both population size and rates of
change. They range in population size from about 200
to over 50,000 (Table 8.1). Seventy to 80 percent had
less than 10,000 residents. The median population of
the local areas was 2303 in 1960, and 3375 in 1980
(Table 8.1 and 8.2).

Six host local areas declined in population during
the 1960s and four declined during the 1970s. At the
other extreme, one increased by more than 10,000 during
the first decade, and three by more than 10,000 during
the second decade.

TABLE 8.1
Aggregate Population and Population Change in Local
Areas Hosting Nuclear Power Stations, 1960-1980 (48
Sites)*

Population	1960	1970	1980
All sites	274,282	334,810	450,372
Median site	2,303	2,682	3,375
Minimum site	197	277	255
Maximum site	34,848	44,899	51,865

Population Change	1960-1970	1970-1980
All sites	60,528	115,562
% All sites	22.1	34.5
Median site	368	765
Minimum site	− 215	− 223
Maximum site	10,540	23,008

*St. Lucie is not included in these totals.

Some population growth is characteristic of nearly
all of the local areas. Forty of the forty-eight (83
percent) increased in population during both decades;
only one site, Beaver Valley 1 and 2, Pennsylvania (383
inhabitants in 1960), decreased during both decades.
Seven decreased during one decade and increased during
the other: Big Rock Point, Michigan; Cooper, Nebraska;
Davis-Besse, Ohio; Farley, Alabama; Hatch, Georgia;
Indian Point, New York; and Point Beach, Wisconsin.
Most grew more during the 1970s than during the 1960s,
as indicated by the fact that the median change during
the second decade, 765 persons, was more than double
the median change during the first decade, 368 persons
(Table 8.1).

The local areas that grew the most during the 1960s
also tended to be the communities that grew most during
the 1970s. A coefficient of correlation (r) of .60
(significant at .001 level) between population changes
of the two decades support this conclusion. But the
association is far from perfect and implies that different growth factors may have been operating in the
two decades. For example, of the following six local
areas in Table 8.3, four had substantial population

TABLE 8.2
Population Change in Local Areas Hosting Nuclear Power
Stations, 1960-1980, by Size Class

Local Host Area Population Size	1960		1970		1980	
	Num.	%	Numb.	%	Num.	%
< 1000	9	19	7	15	6	13
1000-9999	29	60	30	63	28	58
≥ 10,000	10	21	11	23	14	29
TOTAL	48	100	48	101	48	100

Change by Size of Change Increment	1960-1970 Number of Sites	%	1970-1980 Number of Sites	%
Absolute decrease (-225 to -1)	6	13	4	8
0 to +100	8	17	8	17
+101 to +500	13	27	6	13
+501 to +1,000	5	10	8	17
+1,001 to +10,000	15	31	19	40
≥+ 10,000	1	2	3	6
TOTAL	48	100	48	101

growth during both decades but much greater growth
during the 1970s, while the other two had more growth
during the 1960s.

The mathematical model apportioned a total population growth of 176,000 people among national, regional,
county, and local growth components. The regional
component, followed closely by the national, was found
to be the largest during the period 1960-1970, while
local and county components of growth were the smallest. However, in the 1970-1980 period, the local component is larger than all other components (Table 8.4).

The National Component

If the host local areas had grown at the national
rate during 1960-1980, their aggregate population would

TABLE 8.3
Population Change in Six Selected Host Sites

| | Population in 1000s | | | |
| | | Pop. Change | | |
Local Host Area	Pop. 1960	1960– 1970	1970– 1980	First Plant Operating Date
Turkey Point, FL	22.1	6.7	23.0	1972
Crystal River, FL	4.1	4.0	15.2	1977
Pilgrim, MA	14.4	4.2	17.3	1972
Oyster Creek, NJ	1.9	2.7	9.5	1969
San Onofre, CA	34.8	10.0	5.3	1968
Zion, IL	22.5	8.4	1.5	1973

TABLE 8.4
Aggregate Components of Population Change in 48 Local Areas Hosting Nuclear Power Stations, 1960–1980

Population at Start of Period	274,282		334,810	
	1960–1970		1970–1980	
Components	Population	%	Population	%
National	36,895	61	38,268	33
Regional	52,353	86	37,206	32
County	-31,821	-53	1,247	1
Local	3,101	5	38,841	34
Net Total Change	60,528	99*	115,562	100
Population at End of Period	334,810		450,372	

*Error due to rounding.

have increased from 274,000 to 347,000, an increase of
73,000, or only 41 percent of their actual growth
during the two decades. The national component is much
more important in the first decade, during which it
contributed 61 percent of the total growth, than during
the second decade (Table 8.4). Over the two decade
period, national population growth trends account for
the second largest share of growth after the regional
component in the host local areas.

The Regional Component

The region, composed of those counties adjacent to
the host county, is the largest growth component over
the full twenty-year period. Even after subtracting
the national component, almost 90,000 people are
estimated to have been added to the host local areas
due to regional trends. The regional component is also
relatively constant in strength among the forty-eight
regions, as indicated by a correlation (r) of .82
between the regional components during 1960-1970 and
1970-1980. However, a greater regional effect was felt
during the 1960s when the regional component was
52,353, compared to 37,206 during the 1970s (Table
8.4).

The County Component

The county component (host counties minus host
local areas) is negative during the 1960s, and positive
but negligible during the 1970s (Table 8.4). The negative component during the first decade does not mean
that the host counties absolutely declined in population. Indeed, many substantially increased. Rather,
the negative component means that the host counties as
a whole grew more slowly than their adjacent regions.
The 1200 positive population increment during the
1970s means that rates of population change in the host
counties were about the same as those in the host
regions. Had rates in the host counties consistently
exceeded those of their regions, the aggregate county
component would have been much higher. Thus, the
county component does not appear to be an important
source of local growth.

The Local Component

The most interesting results are the local components. During the 1960s, the local component was 3100
or five percent of the total aggregate host local area

increase (Table 8.4). This finding means that the host local areas as a whole grew more rapidly than the host counties, which had a negative aggregate component, and at about the same rate as their host regions. Had they grown far more rapidly than the host regions, the aggregate local component would have been greater.

Further evidence to support the conclusion that the regional component was clearly the dominant force during the 1960s is provided by three statistical measures. The coefficient of correlation between population change in the forty-eight host local areas and the regional components for 1960-1970 was .68 (significant at the .001 level). Neither the county nor local components of change exhibited statistically significant correlations with total population change during the 1960s.

In strong contrast to the findings for the 1960s, the local component during the 1970s becomes the most important contributor to local change. From a five percent share of aggregate local growth, the local component jumped to almost a 34 percent share of the total change, almost 39,000 people (Table 8.4). The correlation between total population change of the host local areas and the local components increased from .11 during the 1960s (not significant at .05), to .68 during the 1970s (significant at .001). The correlation between total local area population change and the regional components remained at .68 during the 1970s, while the correlation with the county components remained insignificant.

The local components during the 1970s are not only more important, but they also have a different geographical distribution; that is, the local areas with the larger local components in the 1960s are not the same ones that had large local components in the 1970s. The correlation between the local components in the 1960-1970 period and the components in the 1970s is -.09 (not significantly different from 0.0 at the .05 level). By comparison, as noted above, the correlation between the regional components of the two decades was .82 and .45 between the county increments, an implication of relatively similar geographic patterns of change over the two decades. In short, local components of change, those population changes which cannot be accounted for by national, regional, or county growth trends, became important during the 1970s and had a different geographic distribution from the much smaller local components of the 1960s. As most of the nuclear plants came on line during the 1970s, the plants may reasonably be considered to have induced this growth. However, other forces, such as local infrastructure development and other industrial growth, are also possible causal factors.

CLASSIFYING THE LOCAL HOST AREAS

One of the objectives of this research was to develop a classification system based on demographic change from which a list of potentially important case studies might be selected. Two demographic change groups were isolated, based on two criteria: (1) the absolute size of population change, and (2) the relative importance of the local component of change. For example, local areas with relatively small absolute population changes of less than about 1,000 probably should not be included because the number of people is so small that random factors could have played an important role in their change. The criteria were applied to isolated places with sharply contrasting local components and produced the following two groups: Group (1): Host areas where the local components of change were very large, accounting for at least two-thirds of total local growth; and Group (2): Host areas where the local components of change were very small (negative), although absolute local change was large and positive.

Sixteen plant sites are in the first group, host local areas with local components of change that are a large percentage of the total population change during at least one of the two decades (Table 8.5). Three places qualify in both decades: Ginna, FitzPatrick, and Rancho Seco.

Eleven sites are in the second group, with negative local components and large, positive total population increases. Three sites qualify in both decades on this criterion: San Onofre, Diablo Canyon, and Calvert Cliffs (Table 8.6).

Two South Carolina sites, Robinson and Oconee, had large negative local components during the 1970s following large, positive local components during the 1960s. Pilgrim, Browns Ferry, and Turkey Point manifested the opposite characteristic--from a negative local component of change during the 1960s to a positive local component of change during the 1970s.

As most of the nuclear power stations opened during the 1970s, the finding that the local components of change were important during the 1970s is consistent with the hypothesis that nuclear power stations might be a cause of local population change. The twelve sites with large, positive local components of change relative to their total population during the 1970s, the six with negative local components and a large, total increase, and the three that switched from a negative component (1960-1970) to a strong positive local component (1970-1980) are likely to be the best cases for further study. Host areas with "small" populations and "small" local components might also yield useful information, if only as controls.

TABLE 8.5
Local Host Areas Where the Local Component of Change is a Large Percentage of the Total Population Change

Host Site	Site Number	Local Component	Total Local Change	Year Plant(s) opened, 19___ or are scheduled to be opened
1960 – 1970				
Robinson, S. Carolina	38	3,200	291	71
Sequoyah, Tennessee	42	1,996	2,541	81, 82
Oconee, S. Carolina	29	1,992	1,576	73, 74, 74
Rancho Seco, California	37	1,804	3,076	75
Ginna, New York	19	1,109	1,755	70
FitzPatrick, New York	16	863	1,130	69, 75, 86
Monticello, Minnesota	27	856	1,152	71
1970 – 1980				
Turkey Point, Florida	46	15,201	23,008	72, 73
Pilgrim, Massachusetts	33	14,853	17,307	72
Oyster Creek, New Jersey	30	6,697	9,545	69
Browns Ferry, Alabama	5	3,195	3,621	74, 75, 77
Rancho Seco, California	37	3,002	4,844	75
FitzPatrick, New York	16	1,921	1,836	69, 75, 86
Ginna, New York	19	1,790	1,466	70
Fort Calhoun, Nebraska	17	1,744	1,938	73
Connecticut Yankee, Conn.	8	1,412	1,449	68
Zion, Illinois	49	1,226	1,533	73, 74
McGuire, North Carolina	25	1,190	1,462	81, 83
Peach Bottom, Pennsylvania	32	1,072	1,268	74, 74

TABLE 8.6
Local Host Areas with Large Population Increases but Negative Local Components

1960 - 1970

Host Site	Site Number	Local Component	Total Local Change	Year Plant(s) opened, 19___ or are scheduled to be opened
Browns Ferry, Alabama	5	-7,192	- 180	74, 75, 77
Millstone, Conn.	26	-2,056	1,836	70, 75, 86
Turkey Point, Florida	46	-1,202	6,708	72, 73
Diablo Canyon, California	13	-1,124	1,153	79
Calvert Cliffs, Maryland	7	-1,039	981	75, 77
San Onofre, California	41	-939	10,051	68, 82
Pilgrim, Massachusetts	33	-792	4,181	72
Crystal River, Florida	11	-597	4,008	77

1970 - 1980

Host Site	Site Number	Local Component	Total Local Change	Year Plant(s) opened, 19___ or are scheduled to be opened
San Onofre, California	41	-11,743	5,310	68, 82
Diablo Canyon, California	13	-3,704	493	79
Robinson, South Carolina	38	-1,567	3,485	71
Calvert Cliffs, Maryland	7	-1,458	3,283	75, 77
Prairie Island, Minnesota	35	-891	887	73, 74
Oconee, South Carolina	29	-69	2,464	73, 74, 74

THE ST. LUCIE SITE: A SPECIAL CASE

The St. Lucie site lies in eastern Florida on the Atlantic coast in one of the most rapidly growing areas of the United States. The host local area boundary assigned to the station is unsatisfactory for apportioning to components due to its resemblance to the county; therefore, the county was used as the host local area until a smaller local area can be defined. With respect to the other forty-eight sites, St. Lucie manifests very large regional increments and very large negative local increments of growth. Specifically, the regional (contiguous counties) components for 1960-1970 and 1970-1980 were 16,200 and 40,000, respectively, while the local increments were -9900 and -10,200, respectively. The population of St. Lucie County more than doubled during the two decade study period. However, during this same period, the population of the surrounding counties almost tripled, an increase which led to a negative local component for St. Lucie. Clearly, this unusual site is an important case and deserves careful consideration in any further study.

FACTORS CONTRIBUTING TO LOCAL
POPULATION CHANGE

A number of factors other than the presence of nuclear power stations may explain why some host communities have grown more rapidly than others: regional population and employment trends; degree of urbanization, which spurs growth; coastal location; type of tax structure; the date the plant was built; infrastructure such as water, sewers, highways; the physical geography of the area; and the presence of workers constructing the nuclear plant. A survey is underway to try to explain the findings reported above. As data from this project are not yet available, the following sections should be viewed as initial tests of hypotheses concerning the significance of urbanization and water body location in contributing to local population change.

Urbanization as a Factor

The forty-nine host counties, including the host local areas, vary widely in degree of urbanization. In 1970, the median host county was 44 percent urbanized and the range was from zero to 98 percent. The weak association between host county urbanization in 1970 and local population change is indicated by the four correlations in Table 8.7.

TABLE 8.7
Four Correlations of Urbanization and Population

	Correlations	r	Significance
1.	% urban 1970 with local population change 1960–1970	.36	.012
2.	% urban 1970 with local population change 1970–1980	.14	not signif- icant at .05
3.	% urban 1970 with local com- ponent of population change 1960–1970	-.02	not signif- icant at .05
4.	% urban 1970 with local com- ponent of population change 1970–1980	.11	not signif- icant at .05

We conclude that the extent of urbanization of the host counties is not an important correlate of popula-tion change in the host communities.

Coastal Location as a Factor

The East, West, and Gulf Coasts have attracted many residents both from outside and inside the country. Perhaps the salt water location of the coastal sites has made them more attractive than the remaining host communities. In order to test this hypothesis, we divided the forty-nine locations into twenty-three that had a local component of change during 1970–1980 that was at least half of the total population change during the 1970s and twenty-six that had a less important local component of population change. We also divided the communities into categories of water body location.

The results were inconclusive. Contrary to our expectations, the local component was the most impor-tant factor of population change in only three of the eleven salt water coastal sites. It was, however, extremely high in the three where at least half of the population growth was attributed to the local component (Turkey Point, Pilgrim, and Oyster Creek). Indeed, these three had by far the largest local components among the forty-nine during 1970–1980 (Table 8.5).

The ten Great Lakes host communities did not experience much population growth in comparison to the remaining thirty-nine sites. Seven of the ten, how-ever, had local components exceeding half of their total population growth, and three of the ten had among the largest local components during the 1970s. The remaining twenty-eight sites are adjacent to large lakes, reservoirs, impoundments, canals, and rivers. The local component of population change among these sites was not noteworthy.

In summary, neither of the hypotheses that could be tested with readily available data yielded a positive result. The questionnaire survey hopefully will be more productive in isolating the most important explanatory factors.

SUMMARY

Site selection criteria used by the Nuclear Regulatory Commission emphasize the selection of low population areas in which little growth is anticipated. This research examines population growth after site selection for the period 1960 to 1980 for forty-three operating sites. Substantial increments of population increase were found, only partially explained by national, regional, and host county growth trends impacting local host areas. These local components of change became especially important in the decade of the 1970s, when most of the plants were in full operation. The decade of the 1970s also saw a marked shift from the geographic pattern of growth of the 60s, when few plants were in operation. These larger and different growth components of the 1970s, also unexplained by preliminary analysis of correlation with coastal locations and degree of urbanization, are classified into categories with high potential and interest for further research.

NOTES

1. We would like to thank the Nuclear Regulatory Commission and Brookhaven National Laboratory for providing funds for this research; Dr. William Metz of Brookhaven National Laboratory for providing data and reviewing the research; Donald Cleary and Leonard Soffer of the NRC for their helpful comments; Donald Barrows of Rutgers University for helping to gather the data; and William Dolphin of Rutgers University for assisting with the computer programming.
2. The reference here is to investor-owned utilities that pay real estate taxes directly to local government entities.
3. The specific question asked was: "For each type of plant, please tell me the closest distance such a plant could be built from your home before you would want to

move or to actively protest, or whether it wouldn't
matter to you?"

REFERENCES

Buchanan, J. 1975. "AEC Working Paper on Population
    Density Around Nuclear Power-Plant Sites," Nuclear
    Safety, 16(1):1-7.
Chalmers, J., D. Pijawka, P. Bergmann, K. Branch, J.
    Flynn and C. Flynn. 1982. Socioeconomic Impacts
    of Nuclear Generating Stations: Summary Report on
    the NRC Post-Licensing Studies (NUREG/CR 2750).
    Prepared for U.S. Nuclear Regulatory Commission.
    Springfield, Va.: National Technical Information
    Service.
Council on Environmental Quality. 1980. Public Opin-
    ion on Environmental Issues: Results of a National
    Public Opinion Survey. Springfield, Va.: National
    Technical Information Service.
Edel, M. and E. Sclar. 1974. "Taxes, Spending, and
    Property Values: Supply Adjustment in a Tiebout-
    Oates Model," Journal of Political Economy,
    82/5:941-954.
Friesema, H., et al. 1979. Aftermath: Communities
    After Natural Disasters. Beverly Hills: Sage
    Publications.
Haas, J., R. Kates, and M. Bowden. (editors). 1977.
    Reconstruction Following Disaster. Cambridge,
    Mass.: The MIT Press.
Houts, P. and M. Goldhaber. 1981. "Psychological and
    Social Effects on the Population Surrounding Three
    Mile Island After the Nuclear Accident on March 28,
    1979." In Energy, Environment, and the Economy, S.
    Majumdar eds., Philadelphia: Pennsylvania Academy
    of Sciences, pp. 151-164.
Marsh and McClennan, Inc. 1980. Risk in a Complex
    Society: A Marsh and McClennan Public Opinion
    Survey. New York: Marsh and McClennan, Inc.
Oates, W. 1969. "The Effects of Property Taxes and
    Local Public Spending on Property Values: An
    Empirical Study of Tax Capitalization and the
    Tiebout Hypothesis," Journal of Political Economy,
    77(6):957-971.
Palm, R. 1981. Real Estate Agents and Special Studies
    Zones Disclosure: The Response of California Home
    Buyers to Earthquake Hazards Information. Mono-
    graph #32. Boulder, Colo.: Institute of Behavior-
    al Science.
Policy Research Associates. 1977. Socioeconomic Im-
    pacts: Nuclear Power Station Siting, (NUREG-0150).
    Prepared for U.S. Nuclear Regulatory Commission.

Springfield, Va.: National Technical Information Service.

Price, H. 1968. "Siting of Power Reactors," Nuclear Safety, 9(1):1-4.

Purdy, B., et al. 1977. A Post Licensing Study of Community Effects at Two Operating Nuclear Power Plants (ORNL/NUREG/TM-22). Prepared for U.S. Nuclear Regulatory Commission. Springfield, Va.: National Technical Information Service.

Shields, M., et al. 1977. Socioeconomic Impacts of Nuclear Power Plants: A Paired Comparison of Operating Facilities (NUREG/CR-0916). Prepared for U.S. Nuclear Regulatory Commission. Springfield, Va.: National Technical Information Service.

Slovic, P., et al. 1979. "Rating the Risks," Environment, 21(3):14-20 and 36-39.

Wolpert, J. 1964. "The Decision Process in Spatial Context." Annals, Association of American Geographers, 54:537-58.

Wolpert, J. 1965. "Behavioral Aspects of the Decision to Migrate," Papers and Proceedings, Regional Science Association, 15:159-69.

Wolpert, J. 1966. "Migration as an Adjustment to Environmental Stress." In Man's Response to the Physical Environment, Journal of Social Issues, R.W. Kates and J.F. Wohlwill (eds.), 22:92-102.

APPENDIX

SUPPORTING POPULATION DATA

Site No.	Name	State	Local Area Population			Population Change		Local Component of Change	
			1960	1970	1980	1960-1970	1970-1980	1960-1970	1970-1980
1.	Arkansas	AR	1040	1369	1969	329	600	-9	108
2.	Duane Arnold	IA	729	736	792	7	56	-103	60
3.	Beaver Valley	PA	383	328	255	-55	-73	-3	-11
4.	Big Rock Point	MI	2751	3519	3296	768	-223	162	-1193
5.	Browns Ferry	AL	20532	20352	23973	-180	3621	-7192	3195
6.	Brunswick	NC	3355	4346	6828	991	2482	680	501
7.	Calvert Cliffs	MD	5423	6404	9687	981	3283	-1039	-1458
8.	Connecticut Yankee	CT	3466	4934	6383	1468	1449	466	1412
9.	Cook	MI	2016	2146	2212	130	66	61	136
10.	Cooper	NE	567	440	488	-127	48	-8	160
11.	Crystal River	FL	4055	8063	23275	4008	15212	-597	513
12.	Davis-Besse	OH	1570	1355	1706	-215	351	-234	387
13.	Diablo Canyon	CA	7161	8314	8807	1153	493	-1124	-3704
14.	Dresden	IL	197	439	1236	242	797	207	742
15.	Farley	AL	1590	1513	1702	-77	189	-269	-152
16.	FitzPatrick	NY	2489	3619	5455	1130	1836	863	1921
17.	Fort Calhoun	NE	10500	12000	13938	1500	1938	425	1744
18.	Fort St. Vrain	CO	1785	1851	3123	66	1272	-361	576
19.	Ginna	NY	4259	6014	7480	1755	1466	1109	1790
20.	Hatch	GA	1665	1583	2152	-82	569	147	247
21.	Indian Point	NY	2019	2110	2041	91	-69	-123	396
22.	Kewaunee	WI	1094	1105	1142	11	37	-32	33
23.	La Cross	WI	559	728	787	169	59	246	49

Site No.	Name	State	Local Area Population			Population Change		Local Component of Change	
			1960	1970	1980	1960-1970	1970-1980	1960-1970	1970-1980
24.	Maine Yankee	ME	1800	2244	2832	444	588	424	28
25.	McGuire	NC	1578	1992	3454	414	1462	-65	1190
26.	Millstone	CT	15391	17227	17843	1836	616	-2056	213
27.	Monticello	MN	1088	2240	3599	1152	1359	856	390
28.	North Anna	VA	2283	2372	3096	89	724	-115	92
29.	Oconee	SC	11397	12973	15437	1576	2464	1992	-69
30.	Oyster Creek	NJ	1940	4616	14161	2676	9545	896	6697
31.	Palisades	MI	2323	2659	2706	336	47	-39	-394
32.	Peach Bottom	PA	1325	1424	2692	99	1268	-80	1072
33.	Pilgrim	MA	14425	18606	35913	4181	17307	-792	14853
34.	Point Beach	WI	458	580	489	122	-91	79	-75
35.	Prairie Island	MN	12092	12834	13721	742	887	149	891
36.	Quad Cities	IL	953	1003	1050	50	47	-42	69
37.	Rancho Seco	CA	4927	8003	12847	3076	4844	1804	3002
38.	Robinson	SC	24611	24902	28387	291	3485	3200	-1567
39.	St. Lucie	FL	39294	50836	87182	11542	36346	-9912	-10220
40.	Salem	NJ	1293	1400	1547	107	147	73	102
41.	San Onofre	CA	34848	44899	50209	10051	5310	-939	-11743
42.	Sequoyah	TN	8377	10918	13565	2541	2647	1996	1305
43.	Surry	VA	2567	2704	2982	137	278	471	529
44.	Three Mile Island	PA	3053	3453	5138	400	1685	45	160
45.	Trojan	OR	2632	3469	4202	837	733	94	-106
46.	Turkey Point	FL	22149	28857	51865	6708	23008	-1202	15201
47.	Vermont Yankee	VT	865	1024	1175	159	151	68	47
48.	Yankee Rowe	MA	231	277	336	46	59	34	58
49.	Zion	IL	22471	30866	32399	8359	1533	2978	1226

9

The Pattern of Public Response to Nuclear Facilities: An Analysis of the Diablo Canyon Nuclear Generating Station[1]

K. David Pijawka

During the past five years, as nuclear power has emerged as a national social and political issue, research on public response to nuclear power has proliferated at both local and national levels. These studies have become more persuasive following the Three Mile Island accident. Investigations have included attitudinal surveys (Melber, et al., 1977; Kasperson, et al., 1979; and Flynn, 1980), sociological profiles and analyses of motivations of nuclear opponents (Pijawka, 1982), perceived risk research (Pahner, 1979; Gatchel, et al., 1979), and case histories of community conflict over facility siting or expansion (Ebbin and Kasper, 1974; Purdy, et al., 1977). Few authors, however, have documented the nature of the response to a nuclear facility over time, particularly over a period of nearly two decades. The purpose of this paper is to describe and explain the public response to the Diablo Canyon plant, located in San Luis Obispo, California, from the early 1960s to the post-Three Mile Island period.

The response to the proposed Diablo Canyon facility began in the early 1960s as a local issue over the siting of the plant. It later expanded as environmental concerns surfaced as political issues, and then heightened as the facility became the focal point of the regional anti-nuclear movement. The changing nature of both the nuclear-related issues and the anti-nuclear constituency will be described. It is also important to ascertain the degree to which local residents participated in public activity over the facility, to gauge their level of concern, and to measure the salience of issues provoked by the construction and probable operation of the nuclear station. Despite the fact that the construction of the station was completed in 1975, it has not been issued a full operating license, primarily because of the debate over seismic risk posed to the nuclear facility.

The history of public response to the Diablo Canyon project has been both formal and informal. Formal public responses consist of two major elements: (1) responses by groups through normal governmental channels, such as contentions raised by intervenors during permit hearings; and (2) local governmental responses, as illustrated by resolutions passed by communities to reduce potential hazards. Informal public responses refer to events or activities that take place outside existing political channels and are directed at issue resolution. The emergence of public interest and environmental organizations, as well as organized public protests, are examples of such activities. In addition, changes in the attitudes, concerns, and behaviors of residents are considered important components in the informal public response to the Diablo Canyon plant.

PRE-CONSTRUCTION PERIOD, 1964-70

The initial public response to the siting and proposed construction of a nuclear station in San Luis Obispo County began during the pre-construction period with the general planning efforts undertaken in the early 1960s; it continued until after construction permits for two reactor units were issued in 1968 and 1970. This period was characterized by a shift in responses from a generally favorable climate in the early 1960s, when the public attitude was generally favorable to nuclear facilities, to a less favorable climate where public concern over environmental questions engendered local opposition to the nuclear plant siting. This period is indicative of the changing attitudes of many Americans regarding the impact of nuclear power plants on the natural environment. The nature of the public response to the Diablo Canyon plant during this period was most observable during the early siting controversy by the contentions expressed in the California Public Utilities Commission Hearings and in the Atomic energy Commission's Construction Permit hearings.

Site Selection Issue

The site initially selected for the plant was a 1,120-acre (453 ha) lot near the coastal sand dunes at Nipoma in San Luis Obispo County, approximately 20 miles (32 km) south of the present Diablo Canyon site. The selection of that site was favored by the county supervisors, and there was no significant public opposition at that time. Rather, information from interviews with key informants suggests that the tax revenues from constructing the facility were expected to be large and relatively important to the county economy. The anticipated increased revenues and the expected increased sales in the business community were factors identified by key informants as enhancing local acceptance of the proposed site.

Early concerns about the plant centered on the environmental effects regarding the Nipoma Dunes location. Both the Sierra Club and the California State Resources Agency (parks and recreation) took the position that this part of the coast should be developed as a scientific and recreational area rather than an industrial area. The opposition to locating the plant at Nipomo Dunes was not directed at preventing the establishment of a nuclear power station per se (at that time, the Sierra Club had not taken a position against the development of nuclear power), but was motivated to preserve what was considered to be a unique coastal environment. In fact, the Sierra Club's position was that if the Nipomo Dunes site was selected for the plant, it should be set 4000-5000 feet (1220-1524 m) inland from the beach area. According to the utility, however, constructing the plant so far inland would have entailed a significant loss of economic advantages, especially regarding the cooling design system. Because of the additional costs, the utility rejected the inland siting proposal. A decision was then made by the Pacific Gas and Electric Company to work in concert with the state agencies, the County Planning Commission, and the conservation groups in an effort to find a suitable alternative site. The Diablo Canyon site was selected with the assistance of local conservation groups and the County Planning Department.

The site-selection process took place over a period of two years, 1964-66. Comparisons were made of eleven potential southern coastal sites using seven siting criteria: availability of circulating water, isolation for the exclusion area, land availability, suitable routes for transmission lines rights-of-way, community acceptance, access to transportation, and a seismologically adequate foundation. Three of the sites were rejected because of their location within the Vandenburg Air Force Base exclusion area in Santa Barbara County. The other sites, except for the Avila Beach site, were also rejected on the basis of unfavorable community acceptance, as these locations were near existing or proposed state parks or residential areas. The Avila Beach site was withdrawn from consideration because of privately planned development of the area. In 1966, the Diablo Canyon site was selected, following agreement by the State Resources Agency, the Sierra Club, San Luis Obispo County, and the County Conservation Association.

The California Public Utilities Commission Hearings (CPUC), 1967-69

Prior to issuance of a construction permit by the Atomic Energy Commission, a utility in the State of California must obtain a Certificate of Public Convenience and Necessity. The utility is required to furnish information on cost of power, safety factors, and environmental effects in an effort to support its claim that the nuclear power plant is needed and can operate safely. As part of the CPUC evaluation process, public hearings were held in 1967 for Unit 1, and in 1968 for Unit 2. Unit 1 was approved in November 1968 and Unit 2 in March 1969. The hearings afforded an opportunity to ascertain the topics of public concern over the Diablo Canyon plant.

At the CPUC hearings, public opposition to the Diablo Canyon nuclear plant was limited principally to the environmental effects of the plant related to its coastal location. Contentions were made that the construction and operation of a nuclear power plant would adversely affect the natural coastline, its recreational potential, and the coastal wildlife and fish. The expression of the need for preserving the natural state of the coastal area was spearheaded by the Scenic Shoreline Preservation Conference (SSPC), an environmental interest group formed specifically to oppose the siting of Diablo Canyon plant.

The formation of the SSPC is particularly important to an understanding of the emergence and the duration of nuclear plant-related issues. When the Sierra Club originally approved the plant site at Diablo Canyon, its membership had not reached a consensus regarding the site. In fact, a number of Sierra Club members had taken a position opposing the building of a nuclear plant in any part of the coastal region. The internal conflict within the Sierra Club resulted in some members withdrawing from the organization and subsequently forming the SSPC to oppose the Diablo Canyon project. The SSPC presented its views on environmental effects at the CPUC hearing and participated as intervenors in the later AEC/NRC licensing permit hearings. The group was centered in Santa Barbara County, but some of its members resided in San Luis Obispo County.

Construction Permit Hearings, 1968-70

Pacific Gas and Electric submitted an application to the AEC for a construction permit for Unit 1 in January 1967, and a public hearing was held in San Luis Obispo in February 1968. The permit was issued in April 1968, following a brief public hearing at which a number of environmental concerns were expressed. These concerns did not differ from those expressed during the CPUC hearings. The construction permit application for Unit 2 was filed in June 1968, and a short public hearing was held in January 1970. The main contentions of the intervenors included potential environmental effects of plant operation and the instability of the geological foundation on which the plant would be built. In August 1970, seven months after the January hearing, the hearing was reopened on the basis of additional geological evidence presented by the intervenors. However, the Atomic Safety Licensing Board did not find the evidence persuasive, and in December 1970 the construction permit for Unit 2 was issued by the AEC.

Attitudes Toward the Diablo Canyon Project During the Pre-construction Period.

The Utility. Planning for the nuclear generating facility began in the early 1960s. At that time, Pacific Gas and Electric had made decisions to invest heavily in nuclear power development, and the Diablo Canyon plant was one of five existing or proposed facilities. According to the Pacific Gas and Electric Company, the decision to build nuclear plants in their service area was based on the argument that the development of nuclear power would have economic and environmental benefits over fossil-fueled facilities. The utility company was also cognizant of the importance that state and local agencies placed on preserving the enviromental quality of the coastal area, and it stated that nuclear power would be environmentally advantageous and would not discharge combustion products into the atmosphere. The development of a large fossil-fueled facility in nearby Morro Bay reinforced the utility's argument for an environmentally cleaner alternative power source for the region. Moreover, in the 1960s, the national scientific consensus and political climate favored the development of nuclear power, and this public predisposition was apparent in the region. Nuclear power for generating electricity was viewed as economically and environmentally preferable to conventional generation.

Community and Political Leaders. Between 1962 and 1970, local residents held a generally favorable attitude toward the Diablo Canyon plant, and local politicians and the business community had a highly favorable attitude toward its location in the county. Recent interviews with business leaders indicate that, in general, there was support for the plant when it was announced because large increases in economic activity and tax revenues were anticipated. All of the area

business leaders interviewed agreed that during the
1960s, unemployment was generally high, the tourist
industry was experiencing large seasonal fluctuations,
and industrial employment was needed to increase the
tax base and to reduce unemployment. The nuclear
plant, it was perceived, would aid in meeting these
objectives.

The San Luis Obispo county Board of Supervisors
unanimously endorsed the Diablo Canyon plant. Interviews with the county administrators, supervisors, and
planners who held positions at that time suggest that
the location of the plant was acceptable to the county
because it was relatively isolated and without public
access. The Board of Supervisors publicly endorsed the
nuclear plant, largely motivated by the increased tax
base the plant would generate. In fact, members of the
board fought among themselves, each wanting the plant
to be located in their district. In addition, efforts
were made to redraw school district boundaries at that
time to equalize the tax benefits from the plant.

The planning department, while generally favoring
the plant, was originally interested in issuing a
conditional use permit for Diablo Canyon pending the
completion of a number of zoning studies. However, the
California District Attorney ruled that the state
(CPUC) superseded the county in zoning for utilities.
Despite this setback, the local planning department
continued to monitor the project's development. In
addition to helping locate the plant, the planning
department actively participated in challenging the
utility over a number of transmission line issues. The
department criticized the routing of transmission lines
over valued scenic landscapes and the method by which
towers were built which resulted in erosion problems;
in two cases, it was successful in changing the
routing.

The Public. Pacific Gas and Electric Company
sponsored a public opinion poll in 1967 that questioned
attitudes toward nuclear power. The poll found that 75
percent of the residents of the City of San Luis Obispo
favored the construction of nuclear power plants near
their community, while only 7 percent opposed such
facilities. However, only 40 percent of the residents
in the City of San Luis Obispo were aware of a proposed
plant site in their own area. Of the respondents in
the City of San Obispo, 61 percent did not see any disadvantages in having an atomic plant nearby, and 19
percent indicated that there were disadvantages to
having an atomic power plant in or near their community. However, of those respondents who indicated
there were disadvantages, less than one-third were
concerned about exposure to radiation, and about one-
fifth expressed fear of a plant accident. Only five

percent of those who were concerned about the nuclear power plant identified its potential for changing the nature of their community as the basis of their concern.

The poll indicated that 63 percent of the City of San Luis Obispo sample perceived that a nearby atomic power plant would have advantages. The principal advantage, that nuclear plants would produce more economical electricity, was followed by the view that a nuclear power plant in the area would attract industry or create jobs. In addition, a large number of respondents indicated that nuclear power plants would have air quality advantages over fossil-fueled plants. Almost 10 percent of respondents indicated that one advantage of having a nuclear power plant near their community was that it would reduce property taxes.

In general, it can be said that the plant originally was strongly favored by the major political, business, and agricultural interests in the county and also by most residents, as the City of San Luis Obispo survey suggests. This acceptance may have been related to economic considerations: new job opportunities were identified as the major concern of San Luis Obisop city residents, as shown in the 1967 survey. Despite this predilection, public concerns over environmental quality should not be downplayed; with a large elderly (retired) population to whom environmental conditions were important, and a large, politically active professional population in the City of San Luis Obispo, the preconditions for intervention in the siting process existed.

THE ENVIRONMENTAL HEARINGS

The initial environmental hearings over the Diablo Canyon plant, Unit 2, took place in 1973, but a number of unresolved questions resulted in periodic reopenings of this hearing until 1978, when final decisions were made on the environmental impacts. The 1973 hearings were necessitated because the National Environmental Policy Act (NEPA) required that construction license renewal be accompanied by an environmental assessment. The passage of NEPA in 1970 and its application to nuclear power plants required an environmental impact statement for plants that had not yet been issued an operating license. The NEPA hearings took place while construction of the plant was underway. Consequently, a number of specific environmental effects from plant construction became contentions at the hearing in addition to the more generic environmental issues.

The results of a pre-hearing in 1973 ruled out any discussion of the danger posed by earthquakes to plant safety as part of the environmental hearing. As a consequence, the intervenors charged that it was irresponsible for the regulatory agency not to consider seismic matters at this hearing because of the 1971 San Fernando Valley earthquake which may have altered the geophysical nature of the area. The intervenors argued that the information gained from the 1971 earthquake demonstrated that the geologic data used for the plant's construction permit in 1966 were invalid. According to the intervenors, accelegraphic records showed that ground movement in geologic areas similar to the Diablo Canyon site could be higher than the maximum values that were used to design the Diablo Canyon plant. However, contentions over thermal discharges and their impacts, power requirements, transmission lines, and cooling techniques were approved for inclusion in the hearings.

Concern over Environmental Quality: Regional Background

Historical evidence suggests that the principal concern expressed by the public over Diablo Canyon during the pre-construction period was related to the potential damage to the coastal environment. Expressions of concern, as measured by participation in the hearings, was generally limited to environmental interest groups and a few individuals expressing self-interest contentions. A number of key informants, including intervenors, suggested that the relatively limited opposition to the plant during the pre-construction period may have been the outcome of the Sierra Club's endorsement of the Diablo Canyon site, the County Planning Department's efforts in monitoring the plant's development, and the general public attitude that nuclear power posed fewer air quality problems than fossil fuel plants. However, these individuals also pointed out that the protection of the coastal environment had been a long-time concern of residents in the region and that, by the early 1970s, environmental/conservation issues were of principal interest to residents of the area.

The importance and growth of environmental quality as a major public issue in the region can be illustrated by the following examples. In 1964, the proposal to site a nuclear power plant at Bodega Head, California, was not successful due to public opposition based on seismic and environmental concerns. In the mid-1960s, while San Luis Obispo County experienced substantial growth in its population and in its economic base as the tourist industry expanded, the planning philosophy established by the county was one based on preserving agriculture, preserving the coast for tourism, and protecting natural resources. To this end, the planning department monitored the construction activity of the Diablo Canyon plant and criticized a number of the utility's decisions. In addition, the planning commission defined limits for the growth of urbanization in the county. For example, the agricultural preserve program was a concept instituted to restrict urban growth in the northern and southern parts of the county by reducing certain taxes on farmland under the condition that the land not be sold for development for a set period of time. In addition, by the late 1960s, coastal communities in the area, such as Cambria and Morro Bay, had a strong environmental philosophy. Because of water supply problems in this part of the county, limited growth was planned. Morro Bay passed a moratorium on further development, and the California Coastal Commission passed regulations to prevent growth in communities with water supply problems. The City of San Luis Obispo had also taken strong planning measures to restrict growth by discouraging industrial growth and by passing a law that would refuse annexation without public approval.

Strong support for preserving and protecting the environment manifested itself in the late 1960s and early 1970s in a number of public issues. The proposal to construct the Lopez Dam for county flood control resulted in a clash between government officials and environmentalists who claimed that the project would destroy important and unique natural habitats. Similarly, the proposal to develop the Port San Luis Harbor as a recreational-boating marina was objected to as growth-inducing, and the Hearst Ranch project to develop 28,000 acres (11,336 ha) in the northern part of the county was opposed by a number of the county's conservation groups. Finally, the Lomex project to explore for uranium emerged as a countywide issue as groups expressed concern over the adverse effects uranium mining could have on groundwater quality.

The trend toward conflicts over growth issues resulted in a change in the structure of the County Board of Supervisors. This change resulted in the election of two environmentalists to the five-member board. Since 1972, the board has been split between pro-growth advocates (three supervisors) and environmentalists (two supervisors). Although the evidence indicates that this change did not occur directly as a result of the Diablo Canyon plant, the long controversy over the safety of the plant reinforced the split among the county supervisors. For example, a vote to have the board be part of the intervenors during the licensing hearings was defeated. The split was also apparent on most growth issues. For example, with respect to the Lomex uranium project, a resolution to ask the Bureau of Land Management to place a moratorium on mining failed.

In the early 1970s, the need for environmental protection received strong support statewide, especially
with respect to the coastal areas. The level of environmental activity and concern over growth questions
were particularly intense in the local area. This concern was not initiated by the proposal to build the
Diablo Canyon plant; rather, environmentalists viewed
the decision to construct and operate a nuclear plant
in San Luis Obispo County as another example of encroachment on limited natural space, in addition to its
posing a number of serious environmental impacts. When
the nuclear plant was an issue at the CPUC hearing, the
County Board of Supervisors unanimously endorsed the
utility's plans for its construction. By 1972, the
political climate in the county had changed: environmental protection had become a respectable cause, the
coastline initiative to limit coastal development had
passed, and a "go-slow" development attitude was
reflected in the county election of two avowed
environmentalists.

Thermal Emissions Impact Dispute

The major issue during the environmental hearings
centered on the impact of thermal emissions on the
abalone population in the Diablo Cove. The intervenors
argued that the AEC's final environmental report did
not include complete studies of the thermal effects on
marine life and that it did not provide sufficient
information on the thermal plume. At issue was the
intervenor's objection to the size and shape of the
thermal plume, based on studies provided by the
utility.

In the Final Environmental Statement, the AEC
expressed concern that the thermal discharge at Diablo
Cove could reduce the abundance of kelp, which in turn
would adversely affect the abalone population. The
catch of abalone in the area was estimated to be substantial, amounting to 53 percent of commercial fish
caught between 1965 and 1970. The AEC staff also
argued that additional reductions in algae might occur
and that this, in turn, might further reduce the
abalone population in the area. Based on an assumption
that the algae could be reduced in an area equal to
one-half the cove area, the AEC estimated a total loss
of 110,000 abalone.

Utility consultants disputed the loss estimate and
argued that only a small population near the plant was
in danger; most of the abalone, it was concluded, were
able to tolerate warmer water. The utility also argued
that the area of loss (35 acres (14.2 ha)) was comparatively smaller than the AEC estimate and, additionally, other species of kelp would take over in warmer

water while, at the same time, the abalone would adjust to the thermal changes. Finally, the AEC concluded that the analytical data used to predict the impacts of the discharge were not sufficient. The utility was ordered to continue construction while studying the impacts of the discharge. The AEC stated that any unacceptable degree of damage could result in plant shutdown. The intervenors, however, criticized this ruling because a standard to define "unacceptable" was not determined.

The issues expressed at the hearings focused on two major issues: the destruction of the natural coastline and interference with abalone fishing. The intervenors ultimately failed to halt construction of the plant on the basis of environmental issues, and in June 1978, the Atomic Safety Licensing Board reached its decision that Diablo Canyon had met the federal environmental requirements. It should be noted, however, that the opposition to the plant conformed to the constraints imposed by the rules and regulations of the AEC. Only environmental issues were discussed; concerns over health effects and seismic risk were not addressed at these hearings.[2]

OPERATING LICENSE HEARINGS: THE SEISMIC RISK ISSUE

The application for an operating license and the Final Safety Analysis Report were filed in September 1973. According to utility personnel, key county informants, intervenors, and media coverage, the seismic hazard was the principal issue during the operating license hearing.

The salient issue in the debate over the issuance of the operating license was centered around the risk posed by siting the Diablo Canyon nuclear plant near an existing offshore earthquake fault. In order for the construction permits to be granted, the utility was required to furnish supporting data that the design criteria would meet postulated levels of seismic risk. The utility's assessments were approved by the AEC based on investigations of the site in 1965-66. However, the discovery of an offshore fault (Hosgri Fault) in 1972, at a distance of three to five miles (4.8-8 km) from the plant site, altered the accepted assumptions of the seismic risk in the area and initiated a process of risk assessment and regulatory decision-making over a period of several years (1972 to 1979), when an Initial Partial Decision was made by the NRC following a series of public hearings in 1978. The period was characterized by high levels of uncertainty and conflicting assessments of the risks posed by the offshore fault.[3] For the regulatory agency, it was a particularly difficult period because decisions had to be made under conditions of scientific uncertainty and in a context of growing public opposition and concern over a broad array of generic safety questions over nuclear power. For the utility, the seismic risk issue engendered increased local and regional public concern and opposition to the plant and resulted in costly delay and design retrofitting. For the intervenors, the seismic risk issue was considered to be the most critical area of concern; the principal intervenors who were interviewed held the view that the operation of the plant in itself posed unacceptable risks to the public and that the location near a fault substantially increased the probability of a major accident.

Issues in the Risk Assessment Process

The risk assessment process may be considered to consist of three major components of risk: identification, estimation, and evaluation. Risk assessment efforts over the Hosgri Fault have been characterized by contentions over scientific fact and interpretation inherent within each major component of the assessment process. Components of risk assessment are outlined as follows: identification--the definition of the fault, its extent and physical characteristics, and the nature of the geological structure at the site--in order to assess the magnitude of the seismic disturbance; risk estimation--measurements of the postulated earthquake magnitude and ground acceleration and probability of earthquake occurrence; risk evaluation--the judgment of the level of risk that is acceptable, the balancing of benefits and risks, and the acceptability of design criteria to withstand the postulated seismic hazards.

Contentions in Risk Identification

The utility consistently argued that its 1966 assessment of the seismic hazard was valid in light of the Hosgri Fault discovery and that the plant was designed to withstand a probable Hosgri earthquake. The original plant design was based on two determinations of seismicity. First, according to the utility, extensive site exploration demonstrated that the site was not affected by significant fault movements and additionally, that there was an absence of seismic activity on land such as might result from an extension of an offshore fault. Second, the design of the plant was based on conservative assumptions of a large quake occurring in the region or near the site; a 6.75 magnitude earthquake was the maximum ever expected to occur at the site, and the plant was designed to be able to withstand a force of that magnitude safely.[4]   The

Hosgri Fault was assessed to present a low level of seismic activity that could be accommodated by the plant's original design.

The intervenors' contention was that the utility failed to determine the length of the offshore fault adequately, its relationship to the regional tectonic structure, and the nature of the displacement along the fault, all of which were considered to be critical factors in determining the magnitude of a possible earthquake and the probability of its occurrence. Supported by a number of scientists, the intervenors' position was that the Hosgri Fault was connected or linked to other offshore faults, resulting in a much lengthier fault with a potential for earthquakes of much greater magnitude.[5] The utility, with support from the United States Geological Survey (USGS) and the NRC staff, criticized these views and argued that the Hosgri Fault was not linked, but rather part of a regional fault system characterized by individual, discontinuous faults. Further, the utility characterized the fault as a "capable" fault with a small to moderate magnitude, based on its relatively short length (50-100 miles (80-160 km)) and minor surface displacement--a second or third order fault. The NRC Atomic Safety Licensing Board supported the utility's position that the Hosgri Fault was a distinct unconnected break. The NRC used this finding to deny the request of the intervenors to halt construction work until the seismic risk issue was resolved.

Contentions in Risk Estimation

In 1975, the NRC requested the USGS to evaluate the Hosgri Fault, based on a contention that the ground motion value of 0.5g used by the NRC (the re-evaluated value) was inadequate and the contentions that the 7.3 magnitude earthquake in 1927, thought to have originated in the Lompoc Fault further offshore, could be reassigned to the Hosgri Fault. The USGS studies, based on 1975 and 1976 assessments, resulted in the following findings: (1) the criterion for the design to withstand an earthquake was recommended to be a 7.5 magnitude event based on the fact that the 1927 earthquake with a possible magnitude of 7.3 may have occurred at the Hosgri site. Moreover, the large magnitude of the 1927 earthquake would preclude attribution to the Lompoc Fault because of the fault's relatively short length of eight miles; and (2) a possible 7.5 magnitude earthquake exceeded the plant's seismic design criteria, and the 7.5 value was recommended to determine the safety of the plant in the event of an earthquake.

The utility argued against the USGS findings and recommendations. According to the utility consultants, the Hosgri Fault was not the source of the 1927 earthquake because there was no evidence of any significant offset of the sea floor. The utility consistently argued that the plant's design to meet a 6.5 magnitude earthquake was reasonable.

In 1976, based on the USGS findings, the NRC requested PG&E to reevaluate the seismic structure on the basis of a 7.5 magnitude earthquake and to retrofit the plant to meet the 7.5 design criteria. Of critical importance was whether the Diablo Canyon plant could adequately cope with ground motion generated by a 7.5 magnitude earthquake. The estimation of ground motion acceleration was an area of substantial contention. Based on the NRC staff's estimates, the utility took the position that an unlikely Hosgri event would result in a 0.75g (effective acceleration), which could be met by some modifications to the plant. (The original design was based on a 0.4g peak ground acceleration.) The intervenors, however, argued that, based on information gained in the 1971 San Fernando Valley quake, a 1.15g (peak acceleration) for a 7.5 magnitude earthquake was probable.[6] In contrast to this position, PG&E held the view that a 6.5 magnitude earthquake and a 0.5g surface acceleration were adequate for seismic retrofitting; a 7.5 magnitude earthquake and a 0.75g surface acceleration were extremely conservative values, they contended. Despite this contention, the 0.75g acceleration was accepted by the NRC as a basis for the plant's retrofitting.

Contentions in Risk Evaluation

Based on the assumption of a 0.75g effective ground acceleration, the utility reanalyzed the plant structures and equipment, a process which resulted in a need to modify the structure of the turbine building at the plant to withstand greater ground movement. The intervenors criticized NRC's apparent acceptance of what they considered an unacceptable risk situation. Interviews with intervenors indicate that the seismic issue was not resolved to their satisfaction: (1) they believed that the assumption of a 0.75g acceleration for retrofitting the plant was inadequate to assure safety; (2) retrofitting was limited to noncritical areas of the plant, and the integrity of the reactor and coolant system was not assured in the case of an earthquake of large magnitude; and (3) retrofitting at the relatively late stage of the plant's construction was perceived as problematic--construction activity should have been terminated much earlier--until the seismic issues had been resolved.

Seismic-related Nuclear Fuel Hazard

As construction of Unit 1 neared completion and fuel loading was impending, the intervenors requested that the Nuclear Regulatory Commission prevent the Diablo Canyon plant from receiving the nuclear fuel. The group's position consisted of three arguments: (1) it was not justifiable to receive and store the nuclear fuel on-site prior to the issuance of the operating license; (2) the utility had not demonstrated that the storage facilities at Diablo Canyon would prevent the fuel from being activated during an earthquake; and (3) the utility had not established an on-site security system that would ensure against fuel sabotage.

The outcome of the request was that the nuclear fuel shipment for storage on-site was delayed. The NRC did not have sufficient or timely information to resolve the fuel safety storage issue, and no policy ruling had hitherto been made on the standard for safety with respect to the relationship between earthquake activity and its effect on stored fuel. In August 1975, four months following the request for a delay in the fuel shipment, the NRC ruled that it would presume that the stored fuel at the Diablo Canyon site would be unsafe until the utility could prove that there was no danger posed by the storage.

The public hearing (9-12 December 1975) on the receipt of nuclear fuel for Unit 1 resulted in four additional safeguards, including storage of the fuel in the spent fuel pool and storage in aborated water solution with a concentration of twice the acceptable standard. In December 1975, the NRC approved the storage of fuel at Diablo Canyon. Another Mother for Peace (AMP) appealed the decision on the basis that an earthquake could cause the fuel to go critical. Both the NRC and the utility discounted the risk--probability of occurrence and the magnitude of the consequences--as being of almost negligible probability. In June, 1976, the decision to store the fuel on site was reaffirmed.

THE NATURE OF PUBLIC INTERVENTION

Stages of Intervention

The complexion and nature of public intervention regarding the construction of the Diablo Canyon plant has changed markedly over time. Four distinct phases of public involvement were discerned. The earliest expression of public concern (1964-66) was related to the site selection and the need to protect a unique coastal dune area. The selection of the present Diablo Canyon site ameliorated this concern.

The protection of the coastline continued to be a critical issue during the CPUC hearings (1968-70) as environmentalists became the key participants. However, concerns were also expressed in this second phase by individuals motivated by self-interest--the potential reduction of the abalone population for fishermen and opposition to the routing of transmission lines.

In the early 1970s, the requirement of assessing the environmental impacts of nuclear generating plants had become legitimized through the passage of the 1970 National Environmental Policy Act and the Federal Water Pollution Control Act, enacted in 1972. Subsequently, the environmental hearings, the third phase of public intervention, were characterized by a shift away from concerns over the general encroachment of the coastal area to issues over thermal and radiological effects. Moreover, this third phase of public response was marked by growth in the number of committed intervenors and in their public support. The intervenors had also become more broad-based, representing environmentalists' interests, parochial/private interests, and ideological/nuclear safety concerns.

The fourth phase of public intervention (1975-81) concerned issues over Diablo Canyon's safety, principally with respect to seismic risk. Other issues, mostly generic safety questions, such as the need for emergency planning, plant security, and waste shipment, were long-standing concerns, but became salient following the Three Mile Island accident in Pennsylvania.

The Intervenors

Four petitions were filed in 1973 to intervene in the environmental hearings. The Scenic Shoreline Preservation Conference was recognized as an intervener. The group was based in Santa Barbara County and had members in San Luis Obispo County. This group was formed in 1966 to oppose the siting of the plant following the Sierra Club's acquiescence and participated in the AEC construction permit hearings. The Luigi Marre Land and Cattle Company, a ranching concern in San Luis Obispo County, also appeared as an intervenor and primarily opposed the routing of transmission lines on their ranch land. The company withdrew as an intervenor following the environmental hearings. The third intervenor, California Polytechnic Ecology Action Club, consisted primarily of a small group of individuals from the state university in the City of San Luis Obispo.

The fourth intervenor, Another Mother for Peace, was perhaps the most significant public organization opposing the project, as measured by the size of its membership, its long-term commitment to nuclear safety issues, and its visibility as the most active local opponent. The organization has a long history of political activism; in the early 1960s individuals who later formed or joined AMP had been involved in opposing nuclear weapons-testing because of the radioactive fallout (strontium-90) effects on children. Because of group members' strong ideological position against weapons-testing and war, the group opposed the war in Vietnam. Its concerns over the effects of radiation persisted, and with the termination of the Vietnam war, the group turned its attention to stopping the construction and operation of the Diablo Canyon plant during the operating license hearings.

Nearly all of the active group members are female and college educated, representing a wide range of ages. Although the nuclear plant issue is of major importance to the group, the group is not a single-issue group and has been very active in a broader scale, especially over anti-war issues, the plight of Cambodian refugees, and the mining of uranium in the county. The group's major concern with respect to nuclear facilities included uranium mining as a health issue, just as they had opposed nuclear weapons-testing because of adverse health consequences; the group has therefore opposed such mining in the county. Group spokespersons indicate that the group has been instrumental in broadening the safety issues by introducing questions of risk that were not previously dealt with at hearings. Such questions include the level of retrofitting and the level of risk inherent in stored fuel under conditions of geological instability. On these issues, the group feels it has been successful in delaying construction, in pressing for more stringent standards for storing fuel, in initiating local efforts to develop emergency evacuation plans, and in educating the public on the hazards of nuclear technology. However, the intervenors have taken the position that "no amount of retrofitting or added band-aid adjustments will make Diablo safe."

The concerns of the intervenors and their supporters over the nuclear power plant were rooted in the environmental movement and in the disarmament/antiwar movement. Furthermore, the Diablo Canyon plant was located in an area where environmental quality was already held as an important community value. Of particular and consistent importance to area residents was the preservation of the coast and its planned growth. With respect to the siting impact of the plant on the coastal environment, the early involvements of the Sierra Club, the planning department, and the SSPC were forerunners to the broader environmental issue of the State Coastal Initiative of 1972. Analyses of newspaper reports, interviews with key elected officials, and examination of voting records demonstrate

that environmental quality concerns were important
community issues in the county in general and in the
City of San Luis Obispo in particular. Moreover,
individuals who consider themselves environmentalists
are a powerful political force in the area.

The intervenors have been gaining local and
regional public support since the mid-1970s. The
intervenors are, at present, closely linked with
regional and local anti-nuclear organizations that have
proliferated during the past few years. The accident
at Three Mile Island (TMI) was considered by the intervenors as the critical point in the history of opposition to nuclear power, as the risks of nuclear technology and the probability of a seismic-related accident
became a perceived possibility to them. The intervenors who were interviewed criticized the NRC for its
policy of promoting nuclear plants and not taking more
stringent actions to assure safety. For example, a
particularly disturbing fact, according to the intervenors, was that the NRC did not terminate construction
work until the seismic assessment was complete, a
decision which resulted, according to the intervenors,
in less stringent seismic retrofitting efforts.

Informal Response

The nature of the public response to the Diablo
Canyon nuclear generating facility has been described
in terms of its two major dimensions: participation in
legally constituted political areas (the hearings process) and local governmental response. The third component, the level of public response outside of normal
governmental channels, is more difficult to ascertain.
For the Diablo Canyon case, two social phenomena were
observed: a proliferation of broad-based organizations
in the Study Area between 1975 and 1981, and a change
in the levels of acceptability and attitudes toward
Diablo Canyon by the general population of the Study
Area.

Public Attitudes Toward Diablo Canyon. Available
public opinion polls were scrutinized in order to
ascertain the degree to which residents in the area
favored or disapproved of the plant and the reasons for
their predispositions. It should be noted that the
survey instruments were not consistent in approach or
in specific questions, and one in particular has been
publicly criticized as biased. However, on the whole,
they can be used as crude indicators of perceptual and
attitudinal change.

The 1967 poll demonstrated a highly favorable attitude toward the siting of the Diablo Canyon Plant, with
only 7 percent of the residents opposing the plant.

Moreover, 61 percent of those interviewed found no
disadvantages relating to the nuclear plant. These
findings seem to be consistent with the relatively low
participation rate of active intervenors at that time.

Although local residents generally favored the
plant in 1975, there were some indications that there
were misgivings and concerns, even among those who
generally favored the plant, and the size of the population opposed to the plant had grown. The poll showed
that 75 percent of San Luis Obispo County residents
favored the plant, but 30 percent of them favored it
only "somewhat." Approximately 20 percent definitely
opposed Diablo Canyon; 5 percent were undecided. Of
those favoring the plant, the "need for power" was by
far the number one reason for the positive response.
In 1967, radiation and accidents were the two primary
concerns regarding nuclear power. By 1975, residents'
concerns had expanded and were more focused; they
varied from generic safety issues to concerns that were
specific to the Diablo Canyon site. Of those opposing
Diablo Canyon, the chief expressions of concern were
that they generally considered the plant not to be safe
and that the issue of waste disposal had not been
resolved. In addition, the problem of radiation leaks,
seismic risks, and thermal emissions were also
important reasons for a negative predisposition. The
analysis results of the poll also indicated that favor-
ability toward the plant was directly related to length
of residence (support increased with lengthier residence, age (those 18-30 opposed the plant in greater
numbers than other age groups, and student status
(full-time students were not as likely to favor the
plant as were part-time students or nonstudents).

The analysis of polling results demonstrated that,
in terms of approval by residents, the response to the
Diablo Canyon Nuclear Generating Station varied geographically within the study area. If the county in
which the plant is located is broken down into its five
political districts, residents of District 5 were least
supportive of the plant. District 5 includes the north
section of the City of San Luis Obispo in which the
university is located, the lower part of Atascadero,
and the rural area east of the City of San Luis Obispo.
The social groups in this district--university faculty
and students (a younger population relative to the rest
of the county), and a large professional population--
would have a stronger predisposition to oppose nuclear
power according to sociological surveys of public
response to nuclear power plants. In fact, most of the
public interest anti-nuclear organizations that emerged
in the Study Area had memberships largely concentrated
in this district, particularly in the City of San Luis
Obispo.

Residents of Districts 1 and 4 were most favorable toward the Diablo Canyon facility; 89 percent of the respondents in District 4 and 79 percent of those in District 1 supported the plant. District 4 is located in the southern part of the county and includes the Five Cities Area. In terms of economic gain generated by the construction of the plant, District 4 was affected to a larger degree than the other districts.

Residents of District 2, which includes the northern and western parts of the City of San Luis Obispo, Morro Bay, and the coastal communities of Cambria and Cayucas (an area where residents placed a particularly high value on environmental quality) expressed concerns over the plant to a degree almost equal to those in District 5.

Information on public attitudes subsequent to 1975 was derived from the 1976 California Nuclear Initiative (Proposition 15) voting statistics and from a partially released 1980 utility-sponsored survey. The California initiative to restrict nuclear power development in the state was supported by about one-third of county residents and reflected the state-wide voting pattern. The 1980 survey, according to newspaper accounts, showed that the size of local support for Diablo Canyon was dwindling; 60 percent of San Luis Obispo County residents favored the plant "if licensed by the NRC", but only one-third of these unconditionally favored the operation of the plant. Moreover, a majority of residents in 1980 expressed concern over whether the Diablo Canyon plant was safe; 55 percent indicated that it was not safe; 47 percent said that it was safe, and 8 percent were undecided. The uncertainties in the seismic risk assessments, the inconsistencies of scientific positions, and the claims made regarding insufficient retrofitting have elevated the seismic risk question as a major local concern. In fact, 60 percent of the respondents indicated a preference for reopening the operating license hearings because of the seismic hazard. The fact that the local newspaper shifted positions and attacked the Diablo Canyon plant may have been an important factor in public attitudes.

The dramatic decline in local support for Diablo Canyon, as evidenced by these surveys, may be hypothesized to be: (1) a function of the media's wide coverage of the seismic controversy, (2) the activities of anti-nuclear organizations in the local area, and (3) the slippage in support for nuclear power development in the state as a whole since 1976. With respect to the latter, the California Opinion Index on energy issues was examined. In May 1976, the index found that 69 percent of state residents favored having a greater number of nuclear power plants, while 19 percent opposed such a proposal. In May 1979, two months after the Three Mile Island accident, the index noted that only 37 percent (a decrease of 32 percent) of California residents responded in favor of nuclear plants and that 55 percent were opposed. This was a significant shift and may be partly explained as an effect of Three Mile Island. A direct relationship cannot be made at this time because of incomplete data to interpolate during the 1976 to 1979 period. In April 1980, the percentage favoring more nuclear facilities in the state increased slightly to 45 percent.

The Anti-Nuclear Movement in San Luis Obispo County. Between 1975 and 1980, a relatively large number of anti-nuclear organizations and anti-nuclear activities surfaced in the region. The opposition to Diablo Canyon reached its height in 1978 with a mass demonstration in which 500 persons were arrested. This growth in opposition is a function of four factors: the historical and present commitment of residents to environmental quality, including environmental activism; the growth of the national anti-nuclear movement, which is centered in California; concerns of San Luis Obispo County residents over the safety of the Diablo Canyon plant, particularly concerning the seismic hazard; and the social structure of the area. Most of the area's organizations and their memberships were centered in the City of San Luis Obispo. The preconditions for a viable public response to the Diablo Canyon plant had been firmly established; in terms of social structure, the city consisted of a large professional population (physicians, architects, university professors), a large student population, and an affluent retirement group that did not want to see the coastal environment changed. Moreover, a significant number of residents in the City of San Luis Obispo were historically politically active and, on the whole, were supportive of environmental causes.

SUMMARY OF PUBLIC RESPONSE

The data support the conclusions that, in general, residents of the area around the plant have become more concerned about the Diablo Canyon plant, public acceptability has noticeably declined, and a committed, politically active opposition has grown. Although the opposition is centered in the City of San Luis Obispo, it has become well integrated into the larger regional anti-nuclear movement, for which the Diablo Canyon plant has become a symbol. Following national trends in public attitudes toward nuclear technology, public concerns have shifted from a highly localized interest in preserving environmental quality to a very broad array of generic safety issues. The findings also suggest that the controversy over the seismic risks and

the Three Mile Island accident heightened public concern and awareness of the hazards posed by nuclear
technology. As the opposition became more regional-
ized, intense, and broad-based, the nuclear plant also
became representative of such things as loss of local
control, industrial monopolization, and nuclear proliferation. Regional opposition to the nuclear station
has subsequently grown, both in membership and in
intensity.

The level of public concern varied geographically
within the region. Opposition to the plant was centered in the City of San Luis Obispo and in some of the
small coastal communities; elsewhere, active opposition
to the plant was less observable. Communities having
high levels of public opposition or concern were,
concomitantly, those communities most sensitive to
questions of environmental quality. By and large,
these were the coastal communities with large retired
and affluent populations (i.e., Shell Beach, Cayucos,
Morro Bay, and Cambria).

Interviews with key informants and data gathered
from opinion surveys suggest that the accident at the
Three Mile Island nuclear power plant heightened public
concerns over the Diablo Canyon plant, particularly
over seismic issues and evacuation plans. There is
some evidence that opposition to the Diablo Canyon
plant grew in the area following TMI, but a direct
causal link could not be established. However, since
TMI, a great number of local residents have indicated
that further testing on seismic matters should be
continued. Again, this expression of concern may be a
direct function of recent earthquake events in the
larger region (e.g. Imperial Valley). The TMI accident, according to the key informants who have taken a
position against the plant, has made a seismic-related
nuclear accident more plausible and realistic.

For some of the local political jurisdictions,
environmental concerns have been important issues, and
the pattern of elected officials has long reflected the
split between the growth versus no-growth constitu-
encies. The controversy over the Diablo Canyon plant
has reinforced this dichotomy in the political structure. In addition, there is some evidence that the
growth in political activism over the nuclear plant has
had some "trickling down" effects, as members of anti-
nuclear groups have taken strong positions and actions
regarding other matters of environmental quality and
health in the county.

NOTES

1. This chapter, in slightly different form, was published originally in <u>Energy: The International Journal</u>,
7(8):667-680, 1982. Reprinted with permission.

2.   The intervenors contend they are under the legal constraints imposed by the NRC. Although seismic risk is considered a critical area of concern, one intervenor group argued that the main contention has been, and continues to be, that the routine and nonroutine radioactive emissions from an operating nuclear power plant can cause adverse health effects. The group profoundly disagrees with the assumptions made by the NRC and the nuclear industry in regard to allowable exposures to radioactivity.

3. It is important to note that, except for a pre-hearing conference held in 1974, a five-year delay occurred in the public hearing over the operating permit in order to conduct investigations and to determine the nature of the seismic hazard.   The intervenors (Scenic Shoreline Preservation Conference, Another Mother for Peace, and California Polytechnic Ecology Group) presented a position to ban further construction of the plant until the available data on offshore seismic studies were complete.   Earlier, the Atomic Safety Licensing Board (ASLB) made the decision not to consider the seismic issue during the environmental hearings. In April 1974, the ASLB delayed the decision to terminate construction activity at the Diablo Canyon site; the issue was to be taken up at the Operating License Hearings after an assessment and evaluation of the fault and the risk posed to the nuclear facility.   The intervenors perceived this decision to be highly detrimental to their case.   It meant that construction would continue to proceed, making it more difficult for the intervenors to argue later for the plant's termination, especially once its construction was completed.   Second, the plant's opponents argued that further construction advanced, retrofitting would become less acceptable and less suitable in reducing the potential seismic hazard.

4.   According to the utility, a 6.75 magnitude event directly beneath the plant would result in a plant shutdown following a 0.4g surface acceleration.   Prior to the discovery of the Hosgri Fault, intervenors criticized the plant's design because ground acceleration during the San Fernando Valley 1971 earthquake was substantially higher; the design based on 1966 data was considered invalid by the intervenors. The intervenors argued that the plant originally was designed to withstand a force of 0.2g.   The AEC, to be sure of conservatism in the design, required the plant to use figures expressing a double design earthquake (DDE).   Thus, Diablo Canyon was built to withstand 2 x 0.2g or 0.4g. After the Hosgri Fault was discovered, the NRC used the Safe Shutdown Earthquake (SSE) design criteria.   The intervenors argued that the 0.75g figure underestimated the risk by removing the peak acceleration forces which may reach 2.0g.

5. Some seismologists, for example, have theorized that the length of a fault is directly related to the potential magnitude of the earthquake, but this concept is subject to question. One theory postulated by a scientist who testified for the intervenors was that the Hosgri fault joined with the San Gregorio offshore fault north of the Hosgri fault and further connected with the San Andreas, forming a continuous fault of about 200 miles (320 km). This position was discounted by the NRC safety evaluation.

6. The 1.15g value was derived by the USGS in Circular 672, which estimated near-field ground motions to be used for design of the Alaska Pipeline, and was based largely on data estimated from the San Fernando Valley magnitude 6.4 earthquake in which peak ground motion was recorded at 1.25g on a rock ridge adjacent to Pacoima Dam. The utility criticized this interpretation by arguing that the estimate of acceleration was an anomaly because of the unique geologic structure at the dam site which amplified the seismic movements. According to PG&E personnel, the USGS later modified its position that the PGA (peak ground acceleration) values contained in Circular 672 were not necessarily intended to be used for design purposes, but could be reduced, based on engineering rationale.

REFERENCES

Ebbins, S. and R. Kasper. 1974. Citizen Groups and the Nuclear Power Controversy, Cambridge, Mass: MIT Press.

Flynn, C. 1980. Three Mile Island Telephone Survey, NUREG/CR-1093, Washington, D.C. United States Nuclear Regulatory Commission.

Gatchel, A. Baum and C. Baum. 1981. Stress and Symptom Reporting, Maryland: Human Design Group, Oleny.

Kasperson, R., G. Berk, D. Pijawka, A. Sharaf, and J. Wood. 1979. "Public Opposition to Nuclear Power-- Retrospect and Prospect", Committee on Nuclear Energy and Alternative Energy Systems, Washington, D.C. Background Paper No. 5. National Academy of Sciences.

Melber, B., S. Nealey, J. Hammersala, and W. Ranklin. 1977. Nuclear Power and the Public: Analysis of Collected Survey Research. Seattle, Washington: Battelle Human Affairs Research Center.

Pahner, P. 1979. A Psychological Perspective of the Nuclear Energy Controversy, Research Memorandum 76-67, International Institute for Applied Systems Analysis, Laxenburg, Austria.

Pijawka, D. 1982. Socioeconomic Impacts of Nuclear Generating Plants: Peach Bottom Case Study, Washington D.C.: United States Nuclear Regulatory Commission. NUREG/CR-2749.

Purdy, B., E. Peelle, B. Bronfman and D. Bjornstad. 1977. A Post Licensing Study of Community Effects at Two Operating Nuclear Power Plants, Washington, D.C. United States Nuclear Regulatory Commission, OPNL/NUREG, TM-22.

Introduction

Timothy O'Riordan

There is a school of thought among antinuclear activists who believe that their best friend is the nuclear industry itself. Writing in the now defunct British environmental magazine <u>Vole</u>, Walter Patterson, the Canadian physicist prominent in the Friends of the Earth case against a proposal to reprocess spent uranium dioxide fuel at Windscale in Northwest England, wrote:

> There may once, long ago, have been a British nuclear juggernaut. But in the past decade it has become mired to the axles; and any suggestion to the contrary is simply lurid nonsense, and no basis on which to plan any strategy of opposition. In fact, if you really want to upset nuclear people, don't waste your breath accusing them of crimes against society. Just reflect aloud on the shambles of their industry and tell them you're sorry for them. (Patterson, 1981, 9).

The assertion is that <u>the nuclear industry even with its commercial, military and governmental backers is quite incapable of delivering a product that is either safe or cheap.</u>

Expensiveness and safety are connected. Vociferous and informed groups within the body politic demand higher levels of safety from commercial nuclear power plants, so nuclear costs rise. Charles Komanoff, the American economist who specializes in calculating nuclear plant cost escalation due to the ratchet effect of tougher regulatory requirements, recently gave evidence that additional safety features could add as much as 50 percent to the cost of a modern power station (Kamanoff, 1983). Even a much more cautious estimate by the British Monopolies and Mergers Commission (1981), which critically examined the investment performance of Britain's biggest electricity supplier, the

Central Electricity Generating Board (CEGB), suggested that about a quarter of all nuclear plant cost escalation could be attributed to increased safety requirements. However, the Commission also confessed that it had discovered no official analyses carried out in the UK which attempted to establish relationships between safety and costs, including cost consequences to the general public. This is a serious omission and one which does not appear yet to have been rectified.

No one yet knows the total cost of the Three Mile Island incident. The clean up costs alone have already exceeded one million, the loss of revenue to the utility is calculated to be at least $500 million per year while the cost to the rest of the nuclear industry could be as much as thirty million per PWR in terms of retroactive fitting of safety features and lost power production when design changes were instituted. Should another serious accident befall another reactor for a different sequence of reasons, no doubt there would be even greater economic repercussions.

THE DECLINE OF NUCLEAR PLANT ORDERING

A far more significant cost of the TMI incident is the effect on public faith in the nuclear industry to produce a reliable product, the shock treatment given to the Nuclear Regulatory Commission to tighten up its procedures and surveillance, and the general political antipathy toward nuclear power in general as a major component in a future energy strategy. While a senior General Electric Company official (Bray, 1981) was telling a British parliamentary committee that increased safety regulations were the main cause of cost escalation and construction delay in the US nuclear industry and was urging the British not to copy the regulatory rigidities and procedural hurdles of the American experience, many analysts concluded that the industry itself is largely to blame for its stretched construction periods and cost overruns. These they argue, are mainly due to poor design teamwork, appalling site management during construction, and faulty technology. Every time an important piece of equipment goes wrong, or worse still, a serious design failure is reported, shudders are felt throughout the nuclear community worldwide.

There has been no new nuclear power plant order in the U.S. since 1978. In 1982 eighteen contracts were cancelled, including $2.5 billion worth of reactors in Washington State abandoned by the ill-starred Washington Public Power Supply System (appropriate acronym WHOOPS). It has recently been estimated that of the fifty-seven reactors under construction in the US and the five still on the drawing boards, as many as ten may eventually have to be cancelled. The effect on the major PWR contractors could be devastating for the nuclear industry which flourishes on steady plant ordering.

More serious perhaps is the effect on the longevity of highly specialized and skilled design teams. These people (about 20,000 in the US and 2,500 in the UK) cannot work satisfactorily in a climate of uncertainty, nor are they enthusiastic about the humdrum of design modification caused by what they see as fickle political and regulatory requirements. If a major power plant design program is seriously delayed, morale falls and some of the teams break up as particular individuals seek more secure and more rewarding employment elsewhere. Any resurgence in plant ordering, especially if major design changes were to be involved, would therefore require an expensive and time consuming period of retraining. Design team confidence and stability are amongst the most vulnerable "achilles heels" of the commercial nuclear industry.

THE CHANGING ECONOMIC
WORTH OF ADDITIONAL ENERGY

Energy economics have been transformed over the past ten years. The worldwide recession coupled with the escalatory costs of energy (led by oil) have meant that demand forecasts for energy in general and electricity in particular have slumped by as much as five percent per year (or 75 percent of estimated growth rates). This has meant that power station orders so confidently expected in the early 1970s are now an embarrassment. Furthermore the whole scenario culture (the groups of forecasters who seek to predict energy/electricity growth rates over the coming decades) is in disarray. Nobody nowadays is unchallengeable about their assumptions and predictions. In an industry where 15-20 year lead times are becoming the norm and where forty year investment write off period is not unusual, this highly uncertain economic climate is most unsettling. Furthermore, as the technology of energy conservation has improved rapidly and as public behavior favoring energy conservation has become more predictable, future energy/electricity demand scenarios become even more spread on the graph paper. This invites endless speculation about the competence of specialist advice and major political rows over the merits of conservation over "secure" additions to supply.

The conclusions drawn from this brief overview to date are: (1) that the nuclear industry is trapped in an unfavorable political and economic climate that makes its future economic viability somewhat doubtful;

(2) that a new political awareness favoring reduced energy demand as the primary public good rather than increased energy supply means that the costs and benefits of additional baseload nuclear power are being recalculated; (3) that the drive for higher safety standards has probably just about reached its peak, but at a point where the calculated cost per estimated life saved is orders of magnitude higher for the nuclear industry compared with any other industry; (4) that the antinuclear lobby is making more headway nowadays through its economic critique; and (5) that the nuclear industry itself, because it is beleagured but by no means devoid of friends in financial, commercial, military and political circles, cannot be written off as yet another victim of industrial recession. Indeed it is a mean and impressive fighter with billions of dollars at stake on a number of international stock exchanges and with powerful allies in many nations. The nuclear industry may be down but it is certainly not out.

THE SURVIVAL CAPACITY OF THE NUCLEAR INDUSTRY

This is a paradoxical package of conclusions. The nuclear industry has a powerful capacity to survive. Two points are important in this respect. First, the future of the industry is governed by politics not by economics or even the law, and second, alternatives to nuclear power are also governed by politics and economics. Nearly all politicians harbor illusions that economic growth can continue more or less indefinitely and that, through technological genius, people need work less to become more affluent. The nuclear industry is seen as a symbol of technological inventiveness and part of the forefront of economic expansion. It is not just the nuclear technology that is regarded as so important: it is also the associated technological activity that can promote new industrial growth. This was certainly the view taken by the French government in the 1970s when thirty-two PWRs were ordered almost with militiary precision and is a line controversially endorsed by the present British government.

Likewise the West Germans are in a dilemma. Nuclear power is so disliked that successive West German federal and state governments have shied away from major new commitments. The rise of the anti-nuclear "Green" movement to partiamentary status will make any new nuclear initiatives even more politically contentious. Yet the German coal industry is also under fire because of its sulfur emissions and the growing problem of forest damage caused by acid rain. Admittedly coal station emissions are by no means the only source of oxides of sulfur and nitrogen in West Germany and indiginous production of these gases account for only 60 percent of total deposition. Nevertheless the German coal industry is not environmentally popular thus forcing the Germans to take conservation more and more seriously.

So too are the French. Having invested in 42,000 MWe of nuclear power they find they not only have a surplus capacity but a considerable debt burden, much of which is invested in the US stock market and which is at the mercy of US interest rates. But the French have no real alternatives except for conservation and renewables - hence the Mitterand Administration commitment to a massive long term program of electricity conservation, with renewed R & D investment in renewables.

We now see that environmental and economic objectives begin to confront almost any energy supply option, and certainly electricity supply possibilities. Doubtless there will be difficulties facing a massive investment in renewable sources, though these remain unproven for the most part, and the true economics of a serious commitment especially to conservation in electricity use again await empirical examination.

SOME MORE HICCUPS FOR THE NUCLEAR INDUSTRY

Before we turn to the essays that follow, there are two further features of the nuclear fuel cycle that deserve mention. The first is the renewed suspicion of an explicit connection between the civil and military use of nuclear fuel. The French do not disguise the fact that their fast breeder program is not for civil power so much as a source of weapon grade plutonium. The British and American governments have assiduously denied any direct connection between the civil and military uses of plutonium, but suspicions abound. National and international peace movements are constantly attempting to probe through the protective armour of official secrecy to get at the answer. Though no proveable connection has been found, the suspicions grow, and that can do the civil nuclear industry no good.

The second point is the apparent failure of the nuclear industry to recognize where public anxieties confound scientific assurances. The best example of this lies in the field of radioactive waste disposal, where many scientists believe that they know a safe and reliable answer but where the public and their elected representatives remain "irrationally" suspicious. In Britain, for example, opposition to the drilling of test boreholes into various kinds of rock to discover their geological properties for possible long term subterranean storage of high level wastes was abandoned because of fierce local reactions. It is interesting to note that the United States Supreme Court has recently upheld a California law banning the construction of new nuclear power plants until the federal government determines what to do with the radioactive waste. Five other states have adopted similar statutes and presumably others will follow. Fears over waste disposal, irradiated fuel transport and plant decommissioning are all likely to place additional strains on the nuclear industry just at a time when its public image is not supportive.

Another example of this problem of public distrust of scientific and managerial competence lies in perceptions about serious accident and judgements as to whether the local utility, police and civil defense authorities are sufficiently competent to get the public out or at least to protect them from radiation poisoning during a time of emergency.

There are two features of this question. The first relates to public confidence in secure plant performance, namely that no serious accidents will occur, or at least if they do the radiation can be contained and the plant will cease to operate.

The now famous Rasmussen report was the first major attempt to identify likelihood of fault sequences of various kinds, though it avoided detailed discussion of likely consequences. Following that came a number of other major nuclear risk studies in various countries, none of which was regarded as definitive (see Dooley, et al., 1983). In general these studies showed up the vulnerability of the risk assessment methodologies to assumption and extrapolation, the major difficulty being lack of good case experience. These studies, while giving the nuclear community some confidence about their abilities to predict danger and to make the necessary design changes, failed to assuage public opinion. Indeed, the more there are published risk appraisals the more there are "counter establishment" scientists ready to express doubt and undermine public trust.

The weakening of public confidence in scientific expertise (not confined to nuclear matters) is an important contemporary issue. Current, but as yet unpublished research suggests that as the public begin to perceive that nuclear power is not necessary in electricity supply terms and that there are always problems attached to design and construction that could lead to a fault, so the distrust in safety grows. Hence the points made at the outset of this introduction: public faith in nuclear power plant safety is very much a function of their judgement of how far nuclear power is readily needed. So, even in psychological terms, economics and safety are intertwined.

Just as there is growing alarm over whether nuclear and regulatory authorities are fully competent to handle the "back end" of the nuclear fuel cycle (namely irradiated fuel transport, radioactive waste disposal and decommissioning), so the aftermath of Three Mile Island thus resulted in doubts about the competence of those responsible for emergency preparedness and evacuation. Two points are relevant here. One is that a significant minority in the vicinity of most nuclear plants (especially PWRs) now believe that accidents could happen. The other is that they doubt that they could get out unless there is an extremely well ordered evacuation arrangement.

Of crucial importance is power plant siting strategy, for that will determine the location of potentially vulnerable population and their escape routes. Detailed analysis of demographic changes around existing nuclear power stations and appraisals of potential sites with the lowest vulnerable population are part of nuclear technology assessment. Work on these issues has already begun in both the United States and the UK.

Susan Cutter is particularly concerned with the spatial aspects of risk cognition. She seeks to discover how residents in zones surrounding a plant judge the likelihood and consequence of an accident. The main problems with research of this kind lie in the wording and context in which questions are posed, and in the hypothetical nature of the investigation. In an actual accident there may be no correlation between anticipated and actual behavior. Future research in this area may look at potential behavior of mass hystaria because there is no doubt that in this modern communications age people will get to know about a major accident within hours and on a mass scale. Families split up during the working day could be in a terrible dilemma.

The Johnson-Zeigler study section argues precisely these points. They believe that many more people will spontaneously evacuate in an accident than has generally been predicted. But more important, they assert that studies of evacuation behavior during non-nuclear accidents will not assist in predicting behavior following nuclear accidents because people dread radiation. They fear it because they do not understand it and do not know how or when they will be injured. This analysis, which seems plausible for a really major incident of the TMI variety, indicates that planners are probably unprepared for mass evacuation, and that they should concentrate most on getting people to follow planned procedures and routes. This study also suggests who might be over-reactors (mostly young and with children) and who might stay even when they should go (mostly elderly, without children and working class). This helps to target the education effort to produce the maximum effect.

Evacuation plans themselves are analyzed by Sorensen from the point of view of management and communication. This whole collection suggests the importance of knowledge of spatial movement and behavior to hazard preparedness. However, it remains to be seen how far anticipated behavior is predictable in real life crises. A major incident with clogged evacuation routes and signs of radiation sickness would do more to kill the nuclear industry than any antinuclear protest.

REFERENCES

Bray, P. 1981. "Evidence to the House of Commons Select Committee in Energy: The Government Statement on the Nuclear Power Programme. House of Commons Paper 114 Vol. IV, pp. 1336-1342. London: HMSO.

Bupp, I.C. and J.D. Darien. 1981. The Failed Promise of Nuclear Power, New York: Basic Books.

Dooley, J., B. Hansson, R. Kasperson, T. O'Riordan and H. Paschen. 1983. Putting Risk Analysis into Perspective: A Comparative Review of Major Societal Risk Studies of Nuclear Power. Stockholm: Beijer Institute.

Komanov, C. 1983. Evidence before the Sizewell B Inquiry. Capital Costs, Construction Times and Operating Performance of Westinghouse PWRs. London: Council for the Protection of Rural England.

Monopolies and Mergers Commission, 1981. Central Electricity Generating Board. House of Commons Paper 315, London: HMSO.

Patterson, W. 1981. "Nuclear Shock: Horror Expose." Vole. 4(6) p. 9.

Sweet, C. 1981. A Study of Nuclear Power in France. Energy Paper No. 2. London: Polytechnic of the South Bank.

10

Residential Proximity and Cognition of Risk at Three Mile Island: Implications for Evacuation Planning

Susan L. Cutter

Despite considerable social science research effort, a number of questions remain unanswered concerning emergency response planning and the behavior of individuals as a result of the March 1979 accident at Three Mile Island (TMI). For example, previous research on the consequences of the accident found a clear link between residents' proximity to the plant and their propensity to evacuate (Cutter and Barnes, 1982). The highest percentage of those who evacuated during the accident lived within 10 miles (16 km) of the plant and this percentage decreased with distance. What we do not know, however, is whether the residents' cognition of risk also varied with distance. Little is known about the residents' cognition of risks, both from the accident itself and the clean-up operations. Less is known about the relationship between risk cognition and distance from the plant.

It is for this reason that this paper specifically examines the relationship between risk cognition and distance from the source of the threat or hazard using an experienced population, the residents in the vicinity of the Three Mile Island plant. Survey data from 1980 and 1982 are used to assess the effect of distance from the plant on the cognition of risk. Risk, as used in this paper, is defined as both an estimate of the likelihood of accidents, frequency of accidents and an evaluation of the future use of nuclear power to generate electricity. It is suggested that residents living closer to the plant will be more aware of the risks than those living farther away. It is also suggested that there will be some differences between the cognition of societal risks from the production of power from nuclear sources and the more salient risks associated with the production of power from the Three Mile Island plant.

The importance of understanding the spatial correlates of risk cognition is directly linked to improved emergency response planning for accidents of this type.

247

Since the accident at TMI, emergency response planning for radiological emergencies has been expanded and upgraded. These planning efforts are a recent development, and as such, the explicit theoretical constructs upon which the plans are based are limited to broadly worded guidelines and procedural regulations of the U.S. Nuclear Regulatory Commission.

The few emergency response plans produced to date contain several generic deficiencies (Cutter, 1983). One of these is the size and shape of the planning region addressed in each plan. Few plans incorporate variations in topography, population distribution, transportation access, and other geographic variables when determining the areal extent of the planning region. The plans also assume no spatial variation in the awareness of the risks from the reactor by the local residents. This "awareness space" is akin to geographic models of isotrophic surfaces which assume that movement, information spread, or awareness is possible in all directions. Such models also assume no physical, social, or psychological barriers that would impede information flow, or in this case, the awareness of risks. The end result of these limitations is that all individuals are assumed to have equal access to and knowledge of the information regardless of their location in the region.

The assumption of uniform awareness of risks in a region displays the same weaknesses as isotrophic models. Namely, social, or more importantly, psychological barriers may prevent risk awareness from being distributed evenly in a region. Models of decision-making which assume uniform awareness and employ rational economic approaches often contain unrealistic assumptions about human motivations and behavior, particularly under stressful or uncertain conditions (Tversky and Kahneman, 1974; Simon, 1976; Janis and Mann, 1977). As the success or failure of emergency response planning is dependent upon the behavioral response of the people the plan is designed to protect, such behavioral influences will have a major impact on the government's ongoing planning efforts.

SPATIAL CORRELATES OF BEHAVIOR

The natural hazards literature acknowledges that individuals' cognition of risk is linked to the coping action taken to protect themselves from that risk (Burton, Kates, White, 1978; Perry, Lindell and Greene, 1981). The precise relationship between cognition and response is, however, not fully understood. For the sake of argument, if one assumes that evacuation behavior varies spatially, then one may also assume that cognition of risk which influences the evacuation decision may also vary spatially.

Research exists on the spatial correlates of risk cognition of technological hazards, but it is limited. Diggory (1956) in an early study of health hazards found that individuals farther away from the site of a hazard overestimated the extent of the threat. Reviewing public acceptance of nuclear power plants, Mitchell (1980, 1981) found a higher degree of acceptance of the technology when a plant was proposed (but not yet constructed) or when a plant was located more than five miles from the respondent's community. When residents were asked to personalize the issue, a much higher percentage responded that they were opposed to having the facility nearby. Rogers (1982) examined low-probability hazards (nuclear attack and nuclear power plant accidents) and the role of proximity in determining risk cognition and acceptability. He found that closeness to the hazard decreased the level of cognized risk. Rogers also found no significant association between proximity and the perceived likelihood of an accident. Objective and subjective measures of distance were used, with subjective distance being the more significant factor in risk cognition. The Rogers' study is one of the few to address the relationship between spatial proximity to the threat and cognition of risk from the same threat.

STUDY DESIGN

Residents in the vicinity of Three Mile Island were surveyed by mail in April, 1979. The sampling plan was based on stratified random procedures using distance and direction from the plant. The area was divided into five concentric zones (five-mile (8 km) intervals) and four quadrants (north, south, east, and west). A total of 922 questionnaires were delivered, and 374 were returned for a response rate of 40 percent. Due to financial constraints, no follow-up letter was sent. Sixty-six percent of the 1979 respondents agreed to participate in a follow-up study, and in May, 1980, a second questionnaire was sent to these individuals. A total of 170 questionnaires out of 240 were returned, for a response rate of 71 percent. These same 240 individuals were surveyed for a third time in April 1982 to assess changes in attitudes and their response to ongoing clean-up operations at Unit 2 and the proposed restart of Unit 1. A total of 212 questionnaires were delivered and 141 of these were returned for a response rate of 67 percent.

The sampling design of this three-year study is consistent with other research on the response of residents to the March, 1979 accident (Flynn and Chalmers, 1980; Ziegler, Brunn, and Johnson, 1981). While the sample population is self-selected, possible bias due to self-selection is minimal because the responses are

consistent with and verified by findings of other research groups (Flynn, 1982). The sample may not be
representative of the population of the area; however,
it provides longitudinal data on one segment of the
population and permits assessment of changes in risk
cognition and behavioral response over time.

SOCIETAL AND LOCAL RISKS

Societal risk was measured by two questions using a
fixed-response format. Residents were first asked
which of several statements best described their understanding of the frequency of nuclear power plant accidents in the United States. Choices ranged from no
accidents to one accident per year. Residents were
then asked to check a statement best representing their
views on the future of nuclear power. Choices ranged
from immediate expansion and use of nuclear power
plants without significant modification, through continued use with safety improvements, to immediate
closure of all plants.
It has long been recognized that while the public
will accept risks at a societal level, when these same
risks are localized (for example, building a power
plant nearby) acceptability declines. Residents were,
therefore, asked questions on the future of the TMI
plant and the benefits of TMI compared to its costs and
risks. There were three fixed-response questions and
these constitute the measure of localized risk. The
first two addressed the continued use of each reactor
(TMI 1 and TMI 2). Choices ranged from open immediately, to open after modification and/or clean-up operations completed, to close permanently. The third
question addressed the relative benefits from TMI compared to its potential risks and costs. Again, three
choices were solicited: benefits greater than, equal
to, or less than costs/risks.
The results are only briefly summarized here
because they are discussed in more detail elsewhere
(Cutter, 1983). There were minor shifts in respondents' feelings about the frequency of nuclear power
plants accidents from 1980-82. Forty-one percent of
the respondents in 1980 and 35 percent in 1982 believed
accidents like TMI would occur once or twice in their
lifetime. Nineteen percent of the respondents in 1980,
25 percent in 1980, and 21 percent in 1982 felt the
frequency would be once every three to four years. No
statistically significant changes were evident in the
estimates over the two-year period.
When respondents were asked to select the statement
which best represented their views on the future use of
nuclear power, 70 percent of the 1980 sample and 53
percent of the 1982 sample felt that plants should

continue operating but should be modified with new safety features. Twenty-five percent of the 1980 respondents and 44 percent of the 1982 respondents felt that all nuclear power stations should be permanently closed.

Negative attitudes toward the continued use of TMI are also evident. The majority of respondents in both 1980 and 1982 favor the permanent closure of Unit 2, the damaged reactor. There is a division of opinion regarding Unit 1, the undamaged reactor. In 1980, the greatest number of respondents favored reopening the plant after safety improvements were made with a minority expressing a view of permanent closure. Two years later, the respondents were evenly divided between these two opposing views--44 percent favoring reopening after safety improvements and 43 percent favoring permanent closure. Respondents were also asked to compare the benefits of TMI with its potential risks and costs. A majority of both the 1980 and 1982 sample felt benefits were less than the risks (52 percent in 1980 and 61 percent in 1982). Remaining respondents were evenly split between the other two answers to this question; benefits greater than risks and benefits equal to risks.

MEASURING RISK COGNITION

In order to more fully examine risk cognition, an index of risk was computed. Five questions (discussed in the preceeding section) were used in the construction of the index. These were the frequency of accidents, continued use of nuclear power, status and future use of both reactors at TMI, and the ratio of risks to benefits of TMI. This index was further subdivided into societal and local risk cognition categories.

Mean scores for total, societal, and local risk are shown in Table 10.1. These scores are generally higher in 1982 than in 1980, suggesting a growing level of awareness of risks posed by nuclear power over the two-year period. The difference between the two years is significant for total risk cognition only. This difference can best be explained by the continued controversy over the clean-up of the damaged reactor, the proposed restart of Unit 1, and intense media coverage in the local area of all events related to the plant.

As expected, there are major differences between cognition of societal risks and cognition of localized risks from TMI (Table 10.1). Over the two-year study period, the cognition of local risks is consistently higher than the cognition of societal risks.

TABLE 10.1
Mean Risk Scores

	1980	1982	Change 82/80	T-score[a]
Overall risk[b]	11.92	12.55	.63	2.45*
Societal risk[c]	4.88	5.19	.31	1.89
Local risk[d]	7.04	7.36	.32	1.68
N	170	141		

*p<.05

a Student's T is a measure of the difference between two sample means where the significance is determined by various probability levels associated with the number of degrees of freedom used.
b Scores range from 5 (low risk cognition) to 18 (high risk cognition).
c Scores range from 2 (low risk cognition) to 9 (high risk cognition).
d Scores range from 3 (low risk cognition) to 9 (high risk cognition).

RISK COGNITION AND SPATIAL PROXIMITY

It was assumed earlier that if evacuation behavior varies spatially, then risk cognition may also vary with proximity to the plant. This was tested using both an actual distance measure (derived from the five mile (8 km) zones used in the sampling plan) and a subjective distance measure based on answers to the question: "How far away, in miles, are you from Three Mile Island?" The simple correlation between these two measures is $r = .87$, $p \leq .001$ in 1982, and $r = .86$, $p \leq .001$ in 1980. This close correlation between the two measures permits interchange of data during the analysis, depending on the needed level of measurement (zone is ordinal and subjective distance is interval in scale).

When risk cognition is analyzed by zone, levels of cognition decline with increasing distance from the plant (Table 10.2). While overall risk levels are slightly higher the closer one lives to the plant, there are no statistically significant differences between zones. There is a similar pattern for cognition of societal risks. Mean values by zone are approximately equal and the slope of the line is near zero. Cognition of local risks also follows the gen-

TABLE 10.2
Risk Cognition by Zone
(Mean Scores)

| | ZONES | | | | |
	0-5 miles	5-10 miles	10-15 miles	15-20 miles	>20 miles
OVERALL RISK[a]					
1980	12.57	11.89	11.29	11.87	11.69
1982	13.25	12.78	11.95	12.26	11.66
SOCIETAL RISK[b]					
1980	5.03	4.91	4.72	4.81	4.86
1982	5.48	5.26	4.98	5.03	4.95
LOCAL RISK[c]					
1980	7.54	6.98	6.57	7.06	6.83
1982	7.87	7.52	6.96	7.23	6.71

[a] scores range from 5 (low risk cognition) to 18 (high risk cognition)
[b] scores range from 2 (low risk cognition) to 9 (high risk cognition)
[c] scores range from 3 (low risk cognition) to 9 (high risk cognition)

eral trend and exhibits a slightly stronger association with distance.

Proximity has a greater impact on risk cognition in the zones closest to the plant. As distance increases, residents have the knowledge that they are outside the immediate 5-mile (8-km) zone, and thus distance appears to have a stabilizing effect on risk cognition. In fact, cognition does decrease with increasing distance up to the 15-20 mile (24-32 km) zone.

There is an elevation in mean scores in the 15-20 mile (24-32 km) zone. This might be explained in a number of ways. Residents in this zone live in an evacuation fringe area. As a result, they have a higher degree of uncertainty regarding potential impacts and the likelihood of evacuation than would their counterparts closer to the plant. They might also have a tendency to overestimate the risk because of this uncertainty regarding evacuation. This situation is not unusual and has been documented elsewhere. Diggory (1956) for example, found a greater propensity to overestimate the risk as distance from the threat increased.

Distance is negatively correlated with risk cognition but the strength of the association is weak (Table 10.3). The relationship between risk cognition and distance is not statistically significant in 1980, but is significant in 1982. Distance and cognition on a societal level are also negatively correlated, but the correlation is only statistically significant in 1982. Cognition of local risks appears to be related to distance, but again, the association is a weak one.

DISCUSSION

This analysis of 1980 and 1982 survey data from residents in the TMI area indicates a weak, negative association between proximity to the power plant and cognition of risks. This conclusion is notably different from previous research, which showed a strong relationship between proximity and evacuation of residents during the 1979 accident. Behavioral actions, such as evacuation, may be related to or prompted by proximity to the source of threat, yet an individual's cognition of that same threat may not have a strong spatial component. During the accident, Pennsylvania Governor Thornburgh ordered an evacuation advisory

TABLE 10.3
Relationship Between Risk Cognition and Distance
(Pearson's r)[a]

| | 1980 | | 1982 | |
	ZONE	DISTANCE	ZONE	DISTANCE
Overall Risk	-.095	-.118	-.147*	-.256***
Societal Risk	-.051	-.083	-.099	-.214**
Local Risk	-.115	-.127*	-.173*	-.263***

*p<.05, **p<.01, ***p<.001

a Pearson's r is a statistical measure of association indicating the strength of the linear relationship between two variables. It ranges from +1 to -1 with a zero value indicating no systematic linear relationship.

for pregnant women and pre-school aged children within a five-mile (8 km) radius of the plant. His advisory had a spillover effect with many more people responding to the warning message than the target population (Cutter and Barnes, 1982). The decision to evacuate may also be influenced by social or situational factors, such as the actions of friends and neighbors rather than the cognition of risk. The evidence to date suggests that the behavioral effects of risk estimation or evaluation are secondary to contextual or situational factors.

Attitudes and actual behavior are not always consistent, particularly under times of stress. In this case, the pervasive fear of nuclear power may transcend the effect of distance. Thus, a weak relationship between risk cognition and proximity to the plant may still be possible even though the association between evacuation and distance is strong. The social science literature is replete with studies examining the tenuous links between attitude and behavior, and this study illustrates the poor correlation between the two.

Risk cognition does not have a strong spatial component in a pre-impact context. Distance may, however, provide a cognitive focus for estimating risk during an emergency and may help the individual personalize the hazard. More importantly, distance may serve as a basis for the judgement of the cumulative effects of the risk. If individuals cognize themselves as being a safe distance away, in all likelihood they would also consider themselves less vulnerable to the hazard. The closer individuals are to the threat, the more vulnerable they may feel. Proximity to the threat is perhaps the clearest attribute of the risk that an individual can focus upon during an emergency situation of this type.

The lack of a clear linkage between distance and risk cognition may also be a result of the risk cognition measure used in this study. Survey questions may have been too general, permitting a small disparity in risk cognition levels. No attempt was made in the longitudinal analysis to solicit in detail those attributes of nuclear power which are most influential in determining cognition or attributes of risks. Further exploration of residents' attitudes towards the clean-up operation and the potential restart of Unit 1 may shed some light on this. The 1982 survey incorporated attitudinal measures of risk, and a subsequent analysis of these data may clarify possible spatial relationships.

This paper presents a preliminary analysis of survey data and suggests that risk cognition from a nuclear power station does not significantly vary with distance from the plant. The behavior of residents (e.g., evacuation and other forms of coping activities)

in response to a cognized risk does, however, have a
spatial dimension. The heightened level of risk cognition between 1980 and 1982 may be partially explained
by the continued uncertainty over the fate of both
reactors at the site. Local opposition to the reopening of both reactors is strong, and legal activities
regarding liability and damages are still in the courts
and will not be settled for some time.

There is very little evidence to support the
hypothesized link between distance and risk cognition.
This is true for risk cognition at both the societal
and the local level. While this preliminary analysis
suggests a moderate association between distance and
cognition of local risks, it does not show conclusively
that risk levels are influenced by perceived distance
from the plant. It is, therefore, premature to support
or reject the assumption made by emergency response
planners of a uniform awareness or cognition space.
Data from this preliminary analysis would tend to
support non-uniformity. Further investigation is
needed before the true nature of the spatial correlates
of risk cognition can be determined.

REFERENCES

Burton, I., R.W. Kates, and G.F. White. 1978. The
Environment has Hazard. New York: Oxford University Press.

Cutter, S.L. 1983. "Emergency Preparedness and
Planning for Nuclear Power Plant Accidents."
Applied Geography, forthcoming.

Cutter, S.L. 1983. "Risk Cognition and the Public:
The Case of Three Mile Island," Environmental
Management 7, (forthcoming).

Cutter, S.L., and K. Barnes. 1982. Evacuation
Behavior and Three Mile Island," Disasters 6(2):
116-124.

Diggory, J.C. 1956. "Some Consequences of Proximity
to a Disease Threat," Sociometry 19:47-53.

Flynn, C.B. and J.A. Chalmers. 1980. The Social and
Economic Effects of the Accident at Three Mile
Island: Findings to Date. Washington, D.C.: U.S.
Nuclear Regulatory Commission, (NUPRG/CR-1215).

Flynn, C.B. 1982. "Reactions of Local Residents to
the Accident at Three Mile Island." In Accident at
Three Mile Island, The Human Dimensions. D.L.
Sills, C.P. Wolf, and V.B. Shelanski, eds.,
Boulder: Westview Press, pp. 49-63.

Janis, I.B. and L. Mann. 1977. Decision Making: A
Psychological Analysis of Conflict, Choice and
Commitment. New York: The Free Press.

Mitchell, R.C. 1980. "Public Opinion and Nuclear Power Before and After Three Mile Island," _Resources_ (Resources for the Future) 64:5-8.

Mitchell, R.C. 1981. "From Elite Quarrel to Mass Movement," _Society_ 18:76-84.

Perry, R.W., M.K. Lindell, and M.R. Greene. 1981. _Evacuation Planning in Emergency Management_. Lexington, MA: Lexington Books.

Rogers, G.O. 1982. "Life-experience as a Determinant of Perceived and Acceptable Risk: Residential Proximity, Perceived and Acceptable Risk." Paper presented at the Workshop on Low Probability/High Consequence Risk Analysis, Arlington, VA. June.

Simon, H.A. 1976. _Administrative Behavior: A study of Decision-making Processes in Administrative Organization_. New York: Free Press (3rd Edition).

Tversky, A. and D. Kahneman. 1974. "Judgement under Uncertainty: Heuristics and Biases," _Science_ 185: 1124-30.

Ziegler, D.J., S.D. Brunn, and J.H. Johnson, Jr. 1981. "Evacuation from a Nuclear Technological Disaster," _Geographical Review_. 71:1-16.

11
Evaluating the Effectiveness of Warning Systems for Nuclear Power Plant Emergencies: Criteria and Application

John H. Sorensen

The accident at Three Mile Island Nuclear Power Plant in 1979 was an emergency management disaster. Chief among the problems was ineffective public warning and communications. While it is difficult to assign blame for that condition to any given party or determine if it was due to unique situational factors, the failure led to fairly significant regulatory changes in the arena of public warning and notification. These changes are intended to avoid the problems that arose during the TMI accident.

This chapter reviews these regulations and suggests an alternative set of criteria for evaluating warning systems. The criteria are used to assess the effectiveness of the warning system at the Ft. St. Vrain nuclear power plant in Colorado.[1] The paper concludes with some discussion of the lessons learned from the TMI experience as they apply to warning systems for all nuclear generating stations.

WHAT IS NECESSARY: REGULATORY
CRITERIA FOR WARNING SYSTEM EVALUATION

The accident at TMI raised many serious concerns about the safety of nuclear power. They centered on both technological and human issues. As a result, the U.S. President appointed a 12-member commission to investigate the accident and to recommend appropriate actions on the basis of their findings. As part of their study the commission was requested to examine the adequacy of emergency plans and problems in communicating information to the public. The recommendations of the commission provided the basis for new regulations on warning system capabilities at nuclear power facilities (Kemeney, 1979).

While their recommendations are extensive, they center on two key premises. First, the report stresses that the public requires clear information on nuclear power plant emergencies prior to accidents as well as when they occur. Second it assumes the information must reach the public in a timely fashion in order to be acted upon. These two considerations provide the basis if not the spirit for existing regulations.[2]

The key regulatory requirements for warning systems are as follows: (1) Each organizational level involved with emergency response (licensee, state and local) must establish the administrative and physical means to notify and provide prompt instruction to the public within the emergency planning zone (EPZ) plume exposure pathway;[3] (2) The licensee must demonstrate that the system exists; (3) State and local government is responsible for implementing the system; and (4) A warning system must meet the following criteria (Figure 11.1): (a) It must use an acoustic alerting signal. (b) It must employ commercial broadcasts plus special notification systems (e.g., NOAA radio); (c) The alert signal and information message must be disseminated in the 10-mile EPZ within 15 minutes; (d) The initial notification will reach 100 percent of the population within 5 miles of the site; (e) The system must assure 100 percent coverage within the plume exposure EPZ within 45 minutes. These requirements determine the adequacy of a warning system for regulatory and licensing purposes. They are highly mechanistic in that chief emphases are placed on time and distance of message receipt. Most utilities now have systems in place that meet such requirements. Satisfying the criteria,

FIGURE 11.1
Warning System Criteria

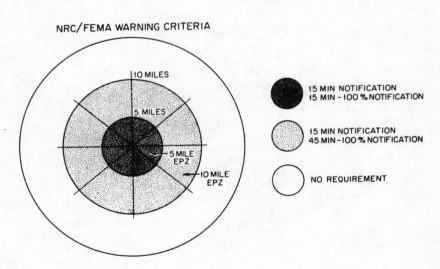

NRC/FEMA WARNING CRITERIA

however, does not guarantee an effective warning system. The following section defines some additional criteria which may help define system effectiveness.

WHAT IS DESIRABLE: SOCIAL CRITERIA FOR WARNING SYSTEM

An Integrated Warning System

The aim of any warning system is to alert as many people as possible to the likelihood and consequences of a potential, impending disaster and to tell them what protective actions to perform. Warning system adequacy can, therefore, be measured by the extent of actions taken that would result in reduced damages and casualties in the event of an emergency and in increased emergency preparedness activities (Mileti, et al., 1981).

An "integrated" warning system (Figure 11.2) performs three basic functions (Mileti, 1975): evaluation, dissemination, and response. "Evaluation" is the estimate of threat from a hazard to people in an area that is at risk. The evaluation processes of importance are detection, measurement, collation, and interpretation of available technical information about

FIGURE 11.2
An Integrated Warning System (Mileti, et al., 1981).

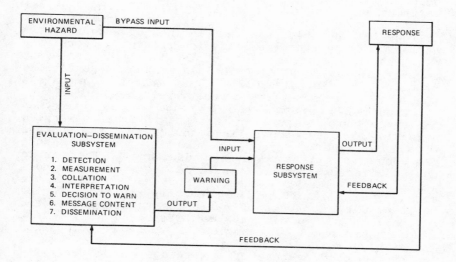

the likelihood of and threat (risks) posed by the
hazard. "Dissemination" of a warning involves deciding
whether or not the risks warrant alerting the public to
the possible danger, explaining the risks, and
suggesting what actions to take. "Response" is the
taking of adaptive action by people receiving the warnings. Once the warning is implemented, however,
actions are influenced by people's interpretations of
warnings. These interpretations are shaped by many
social, economic, psychological, and situational factors which are discussed in the next section.

The adequacy of a warning system is based on having
effective linkages between the three system functions.
It is based neither solely on detection abilities nor
on warning hardware and equipment. The dissemination-
response linkage is vital to achieving adequate and
effective warnings, yet it is the least understood and
the weakest link in most warning systems (Mileti et
al., 1981).

A Summary of Research Findings on
Warning Effectiveness

Numerous behavioral science studies on the effectiveness of human response to warnings have been
conducted by sociologists and geographers (Mileti,
1975; Perry, et al., 1981; Leik, et al., 1981;
McLuckie, 1974; Sorensen and Gershmehl, 1980; Baker,
1979; Gruntfest, 1977). Most of these studies have
focused on warnings of impending natural disasters,
such as floods, hurricanes, volcanoes, tsunamis, or
earthquakes. Several have dealt with the TMI nuclear
power plant accident (Flynn, 1981). The synthesis that
follows is generated from the findings of such investigations. While the generalization presented represents
a consensus of facts, they may not hold true for every
warning situation. Nevertheless, these observations
provide a reasonable basis for evaluating warnings
system effectiveness from a behavioral perspective.

Figure 11.3 summarizes a range of factors that
influence human response to warnings. Because warning
effectiveness has been defined in terms of the success
of the warning in prompting adaptive behavior, such
factors also influence system adequacy. The single
dominant factor that influences people's response to a
warning is, simply, whether or not they believe the
message. Believability, however, has been found to be
influenced by a number of variables both internal and
external to the warning system. Starting at the left
side of Figure 11.3, examples of these relationships
can be provided.

As depicted, nature of a warning can be described
by the following dimensions: (1) its source; (2) the

FIGURE 11.3
Behavioral Considerations for Warning System Evaluation

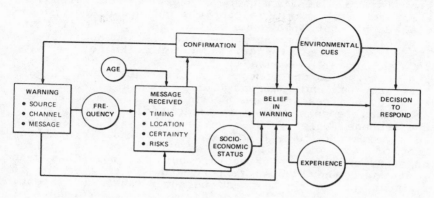

mode or channel by which it is communicated, and (3) the contents of the message. Studies have shown that warnings from an "official" emergency management source are generally more effective than warnings from an un-official source. Warnings issued by local official sources, such as police, are more effective than warn-ings from the federal government (Mileti et al., 1981). Minorities or low-income groups, however, are more skeptical of warnings from a government source (Perry et al., 1980). One problem is that people often con-fuse source with channel; a message coming over TV or radio may not be seen as official despite its origin (Drabek and Stephenson, 1971).

Face-to-face or direct personal transmittal is the most effective mode of communicating a warning. The personal contact conveys the urgency of the situation (Gruntfest, 1977). Sirens, because they are ambiguous, are usually the least effective. Mass media receives mixed reviews; it may be effective in some situations and unheeded in others (NAS, 1980). Automated systems such as tone-alert radios are largely untested but hold great promise as they combine desirable features of personalized and automated systems.

Message content can vary tremendously, and there are no magic formats or words that characterize an effective message. The chief stylistic parameters influencing believability are whether the message is precise, accurate, consistent and stated clearly in simple language. Features of the message content that relate to believability are the length of time to the disaster impact, the location of the projected impact, the projected certainty of the disaster, and the projected magnitude of the risks. These features of content form a dimension that relates to "fear arousal." Messages should maximize people's definitions of potential danger but should not elicit high levels of fear, a condition that may hamper effective response (Mileti, 1975). A critical factor, for example, in the decision to evacuate in response to flood warning is a message that conveys to the listener perceptions of the threats that are real and imminent (Perry, et al., 1980).

Previous research has shown that an almost universal human response to an initial warning is to confirm the message (Mileti, 1975; Perry et al., 1981). People rarely respond to a single warning without clarification or reinforcement. They want to know what neighbors, friends, or relatives are going to do. They want to be sure they should respond and, if they decide to respond, to do the correct thing. Thus, consistency and accuracy in confirmation are strong determinants of response (Leik et al., 1981). Similarly, the frequency or number of times the warnings are received shapes its believability (Perry et al., 1981).

Several factors external to the warning system design are also significant. First, the presence of environmental cues (or visual confirmation) is often important. People evacuate floodplains when they see rising water (Gruntfest, 1977). Second, experience influences behavior. Those people who have been in a similar situation are more responsive to a warning than the uninitiated. Finally, age and socioeconomic status influence response. Elderly people are less likely to hear a warning, believe it when they do hear it, or respond to it after hearing it (Friedsam, 1961). Likewise, warnings are usually less effective in reaching and influencing people of low socioeconomic status.

Table 11.1 summarizes current knowledge about factors which make a warning believable. Other factors as well can shape warning effectiveness based on other social and behavioral processes. These include the level of public information prior to a warning, situational factors such as time of day, experience with false alarms, conflicting information, the gender of the respondent and ability to confirm messages. Together these factors comprise an additional or alternative set of criteria on which to evaluate warning systems for nuclear power plants.

EVALUATING A WARNING SYSTEM:
THE FT. ST. VRAIN POWER PLANT

On the basis of the criteria presented in Table 11.1, it is possible to evaluate the effectiveness of an existing warning system. Of course, several factors

limit the utility of this type of analysis. First, a
warning system will never be perfect, chiefly because
the recipients of the warnings will not always hear the
same things, despite the fact that they are receiving
the same message (Mileti, 1975). Second, an appropriate level of effectiveness for a warning system cannot
be addressed on technical grounds alone. Essentially,

Table 11.1
What Makes a Warning Believable

Factors determining warning belief	Relationship
Factors internal to the warning system	
Number of warnings	As the number of warnings received increases, the believability also increases
Warning source	Official sources are more believable
Warning channel	Direct personal contact is more believable than impersonal channels
Warning message	Accurate, clear, and consistent messages are more believable
Timing	As the length of time to impact decreases, the believablility increases
Location	The closer the recipient is to the impact location, the greater the believability
Certainty	As the likelihood or probability of the event increases, the believability also increases
Risk	As the forecasted consequences become larger, the believability increases
Rumor	As the number of conflicting rumors increases, the believability decreases

TABLE 11.1 (Continued)

Factors external to the warning system

Socioeconomic status	As socioeconomic status increases, people are more likely to believe
Experience	People who have experienced a disaster are more likely to believe again
Environmental cues	If visually or audibly confirmed, the warning is more believable

such effectiveness is a subjective question based on values, acceptable risks, and human preferences. Such factors can be extremely difficult to measure and assess (Burton and Whyte, 1980). Finally, a warning system can be appraised by using some type of economic criteria, although this is problematic for the same reasons. Large per capita expenditures on a warning system design and maintenance may not be viewed favorably in light of low societal benefits, although this is muddled by the right of individuals to be protected from undue harm and by social debate over the value of life.

Ft. St. Vrain

The Ft. St. Vrain power plant is located about 35 miles (56 km) northwest of Denver, Colorado, and approximately 3.5 miles (5.6 km) northwest of the town of Platteville near the confluence of the St. Vrain and South Platte Rivers. The surrounding land use is predominantly rural residential and agricultural; about 2,000 people reside within a 5 mile radius of the plant, and 14,000 within 10 miles. Greeley is the nearest population center, about 14 miles from the plant site. The reactor is a 330 MWe high temperature gas cooled reactor (HTGR) designed by General Atomic. The reactor utilizes a uranium-thorium fuel-cycle, a graphite core structure and helium as a primary coolant. The reactor is housed within a prestressed concrete reactor vessel. While it is possible to achieve an off-site release of radiation from a light-water reactor in 15 to 30 minutes after a transient, the Final Safety Analysis Report (FSAR) on Ft. St. Vrain concludes that a 20 hour time period would elapse before an off-site release would occur at the Ft. St. Vrain facility. Inherently the HTGR is considered a "safer" reactor and accordingly the EPZ has a smaller, 5-mile radius. The remainder of NRC warning requirements apply to this plant.

The Ft. St. Vrain Warning System

Table 11.2 provides a summary of the Ft. St. Vrain warning system according to our evaluation criteria. Information in the table was compiled from a review of utility and state emergency plans and from interviews with emergency officials from the power plant and in the Colorado Department of Emergency Services. Each factor is discussed in turn.

Number of Warnings. The Ft. St. Vrain warning system procedures do not specify how often warnings will be issued or how frequently updated information will be incorporated in the message. It can be assumed that door-to-door notifications will be made only once. The extent of media coverage reaching individuals is difficult to predict. Hence, frequency of warnings is an unknown factor in evaluating effectiveness.

Warning Channel. The Ft. St. Vrain system chiefly relies on tone alerts and door-to-door contact for evacuation notification and on media (television and radio) as the general channel. The personalized contact will increase effectiveness. The impact of warnings received through the news media is not well understood (National Academy of Science, 1980). The avoidance of siren systems will increase effectiveness of the notification, although it may delay the alert. For example, several studies have shown that sirens are confusing and often misinterpreted; a major cause of fatalities in the 1960 Hilo, Hawaii, tsunami has been considered ambiguity in the interpretation of sirens (Lachman, et al., 1961).

Warning Message. The standard messages found in the emergency plans could create some confusion, because of certain contradictory elements in the messages. For example, the messages first state that there is a reason for public concern in that a release of radioactive materials has taken place. Messages then proceed to denegate that by stating people should not be concerned--"there is no cause for alarm"-- or "no serious hazard." This is a confusing message.

Obviously, there is cause for concern, and people will
be concerned. They will probably want far more information than is provided in those sample messages. The
people will also demand much more precise reasons for
being or not being concerned. More detailed instructions should be presented. The failure to do so will
likely result in the same type of confusion experienced
at TMI.

TABLE 11.2
Ft. St. Vrain Warning System Effectiveness
in Light of Social Science Criteria

Factors Determining Belief	Findings
Factors Internal to the Warning System	
Number of warnings	No guidelines on frequency of warnings are provided
Warning source	Variable
Channel	Door-to-door is highly effective although prone to contain errors; Tone-alert notification is largely untested
Message	Official messages need improvement because of ambiguity and confusing contents
Timing	More than adequate, although some-times too much lead time reduces effectiveness
Location	Specificity increases effectiveness
Certainty	Messages appear to be uncertain, reducing effectiveness
Risk	Risks are not outlined clearly, reducing effectiveness
Rumor	Situation specific effects; No provisions for rumor control in warning procedure

TABLE 11.2 (Continued)

Factors External to the Warning System

Socioeconomic status	Not estimated in this study
Experience	Lack of incidents will decrease effectiveness, although TMI provides a surrogate
	Public involvement in testing will increase effectiveness
Environmental cues	There may be "false" environmental cues

Other Factors

Public information	Will increase effectiveness
Situational-- time of day	Will vary the effectiveness
False alarms	A few will not hurt the effectiveness
Conflicting media reports	Will greatly decrease effectiveness
Gender	Women may be more likely to listen to warning
Confirmation	Depends on individual and mechanism to confirm

Specification of location. The messages are very precise as to the small areas likely to be affected by radioactive release. This is desirable. Attention should also be given to specifying more detailed information for other geographical areas as well. Information on safe locations is also valuable. Maps and other graphic means are usually an effecive way of communicating the spatial variability of risks to the public.

Certainty. The sample messages tend to convey some uncertainty about what is occuring. This stems from the conflicting depictions of appropriate levels of public concern, from the lack of details, and from the use of conditional phrases such as "may occur," "not expected," or "however." This will decrease effectiveness.

Risk. In a similar manner, risks are not clearly stated in the message, despite the cognitively fearsome statements of radioactive releases. It cannot be assumed that the population has the same knowledge about risks as radiological experts. Second, individuals vary in their willingness to accept or tolerate risks (Slovic, 1980). Thus, overresponse and lack of response should be expected to occur because of the lack of ability to interpret risks and different risk-acceptance levels. This suggests that more background information and more details on consequences are needed in the warning message.

Rumor. The extent of rumor in any emergency is difficult to predict. Rumor control is an important part of enhancing warning system effectiveness. A predetermined mechanism for controlling rumor is not integrated into the Ft. St. Vrain warning system.

Socioeconomic status. While no data have been collected on the characteristics of residents of the EPZ, research shows that people in rural areas are less likely to heed warnings than urbanites (Foster, 1980). This would create a need for extra effort in warning the rural residents surrounding the Ft. St. Vrain site.

Experience. Numerous studies have shown experience of emergency managers and the public to be a major factor in making a warning system work efficiently (Sorensen and Gershmehl, 1980). Obviously, there has been little experience with nuclear power plant accidents nationwide and no previous accidents at the Ft. St. Vrain plant. To some extent, the TMI incident may educate people about nuclear accidents, thus benefiting the warning process for all power plants. In a more programatic sense, test exercises that involve the public in the warning and response aspects of the emergency may provide simulated experience that increases warning system efficiency.

Environmental cues. There will not necessarily be any environmental cues in a nuclear accident. There may be false cues, such as a plume of steam rising from a cooling tower, which would be, in fact, harmless. In certain instances, smoke from a fire or noise from a rupture of a pressurized container could occur.

Situation. Situational circumstances, such as time of day or season, may have considerable significance for warning system efficiency. It is more difficult to alert people at night than in the daytime. Special provisions for warning during commuting periods may be necessary. In addition, the time of year may pose specific problems. Gruntfest (1977) found that vacationers and campers were difficult to warn prior to flash flood in the Big Thompson Canyon, Colorado. Problems with transients may exist regardless of the type of warning system utilized.

False alarms. Emergency managers are frequently concerned about the "cry wolf" syndrome. If the warning proves to be a false alarm, people will be less prone to respond to subsequent warnings. Practical experience with natural hazard warnings, such as hurricanes, shows that people are not subject to this phenomenon when false alarms are infrequent. In fact, an occasional false alarm may increase system efficiency, because it serves as an educational device. The precise relationship between false alarms and warning efficiency has not been ascertained.

Conflicting media reports. Any well-designed and integrated warning system can be undermined unintentially by the news media during an emergency. Conflicting information, sensationalism, and misinterpretations by journalists are significant problems. For example, ridicule of government instructions for protection against volcanic risks detracted from official efforts to educate and warn the public during the early stages of the eruption of Mt. St. Helens volcano (Sorensen, 1981).

Gender. Previous research has shown that women are more concerned about nuclear risks than men (Hohenemser, Kasperson, and Kates, 1977). Sex differences also influence warning beliefs and response (Mileti, 1975). From these findings, one would postulate that women would be more likely to listen to and respond to a warning regarding a nuclear power plant emergency.

This discussion has highlighted some of the major implications of warnings research for evaluating the effectiveness of current warning procedures at the Ft. St. Vrain nuclear reactor. The reader should be cautioned that they are only implications and are not derived from an empirical study of the Ft. St. Vrain situation itself. This review points out several ways in which the warning system can be improved and its general level of effectiveness enhanced, although changes would be minor. Overall, the Ft. St. Vrain system appears to meet what is dictated by Federal regulations through the use of the tone alert radio

system. Furthermore, no major flaws exist when evaluated in light of existing social science knowledge
about warnings.

CONCLUSION: SOME LESSONS FROM TMI

The accident at TMI provided a unique opportunity
to observe and study public response to an actual
emergency. The ensuing studies can be helpful in
evaluating the Ft. St. Vrain warning plan. To date, at
least seven major surveys of public attitudes and
behavior in relationship to TMI have been made (Flynn,
1981). Several reviews, summaries, and critiques of
these studies have been attempted (Flynn, 1979; Flynn
1981; Dynes, et al., 1979; Dohrenwend, et al., 1981).
Together with a number of non-empirical studies,
reviews, and opinion papers that have been published
since the incident, these studies provide an important
set of lessons for radiological preparedness planning
(Fisher, 1981; Hull, 1981b; Marrett, 1981). Some of
these lessons have resulted in policy changes and new
planning criteria. Others have been overlooked or have
not been incorporated into guidelines or regulations.
Several criticisms have been leveled at changes in
emergency planning caused by the TMI experience (Hull,
1981a; Olds, 1981). In this section such findings are
brought to bear on emergency warning systems.

Individual and Family Response to Warnings

Surveys of individuals following the accident indicated most felt they lacked information on the accident
situation and how they should respond. People received
a great amount of conflicting information and rumors
that lacked validity and were unconfirmable. People
also lacked background knowledge to understand the
events taking place and their possible consequences.
As a result great uncertainties existed over such
critical topics as negative health effects and
radiation exposure. Confusion over appropriate
adaptive response abounded.

The situation was characterized by a slowly developing and unfolding set of events. People had several
days to take in information, form images of what was
happening and what the consequences might be, and make
decisions. As no official evacuation orders were
issued, the decision to respond to warning by leaving
were largely individual and family choices. Despite
these conditions, an estimated 55-62 percent of the
population within a 5-mile (8 km) radius evacuated,
while about 44-54 percent within 5 to 10 miles (8-16
km) also left. Most did so because of fear and per-

ceived risk of harm from the accident. The warning system had failed to convey precise messages about risks, and, the resultant uncertainties were likely responsible for this level of activity despite no evacuation order. Many, in retrospect, have charged that people behaved erroneously by leaving; their response, however, is understandable given the way the warning system functioned.

Emergency Management and Warning Dissemination

Numerous agencies and organizations played roles in the emergency-response effort at TMI. In addition to the licensee, ten federal agencies/organizations, eight state agencies/organizations, five county civil defense groups, and numerous local government departments and groups participated in managing the crisis (Dynes, et al., 1979). Post-accident assessments have identified some reasons the warning system did not function efficiently.

Not surprisingly there was a distinct lack of plans to guide the warning process or the emergency response in general. This resulted in a lack of coordination among the various personnel who formally or informally became part of the warning network. Communication either never existed, broke down or was insufficient. Moreover, those who might have issued effective warnings lacked information because it did not exist, was not available, or could not be understood. The resultant uncertainty and confusion of emergency managers was systematically and graphically conveyed to the public as part of the warning. Subsequently, the warning system "broke down" because it lacked authority and contained numerous contradictions.

The warning process also violated the principles of the "integrated warning system" (Figure 11.2). Little attention was given to the concept of feedback and response. While officials were concerned about the public and what they should do, they were not very cognizant of what they were doing in response to warnings and why, at least until it was too late in the course of events to matter.

General Lessons

While it is difficult to generalize from one experience such as TMI to nuclear emergencies as a generic topic, some attempt is warranted. First, segments of the population will want to avoid the possibility of nuclear incidents regardless of the technical basis of the threats presented. If informed of a danger some may evacuate or take protective action whether or not it is recommended officially in the warning. All people, public and expert alike, will be confronted by uncertainty in a radiological emergency. This makes decisions more difficult and time consuming and may be a delaying factor and a cause of problems in the warning process. Because of inherent complexities of nuclear technologies and because of public inexperience with accidents, it will be difficult to issue warnings that reflect appropriate levels of risk and that will determine appropriate public concern and response.

We know as well that there is no ideal or perfect system by which to judge the performance at TMI or the structure of the Ft. St. Vrain system. Despite the nature of the system and the contents of the warning, people will continue to hear a variety of different messages due to selective perceptions and information processing biases. Some of this can be reduced by pre-accident planning and education.

Emergency planning and hence warning systems for nuclear power plant accidents have improved significantly since TMI due to increased regulatory requirements and planning. Current regulations are not based on behavioral criteria despite the intent of the post-TMI President's Commission recommendations. The increased utilization of social science knowledge in designing warning systems can improve them even further. This is demonstrated by the review of the Ft. St. Vrain system. Should another TMI type accident occur the public would be better informed and better able to respond. Warning systems, however, can still be improved without great expenditures of time and money. Whether another emergency response disaster would be avoided by such improvements still remains a speculative issue.

NOTES

1. Ft. St. Vrain was chosen as the study site because the author was requested to review the warning system as part of a larger project to determine the applicability of the TMI action plan to the Ft. St. Vrain power plant. The TMI action plan was developed for light water reactor technology and Ft. St. Vrain is a high temperature gas cooled reactor. This provided the opportunity to develop the alternative set of evaluation criteria discussed in this chapter.
2. See the Clarification of the TMI Action Plan Requirements (NUREG 0737). From this document the NRC and the Federal Emergency Management Agency have developed 16 standards for emergency planning (Final Regula-

tions on Emergency Planning, <u>Federal Register</u>, Vol. 45, No. 162, August 19, 1980). Implementation details have been outlined in NUREG-0654/FEMA-REP-1 (Rev. 1), <u>Criteria for Preparation and Evaluation of Radiological Emergency Response Plans and Preparedness in Support of Nuclear Power Plants</u>.

3. A 10 mile (16 km) radius. See Olds (1981) for a review and critique.

REFERENCES

Baker, E.J. 1979. "Predicting Response to Hurricane Warnings," <u>Mass Emergencies</u>, 4:9-24.

Burton, I., and Whyte, A. 1980. <u>Environmental Risk Assessment</u>. New York: Wiley.

Dohrenwend, B., et al. 1981. "Stress in the Community: A Report to the President's Commission on the accident at Three Mile Island". In <u>The Three Mile Island Nuclear Accident: Lessons and Implications</u>. T. Moss and D. Sills, eds., New York: Annals of the New York Academy of Science.

Drabek, T., and Stephenson, J. 1971. "When Disaster Strikes," <u>J. Appl. Soc. Psychol.</u>, 1:187-203.

Dynes, R., et al. 1979. <u>Staff report to the President's Commission of the Accident at Three Mile Island, report of the emergency preparedness and response task force</u>. Washington, D.C.: U.S. Government Printing Office.

Fisher, D. 1981. "Planning for Large-scale Accidents: Learning from the Three Mile Island Accident," <u>Energy</u>, 6:93-108.

Flynn, C. 1979. <u>Three Mile Island Telephone Survey: Preliminary Report on Procedures and Findings</u>. (NUREG/GR-1093). Washington, D.C.: U.S. Nuclear Regulatory Commission.

Flynn, C. 1981. "Local Public Opinion." In <u>The Three Mile Island Nuclear Accident: Lessons and Implications</u>. T. Moss and D.Sills, eds., New York: Annals of the New York Academy of Science.

Foster, H. 1980. <u>Disaster Planning</u>. New York: Springer-Verlag.

Friedsam, H. 1961. "Reactions of Older Persons to Disaster Caused Losses," <u>The Gerentologist</u>, 1:34-37.

Gruntfest, E. 1977. <u>What People Did During the Big Thompson Flood</u>, Natural Hazard Working Paper No. 32. Boulder: University of Colorado, Institute of Behavioral Science.

Hohenemser, C., Kasperson, R., and Kates, R., 1977, "The Distrust of Nuclear Power." <u>Science</u>, 196:25-34.

Hull, A.P. 1981a. "Critical Evaluation of Radiological Measurement and of the Need for Evacuation of the Nearby Public During the Three Mile Island Incident." In Current Nuclear Power Safety Issues, Vol. 2. Vienna: International Atomic Energy Agency.

Hull, A.P. 1981b. "Emergency Preparedness for What?" Nuclear News, April, 61-67.

Kemeney, J. 1979. Report of the President's Commission on Three Mile Island. Washington: U.S. Government Printing Office.

Lachman, R., Tatsuoka, M., and Bonk, J. 1961. "Human Behavior During the Tsunami of May 1960," Science, 133:1405-09.

Leik, R., Carter, T., and Clark, J. 1981. Community Response to Natural Hazard Warnings. Final Report to the National Science Foundation. Mineapolis: University of Minnesota.

McLuckie, B. 1974. Warning - A Call to Action. Washington, D.C.: National Oceanic and Atmospheric Administration, National Weather Service.

Marrett, C.B. 1981. "The Accident at Three Mile Island and the Problem of Uncertainty." In The Three Mile Island Nuclear Accident: Lessons and Implications. T. Moss and D. Sills, eds., New York; Annals of the New York Academy of Science.

Mileti, D.M. 1975. Natural Hazard Warning System in the United States: A Research Assessment. Boulder: University of Colorado, Institute of Behavioral Science.

Mileti, D., Drabek, T., and Haas J. 1975. Human Systems in Extreme Environments: A Sociological Perspective. Boulder: University of Colorado, Institute of Behavioral Science.

Mileti, D., Hutton, J., and Sorensen, J. 1981. Earthquake Prediction Response and Public Policy. Boulder: University of Colorado, Institute of Behavioral Science.

National Academy of Science 1980. Disasters and the Mass Media. Washington, D.C.: National Academy of Science.

Olds, F.C. 1981. "Emergency Planning for Nuclear Plants," Power Engineering, August:48-56.

Perry, R., Greene, M., and Lindell, M. 1980. "Enhancing Evacuation Warning Compliance: Suggestions for Emergency Planning," Disasters, 4:433-49.

Perry, R., Lindell, M. and Greene, M. 1981. Evacuation Decision-Making and Emergency Planning. Lexington, Massachusetts: Heath.

Slovic, P. 1980. "Judgment, Choice, and Societal Risk Taking." In Judgment and Choice, Public Policy Decisions. K. Hammond, ed. Boulder: Westview Press.

Sorensen, J., Gershmehl, P. 1980. "Volcanic Hazard Warning System: Persistence and Transferability," *Environ. Management*, 4:125-36.

Sorensen, J. 1981. *Emergency Response to the Mt. St. Helens Eruption*. Natural Hazard Working Paper No. 43. Boulder: University of Colorado, Institute of Behavioral Science.

U.S. Nuclear Regulatory Commission. 1980. "Final Regulations on Emergency Planning," *Federal Register*, Vol. 45, No. 162, Part VIII (August 19).

U.S. Nuclear Regulatory Commission, Office of Nuclear Regulation, Division of Licensing 1980. *Clarification of TMI Action Plan Requirements*, NUREG-0737. Washington: U.S. Nuclear Regulatory Commission.

U.S. Nuclear Regulatory Commission and Federal Emergency Management Agency (1980), *Criteria for Preparation and Evaluation of Radiological Emergency Response Plans and Preparedness in Support of Nuclear Power Plants*, NUREG-0654/ FEMA-REP-1 (Rev.1). Washington: U.S. Government Printing Office.

12

A Spatial Analysis of Evacuation Intentions at the Shoreham Nuclear Power Station[1]

James H. Johnson, Jr. and Donald J. Zeigler

Planning for a radiological emergency at a nuclear power plant requires that procedures be identified for protecting the public in the event of an actual or possible release of radioactive material. Sheltering and evacuation are two primary courses of action which can be taken to effect "dose savings," that is, to reduce whole body, thyroid, and lung exosure to radioactive iodine and other fission products. An announcement of a general emergency will precipitate action in two different decision-making arenas: (1) the public arena in which government officials must decide whether to order evacuation or sheltering, and (2) the private arena in which individuals and families must decide whether to follow the directives of public officials or to act on their own. Experiences with natural disasters and some technological disasters have led to the assumption that there will be a reasonably strong agreement between the decisions of government officials who are carrying out their charge to protect the public and the decisions of individuals who are trying to protect themselves and their families. In other words, governmental directives during an emergency will generally result in a large proportion of the population doing what they have been told to do.

Surveys of south central Pennsylvania area residents immediately after the March 28, 1979, nuclear accident at Three Mile Island (TMI) has demonstrated that planning for a nuclear emergency may require a dramatically different assumption (Brunn, Johnson, Zeigler, 1979). At TMI, the gap between governmental advice and warnings and public response was much larger than one would have anticipated from previous experiences with nonradiological emergencies (Cornell, 1982; Lagadec, 1982; White, 1974; White and Haas, 1975; Burton, Kates, and White, 1978; Hans and Sell, 1974). All preschool children and pregnant women living within five miles (8 km) of the disabled reactor were advised to evacuate until the immediate crisis was over; others within ten miles (16 km) were advised to take shelter indoors. This directive should have resulted in approximately 2500 individuals leaving the area. Instead, an estimated 144,000 residents within fifteen miles (24 km) of the reactor (Flynn, 1979) and thousands more beyond this distance left the area (Brunn, Johnson, and Zeigler, 1979). We are unaware of any other disaster - natural or manmade - in which the ordered evacuation of so few resulted in the evacuation of so many. In fact, the reverse has occurred in some disaster areas. In an earlier paper (Zeigler, Brunn and Johnson, 1981) we termed this tendency for widespread unexpected evacuation the "evacuation shadow phenomenon." Elsewhere the process has been referred to as "spontaneous evacuation" (Chenault, Hilbert, and Reichlin, 1979). The basis for this gap between observed and expected behavior seems to be related to the fear or dread of the disaster agent - ionizing radiation - which, although not sensible except in very large doses, is potentially lethal, carcinogenic, and mutagenic (Hohenemser, Kasperson, and Kates, 1977; Nelkin, 1981; Otway, 1977; Slovic and Fischhoff, 1978; Slovic, Fischhoff and Lichtenstein, 1979; Cook, 1982; Erikson, 1982). Such fears were heightened by the conflicting reports issued by both government and utility company officials during the TMI accident (Farrell and Goodnight, 1981; Rubin, et. al., 1979).

Inasmuch as our knowledge of actual human responses to crises involving radiation is limited to a single incident (Barnes, et. al.; 1979; Brunn, Johnson and Zeigler, 1979; Flynn, 1979; 1981), it is our contention that radiological emergency preparedness and response plans should be based on an understanding of the potential behavior of the population which is at risk during a nuclear accident. To assume that people's behavior will necessarily be consistent with official protective action advisories may result in underplanning and ineffective responses.

In order to augment our knowledge of evacuation behavior, a telephone survey was administered to 2595 households in New York's Suffolk and Nassau Counties. The objective was to determine how the population on Long Island would respond to a general emergency at the Shoreham Nuclear Power Station (SNPS; Figure 12.1) and to explicate those factors that would account for behavioral response (Social Data Analysts, Inc., 1982). Information obtained in the survey will be used to estimate the magnitude and spatial extent of spontaneous evacuation in the event of an accident at SNPS, and to determine how accurately intended evacuation rates at the zip code zone level on Long Island can be predicted using a set of social, attitudinal, and locational attributes which previous research has shown to be related to emergency behavior.

FIGURE 12.1
The Study Site

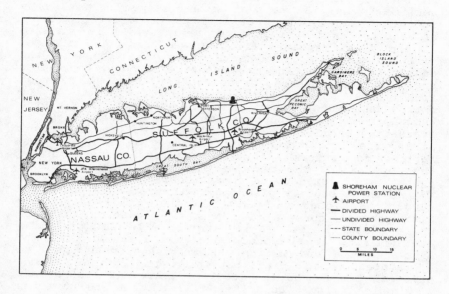

FEDERAL EMERGENCY PLANNING REGULATIONS

Prior to the TMI accident nuclear power plants licensees were required to develop only on-site emergency preparedness plans covering the two-to-three mile (3.2-4.8 km) low population zone around their facilities. This was the case primarily because a major accident involving the release of radiation into the environment was assumed to be a highly unlikely event (NRC, 1975). However, the accident at TMI revealed the high degree of uncertainty which exists with respect to the frequency of major reactor accidents, and in August 1980 the NRC (1980) issued revised regulations which require all licensees to augment on-site plans by devising, in conjunction with state and local governments, off-site plans for two emergency planning zones (EPZs): a ten mile (16 km) plume exposure pathway zone and a 50 mile (160 km) ingestion exposure pathway zone. The NRC has determined, however, that detailed planning encompassing a full range of protective actions, including evacuation, is required only within the ten mile (16 km) EPZ because the "probability of large doses [of radiation] drops off substantially at about ten miles from the reactor" (NRC, 1978, p. I-37). For example, given a core melt accident, the probability of exceeding Protective Action Guide (PAG) doses (25 rems to the thyroid and 5 rems to the whole body) would decline to only 30 percent at a distance of ten miles from the plant. [2] Even if this is an acceptable level of risk for society to assume, a plan which is confined to the plume exposure EPZ may well underestimate the time it would take to vacate the area because the magnitude of spontaneous evacuation beyond ten miles could clog the avenues of egress for those at greatest risk closest to a site (Johnson, 1982). The evacuation shadow phenomenon observed at TMI suggests that, even though it may never be necessary to officially order an evacuation extending more than ten miles from the plant site, the evacuation planning region must extend farther to conform with the behavioral intentions of the population. The ten mile emergency planning zone cannot be presumed to be operationally effective.

As the emergency planning regulations are presently structured, locational and distributional factors may also present constraints on the evacuation process. While the ten mile (16 km) and 50 mile (160 km) EPZs are generic, local conditions vary from one site to another. The evacuation shadow phenomenon may be aggravated by the distribution of population and the physical feature of the local environment. The number of people living within 10 and 50 miles of plant sites varies widely. Whereas the plume exposure pathway around the Indian Point, New York, site encompasses a population of over 200,000, less than 6,000 reside within ten miles of the Cooper, Nebraska plant. Fifty seven thousand persons reside within ten miles of the Shoreham site. In addition to static population densities, diurnal and seasonal variations in settlement patterns also may hamper the evacuation process. Under ideal conditions an evacuation order would find families at home. Only during nighttime hours are such conditions approximated, however. During the daytime, families are commonly split, with some members at work, some at school, and some at home. Conditions such as these promise an increase in crosstown traffic and congestion of telephone exchanges. Population may also swell during the tourist season and on weekends around some sites.

In sum, population density, seasonal and diurnal variations in settlement patterns, and, most important- ly, the potential effects of the evacuation shadow must all be considered if effective evacuation plans are to be developed. In the section that follows, we discuss the extent to which the evacuation shadow phenomenon is likely to exist in the event of an accident at the Shoreham Nuclear Power Station.

THE SHOREHAM EVACUATION SHADOW

In the Shoreham Evacuation Survey, residents were asked to indicate how they were likely to react in the event of a variety of protective action directives issued in the wake of three hypothetical radiological emergencies which are summarized in Table 12-1. It is from the survey respondents' behavioral intentions in response to these three accident scenarios that we derive estimates of the magnitude and geographic extent of spontaneous evacuation.

Magnitude of Spontaneous Evacuation

The first accident scenario recommended only sheltering within five miles (8 km) of the plant; no one was advised to evacuate. However, 215,000 families (one fourth of the population of Suffolk and Nassau

TABLE 12.1
Accident Scenarios and Behavioral Options

Scenarios:

1. Suppose that you and your family were at home and there was an accident at Shoreham. All people within five miles (8 km) of the plant were advised to stay indoors. Do you think you and members of your family would:

2. Suppose that you and your family were at home and there was an accident at the Shoreham Nuclear Power Plant. All pregnant women and preschool children living within five miles of the plant were advised to evacuate and everyone else living between within ten miles (16 km) from the plant was advised to remain in doors. Would you and the other members of your family:

3. Suppose that you and your family were at home and there was an accident at the Shoreham Nuclear Power Plant. Everyone living within ten miles of the plant was advised to evacuate. Would you and other members of your family:

Behavioral Options: (a) Go about your normal business,
(b) Stay inside your home, or
(c) Leave your home and go somewhere else.

Counties) indicated that they would be likely to leave.
The second was modeled after the TMI case where a
partial evacuation within five miles of the plant and
sheltering within ten miles (16 km) was recommended.
Such an advisory should result in the evacuation of
only 2700 families, if only the people advised to
evacuate were to leave. Instead 289,000 families, or
one third of the population of Long Island, indicated
that they were likely to evacuate. The third scenario
recommended a complete evacuation within ten miles.
Such an avisory should result in the evacuation of the
31,000 families living within ten miles. Instead,
based on the survey, approximately 430,000 families,
fully one half of the population of Suffolk and Nassau
Counties, indicated they would evacuate (Table 12.2).
When averaged (across the three scenarios) the stated
behavioral intentions of the survey respondents suggest
that an estimated 39 (approximately 311,000 families)
percent of the population of Long Island is likely to
evacuate spontaneously in the event of an emergency at
the SNPS in which sheltering and/or evacuation is
advised. Thus the intended behavior of Long Island
residents parallels the actual behavior of TMI area
residents and reaffirms the reality of the evacuation
shadow phenomenon. The plume exposure EPZ around Shor-
ham should therefore be extended to conform with the
behavioral intentions of the population.

TABLE 12.2.
Advised and Intended Evacuation in Response to Three
Scenarios.

	Advise to Evacuate	Percent of Total Population	Intend to Evacuate	Percent of Total Population
Scenario 1	0	–	215,000	24.8
Scenario 2	2700	.3	289,000	33.3
Scenario 3	31,000	3.6	430,000	49.6

Source: Compiled by Authors from Shoreham Evacuation
Survey, June, 1982.

Geographic Extent of Spontaneous Evacuation

The evacuation intentions were found to vary by distance and with direction of the residents' homes from the SNPS. Insofar as planning for radiological emergencies is concerned, these spatial correlates of evacuation have been largely ignored. We, therefore, discuss each in considerable detail here, whenever possible comparing intended spatial behavior at Shoreham with actual evacuation behavior at TMI.

Evacuation as a Distance Decay Phenomenon. One of the few undeniably important variables in explaining the pattern of evacuation response is distance from a threatening nuclear reactor. Families living closer to the hazard agent are more likely to evacuate than families living farther away (Figure 12.2). Within 10 miles (16 km) of the Shoreham plant, however, there was little distance decay in the proportion of the population indicating they would evacuate. In this respect, the results of the Shoreham survey parallel the results of our Three Mile Island survey which charted actual rather than hypothetical evacuation rates.

In response to the first scenario in the Shoreham survey, approximately 40 percent of the sampled families in both the five mile (8 km) and six-ten mile (9.6-16 km) zones indicated they would evacuate. In response to the second scenario, 57 percent of the respondents within five miles and 52 percent within six-ten miles of the plant said they would evacuate. In response to the third scenario, 78 percent of the population within 10 miles of the Shoreham site would evacuate.

FIGURE 12.2
Distance Decay Evacuation Curves.

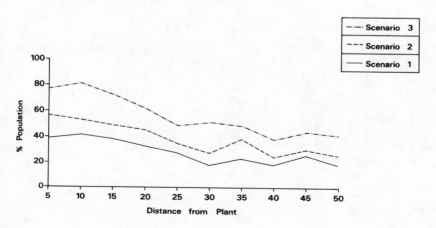

Beyond the ten mile (16 km) zone, the distance decay curve does not register a marked discontinuity until approximately 25 miles (40 km) from the plant. This finding suggests that an expanded evacuation planning zone would approximate more precisely the zone of behavioral intentions to evacuate. By extending the evacuation planning region, however, the perceived zone of danger is also likely to increase, thus further encouraging an expansion of the zone of spontaneous evacuation.

A comparison of the results of the Shoreham evacuation survey with the findings of Three Mile Island evacuation behavior studies (Flynn, 1979) is presented in Figure 12.3. Expected evacuation rates in the event of a Three Mile Island-type accident at Shoreham are compared with actual evacuation rates at Three Mile Island. Within ten miles (16 km) of the respective power plants, the results are amazingly similar and seem to indicate that, while the Shoreham survey solicited only behavioral intentions to evacuate, these intentions will most likely be acted upon in the event of a real accident. Beyond ten miles, however, the disparity between the Shoreham intended evacuation rate and the Three Mile Island evacuation rate increases with distance (Figure 12.3). Several factors may account for the anticipated higher evacuation rate at

FIGURE 12.3
Comparison of Actual Evacuation Behavior at TMI and Intended Evacuation at Shoreham.

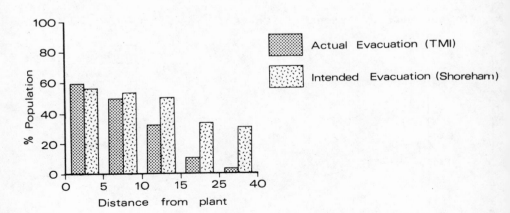

Shoreham: (1) the experiences and images of the Three Mile Island accident; (2) the long- standing controversy over the Shoreham plant; (3) a basic distrust of the Long Island Lighting Company which has been brewing for years; and (4) the fear of being trapped on Long Island (a cul de sac) in the event a more extensive evacuation is ordered.

Evacuation as a Directional Phenomenon. On Long Island there is likely to be a directional bias in the pattern of evacuation in the event of an accident at the Shoreham plant. As the data in Figure 12.4 indicate, residents west of the plant are more likely to evacuate than their counterparts to the east. Beyond the ten mile (16 km) radius, the distance decay effect is much more pronounced to the east than to the west. The population of eastern Suffolk County is not nearly as likely to choose evacuation as the population of western Suffolk County. The differences between their evacuation rates grows larger with each increasingly serious accident scenario.

The people to the east of the plant are in an unenviable position. Despite the fact that they are downwind from the plant, they cannot evacuate to the west without passing through the ten mile (16 km) hazard zone. For some, this may become a perceptual barrier to evacuation. Among those in eastern Suffolk

FIGURE 12.4
Evacuation by Distance and Direction from the Plant.

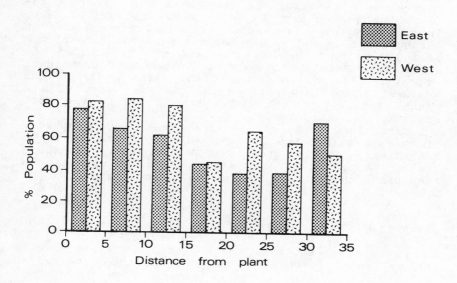

who said they would evacuate, between 15 percent and 19 percent, indicated they did not know the direction they would take. Only Nassau County residents (responding to the second scenario) expressed a proportionately greater amount of indecision. Also, indicative of eastern Suffolk's quandary is the finding that more respondents (12-15 percent) in this part of Long Island said they would go to a public shelter. In addition, between 9 percent and 13 percent of eastern Suffolk residents stated they would evacuate but remain in Suffolk County. It may be assumed that these families would move eastward toward the points, suggesting that preparations need to be made to handle them there (Figure 12.1). On the other hand, between 69 percent and 76 percent of the potential evacuees from eastern Suffolk County indicated that they would head for shelter in Nassau County, New York City, or beyond. This suggests that there will be a sizeable influx of traffic through the 10 mile zone, the area in which evacuation will have to be effective and rapid. South of Shoreham, the island is barely wider than 10 miles and all east-west roads, with the exception of one county route, pass through the plume exposure pathway EPZ (Figure 12.1). What to do about the population of eastern Suffolk County presents a problem for both residents and planners since many of these people feel they must leave their homes, yet have only a few options available.

Overall, more than seven out of ten survey respondents who said they would evacuate, identified New York City or a destination beyond as the place which they would choose for their temporary quarters. This strong directional bias means that most evacuees will not be content with destinations in Nassau and Suffolk Counties, but may attempt to leave the island completely. In the case of Long Island, this directional bias is strongly influenced by the shape and orientation of the island itself. Even in the case of the accident at Three Mile Island, however, a pronounced directional bias was charted. We found that almost half of the evacuees chose destinations to the northwest of the plant (Zeigler, Brunn, and Johnson, 1981), the most sparsely populated quadrant.

Directional bias may result from a lack of options or it may reflect the area perceived by evacuees to be the safest. In all cases it is a phenomenon that needs evaluation and study as part of emergency planning so that plans can be made to include directional predispositions.

PREDICTING EVACUATION INTENTIONS

Given the results of the Shoreham survey, how well can we predict evacuation behavior in the event of a radiological emergency? To answer this question, we aggregated data at the zip code zone level with the percentage of the population who intend to evacuate from each zone (or set of contiguous zones) as the dependent variable (Figure 12.5). The independent variables selected for this analysis were based on the previous research which has shown that the type and timing of an individual's behavioral response in crisis situations depends in large part on the following: socioeconomic status and stage in the life cycle; level of family cohesiveness; prior disaster experience; fear of the impending crisis and perception of the likelihood that it will materialize; distance and direction from the source; level of faith or trust in emergency response officials; and the type, content, and frequency of warnings (Benedict, et al., 1980; Cohen and Ahearn, 1980; Erikson, 1982; Cornell, 1982; Hudson, 1954; Janis, 1954; 1962; 1977; Kuklinski, Metlay and Kay, 1982; Lang and Lang, 1964; Manning, 1982; Melber et al., 1977; Otway, 1977; Perry, 1979; Perry and Green, 1982; Slovic and Fischhoff, 1978; Williams, 1964; Withey, 1962; Zeigler, Brunn and Johnson, 1981).

FIGURE 12.5
Percent Intending to Evacuate

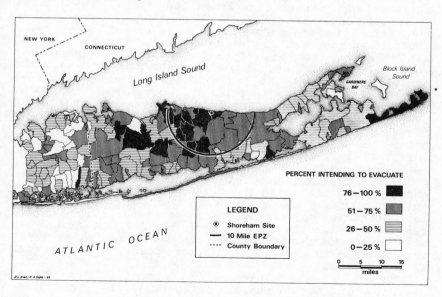

The analysis consisted of three steps. Principal components analysis was used in the first step to reduce the thirty social and attitudinal attributes of the zip code zones to a smaller number of "principal components" representing the underlying structure of the original variables. Principal components are linear combinations of the original variables which emerge in order of their relative importance in explaining the variance of the original data set. In this analysis, only those principal components with eigenvalues greater than one were considered to be significant in summarizing the social and attitudinal attributes of zip code zones. Four factors met this criterion and were rotated in turn using a varimax solution to identify the cluster of variables loading on each factor. After the four principal components were derived, composite scales were built reflecting the "theoretical dimensions associated with the respective factors" (Kim, 1975). By applying the four composite scales to the individual cases within each zip code zone, factor scores were assigned to all zones reflecting their relative position on the factors. These factor scores became the input for the correlation and stepwise multiple regression analyses.

A bivariate correlation model was employed in step two to measure the strength and direction of relationships between the percent intending to evacuate and (1) the four social-attitudinal dimensions identified in the principal components analysis and (2) the locational characteristics, distance and direction, of the zip code zones on Long Island. A stepwise regression model was used in step three to determine how well the social attitudinal and locational variables, individually and collectively, explain evacuation intentions, that is, the percentage of people indicating they would evacuate in the event of an emergency. The maximum r-square selection criterion was used to determine the order in which the variables entered the analysis (Barr, et al., 1976).

Factorial Structure of Long Island. Four factors emerged from the principal components analysis. Two of the factors (1 and 2) explain spatial variations in socioeconomic and demographic characteristics, while the latter two (3 and 4) account for variations in attitudes, opinions, and beliefs about issues pertaining to nuclear power. A brief description of each factor follows:

Factor 1, explaining 32.5 percent of the total variance in the original set of social-attitudinal variables, is an index of socioeconomic status. Among other social characteristics, the zip code zones with positive scores on this factor contain well educated and fairly high income households, whereas the ones with negative scores contain less well educated and less affluent households (Figure 12.6). Accounting for 27.3 percent of the total variance, Factor 2 is a stage-in-the-life-cycle index. Zip code zones with positive scores on this dimension contain either a significant number of elderly (over 65) households or young (under 35) families with preschool and school age children. In contrast, in zip code zones with negative scores on this component, most of the households are in their middle years (35-64) and may or may not have children living in the home (Figure 12.7).

Factor 3 is an index of nuclear opposition and explains a little over one fourth (25.8 percent) of the total variance. Most of the households in the zip code zones with positive scores on this component oppose nuclear power in general, and the completion of the Shoreham plant in particular, and have little or no faith in technical experts' ability to evaluate the risk of nuclear power accurately. By contrast, zip code zones with negative scores contain individuals and families who are less opposed to, and perhaps even favor, both nuclear power and the completion of the Shoreham plant, and who believe technical experts' assessments of the risk of nuclear power (Figure 12.8). Factor 4 is a distrust index and accounts for approximately 15 percent of the total variance. In zip code zones with positive scores on this factor, households would trust neither the utility nor public officials at all to tell them the truth about an accident at the plant, whereas in the zip code zones with negative scores the individuals are more inclined to trust emergency response officials (Figure 12.9). The extent to which these factors and the locational variables are related to evacuation intentions is addressed in the next section.

   Correlates of Evacuation Intentions. The results of the bivariate correlation analysis appear in Table 12.3. Distance from the Shoreham plant appears to be the most important correlate of intended behavior at the zip code level. The fact that it is a moderately strong inverse relationship ($r = -.52$) further confirms the distance decay curves in Figure 12.2 which show the proportion of the population indicating they would evacuate declining with increasing distance from the Shoreham plant (also see Figure 12.5). Intended evacuation is directly related to stage-in-life-cycle ($r = +.35$), suggesting that zip code zones in which young families with pre-school and school age children are concentrated, are more likely to experience higher rates of evacuation than those in which older families without children living at home predominate. Also there is a moderately strong correlation between percent intending to evacuate and nuclear opposition

FIGURE 12.6
Socioeconomic Status Levels

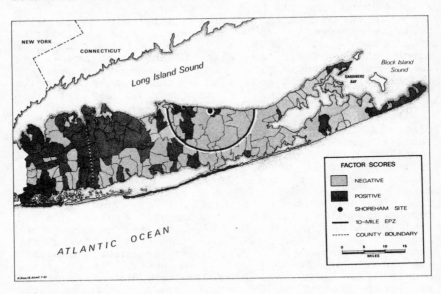

FIGURE 12.7
Stage-in-Life Characteristics

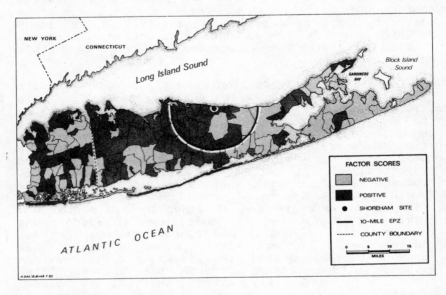

FIGURE 12.8
Nuclear Opposition

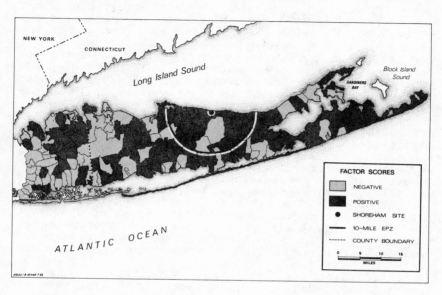

FIGURE 12.9
Distrust of Public Officials

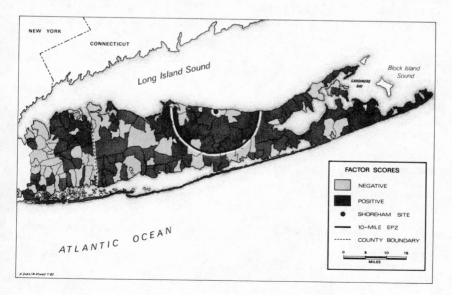

(r = +.35). That is, the higher the proportion of the population within each of the zip code zones who oppose nuclear power and completion of the Shoreham plant, and who have no faith in the ability of technical experts to evaluate the risk of nuclear power, the higher the rate of intended evacuation (Figures 12.5 and 12.8). The relationship between evacuation intention and the distrust of public officials (Factor 4), socioeconomic status (Factor 1), and direction of the zip code zones with respect to the plant were not statistically signi-ficant (Table 12.3).

In short, individually, only three of the social, attitudinal, and locational variables are statistically significant correlates of evacuation intentions at the zip code zone level. In the real world, however, the decision to evacuate is likely to be a function of the interaction or inter-relationships among all of the variables considered in the correlation analysis and perhaps others. The results of the regression analysis reveal the extent to which the six variables measuring the social, attitudinal, and locational attributes of the zip code zones comprising Long Island together explain spatial variations in evacuation intentions.

Determinants of Evacuation Intentions. Once the interactions among the six independent variables were taken into consideration, the maximum r-square regres-

TABLE 12.3
Correlation Coefficients

Variables	R-Value	Significance
Distance from Plant	−.52	.0001
Stage in Life Cycle (Factor 2)	.35	.0001
Nuclear Opposition (Factor 3)	.35	.0001
Distrust Public Officials (Factor 4)	−.11	.2364*
Socioeconomic Status (Factor 1)	−.01	.8754*
Direction from Plant	.04	.6714*

Source: Compiled by Authors from Shoreham Evacuation Survey, June, 1982.

*Not statistically significant

sion technique revealed that the model that "best" explains evacuation intentions consists of five independent variables. As Table 12.4 shows, distance from the plant is the most important determinant, accounting for 27 percent of the total variance in the dependent variable - evacuation intentions. The second most important variable is nuclear opposition (Factor 3), explaining six percent of the total variance. The third most important variable, contributing nearly five percent to the explained variance, is direction from the plant. Stage-in-life-cycle is the fourth most important variable, and it accounts for two percent of the total variance. The fifth most important variable, distrust of public officials, explained approximately one percent of total variance. Together these five variables accounted for 42 percent of the total variance in evacuation intentions among the zones comprising Long Island.

These results, together with the findings of the analyses presented in the preceding section of this chapter, suggest that evacuation is most likely to occur from zones where most of the households are young and contain children and where opposition to nuclear power is high. A significant proportion of these zones lie west of the plant. However, the fact that we are able to account for only 42 percent of the variance in the dependent variable (evacuation intentions) also indicates that some important explanatory variables

TABLE 12.4
Coefficients of Determination: Stepwise Regression
Model (N=120)

Variables	Beta Weights	R-Square	R-Square Change
Distance from plant	-.0107	.273	-
Nuclear Opposition (Factor 3)	-.0583	.337	.064
Direction from plant	.1066	.384	.047
Stage in life cycle (Factor 2)	.0793	.407	.023
Distrust Index (Factor 4)	.0309	.423	.016

Source: Compiled by Authors from Shoreham Evacuation Survey, June, 1982.

have been ignored in specifying the model. More importantly, the results suggest that the determinants of evacuation identified in previous studies of non-radiological emergencies are not strong predictors of who is most likely to evacuate from a nuclear accident. In this regard, the results of the regression analysis support our proposition that radiological emergencies are likely to prompt public reactions (i.e., more extreme behavior) that differ significantly from the way previous studies have shown humans respond to other kinds of crises. Further, the findings also tend to support Firebaugh's (1981, p. 150) view that:

> There is clearly something more to the public's perception of nuclear risk than the perceived actuarial risk of fatality. Risks from nuclear energy are not, in fact, limited to death and injury but include hard to quantify but very real disruptive social effects such as evacuation, loss of income, emotional stress, real estate devaluation, and potential for land despoliation ... Since the real risk of nuclear power must include such deleterious effects, the intuitive "folk wisdom" perception of risk may be a more accurate, operational measure than actuarial statistics.

In short, the variance in evacuation intentions unaccounted for by the regression model employed above (58 percent) most likely is attributed to the residents' perceptions of those "hard-to-quantify, deleterious" social effects.

SUMMARY, CONCLUSIONS AND IMPLICATIONS

We have presented empirical evidence which supports the proposition that nuclear accidents are likely to give rise to higher levels of extreme behavior than has been reported in studies of non-radiological emergencies. More specifically, drawing on data gathered in a survey of 2595 Long Island households and using descriptive and multivariate statistical analyses, we have demonstrated the extent to which people are likely to ignore official instructions and "spontaneously evacuate" in the event of a radiological emergency at the SNPS.

Our analyses suggest strongly, as in the local public reactions to the TMI accident, that spontaneous or voluntary evacuation is also likely to be the dominant type of behavioral response in the event of an accident at the SNPS. Depending on the scenario, 25 percent to 50 percent of the Long Island population is likely to evacuate spontaneously, casting an "evacuation shadow" extending at least 25 miles (40 km) beyond and primarily west of the plant site (Figures 12.2 and 12.5). Beyond the fact that a high degree of spontaneous evacuation can be expected, our analyses suggest that the pattern of evacuation is largely unpredictable in part because many of the social concerns the public possesses about nuclear power are hard to quantify. Factors that explain evacuation from other kinds of emergencies account for only 42 percent of the total variance in evacuation intentions of Long Islanders, leaving over half (58 percent) unexplained (Table 12.4). This finding lends additional support to our contentions that radiological emergencies differ significantly from nonradiological emergencies.

The results of this study, combined with the evidence from TMI, lead us to conclude that it is unrealistic for federal, state and local emergency response planners to assume that people will follow instructions during a radiological emergency. Because of the potentially catastrophic consequences of a nuclear accident, individuals are likely to ignore governmental directives and leave their homes even though they are located well beyond the specified evacuation zone. Given that a flood of "voluntary evacuees" are likely to originate from well beyond the designated zone of evacuation, these peoples' travel patterns must be taken into consideration in planning for an orderly exodus of the people resident in the designated danger zone, however it is defined.

The evidence presented in this chapter suggests that decisions regarding the most appropriate actions to take during a radiological emergency are likely to be made at the individual household level and cannot be imposed on the population by governmental fiat. In light of this fact, two strategies are likely to enhance evacuation planning for radiological emergencies. First, federal, state and local emergency response planners should (1) accept the fact that human behavior cannot be controlled; (2) try to determine how the population is likely to react, and (3) capitalize on these behaviors by building them into the emergency response plans. Results of the Shoreham Evacuation Survey suggest, for example, that the ten mile (16 km) EPZ currently required by the NRC and FEMA should be extended to 25 miles (40 km) at the SNPS, in part to accommodate the likely high degree of "spontaneous evacuation," and in part to ensure that public exposure to radiation is minimal, in case of a radiological emergency at the plant. Second, emergency response planners also need to educate people about the importance of following directions in the event of a nuclear power plant accident. In the case of the SNPS, for example, the education program should be aimed not only at the zip code zones further from and west of the plant in which families (primarily young and with children) are likely to spontaneously evacuate (over-reactors), and also at those zones closer to and east of the plant in which the residents (primarily older and without children) may be inclined to ignore official orders to evacuate (under-reactors). These are the groups that either endanger society by con-gesting travel arteries (over-reactors) or themselves by remaining in place (under-reactors).

NOTES

1. Financial support for the Long Island evacuation survey was provided by Suffolk County, New York, Peter F. Cohalan, Chief Executive.
2. PAGs are projected levels of radiation exposure at which action should be taken to protect the general population.

REFERENCES

Barnes, K., et al. 1979. Responses of impacted popula-
    tion to the Three Mile Island Nuclear Reactor
    Accident; An Initial Assessment. Discussion Paper
    No. 13. New Brunswick, N.J.: Department of Geog-
    raphy, Rutgers University.
Barr, A.J., et al. 1976. A Users Guide to SAS.
    Raleigh, N.C.: SAS Institute.
Benedict, R., et al., 1980. "The Voters and Attitudes
    Toward Nuclear Power: A comparative Study of
    "Nuclear Moratorium Inititatives," Western
    Political Quarterly, 33:7-23.
Brunn, S.D., J. Johnson, Jr., D.J. Zeigler. 1979.
    Final Report on a Social Survey of Three Mile
    Island Area Residents. East Lansing, MI:
    Department of Geography, Michigan State University.
Burton, I., R. Kates, and G.F. White. 1978.
    Environment as Hazard. London: Oxford University
    Press.
Chenault, W.W., G.D. Hilbert, S.D. Reichlin. 1979.
    Evacuation Planning in the TMI Accident.
    Washington, D.C.: Federal Emergency Management
    Agency.
Cohen, R.E., and F.L. Ahearn. 1980. "Applied Concepts
    in Understanding Disaster Behavior". In Handbook
    for Mental Health Care of Disaster Victims, R.E.
    Cohen and F.L. Ahearn, eds. Baltimore and London:
    The Johns Hopkins University Press, pp. 29-41.

Cook, E. 1982. "The Role of History in the Acceptance of Nuclear Power," Social Science Quarterly, 63:3-15.

Cornell, J. 1982. The Great International Disaster Book. New York: Charles Scribner's and Sons.

Erikson, K.T. 1982. "Human Response in a Radiological Accident," In The Indian Point Book: a Briefing of the Safety Investigation of the Indian Point Nuclear Power Plants, Cambridge, MA: Union of Concerned Scientists; and New York: New York Public Interest Research Group, Inc., pp. 55-59.

Farrell, T.B. and T.Goodnight. 1981. "Accidental Rhetoric: The Root Metaphors of Three Mile Island," Communication Monograph, 48:271-300.

Firebaugh, M.W., 1981. "Public Attitudes and Information on the Nuclear Option," Nuclear Safety.

Flynn, C.B., 1979. Three Mile Island Telephone Survey: Preliminary Report on Procedures and Findings. Washington, D.C.: U.S. Nuclear Regulatory Commission (NUREG/CR-1093).

Flynn, C.B., 1981. "Local Public Opinion." In The Three Mile Island Nuclear Accident: Lessons and Implications, Vol. 365, Annals of the New York Academy of Sciences, T.H. Moss and D.L. Sills, eds. New York: The New York Academy of Sciences, pp. 146-158.

Hans, J.M., Jr. and T.C. Sell. 1974. Evacuation Risks - An Evaluation. Las Vegas, NV: U.S. Environmental Protection Agency.

Hohenemser, C., R. Kasperson, and R. Kates. 1977. "The Distrust of Nuclear Power," Science, 196, (April 1):25-34.

Hudson, B.B., 1954. "Anxiety Response to the Unfamiliar," Journal of Social Issues, 10:53-60.

Janis, I.L., 1954. "Problems of theory in analysis of stress behavior," Journal of Social Issues, 10:12-25.

Janis, I.L., 1962. "Psychological Effects of Warnings," In Man and Society in Disaster, G.W. Baker and D.W. Chapman, eds., New York: Basic Books, Inc., pp. 55-92.

Janis, I.L., 1977. "Emergency Decision Making: A Theoretical Analysis of Responses to Disaster Warnings," Journal of Human Stress, 3:35-48.

Johnson, James H., Jr., 1982. Testimony on Behalf of Joint Intervenors. In the matter of emergency planning for Pacific Gas and Electric Company's Diablo Canyon Nuclear Power Plant (Units 1 and 2).

Kates, R., (ed), 1977. Managing Technological Hazards: Research Needs and Opportunities. Program on Technology, Environment, and Man, Monograph No. 25. Boulder: Institute of Behavioral Science, University of Colorado.

Kim, J.O. 1975. Factor analysis. In _Statistical Package for the Social Sciences_, N.H. Nie, et al. eds., New York: McGraw Hill, pp. 468-514.

Kuklinski, J.H., D.S. Metlay, W.D. Kay. 1982. "Citizen Knowledge and Choices on the Complex Issue of Nuclear Energy," _American Journal of Political Science_, 26:615-642.

Lagadec, P. 1982. _Major Technological Risks: An Assessment of Industrial Disasters_. New York: Pergamon Press.

Lang, K. and G. Lang. 1964. "Collective Responses to the Threat of Disaster." In _The Threat of Impending Disaster: Contributions to the Psychology of Stress_, G.H. Grosser, H. Weschsler, and M. Greenblatt, eds. Cambridge, MA: The M.I.T. Press, pp. 58-75.

Manning, D.T., 1982. "Post-TMI Perceived Risk from Nuclear Power in Three Communities," _Nuclear Safety_, 23:379-384.

Melber, B., et al., 1977. _Nuclear Power and the Public: Analysis of Collected Survey Research_. Seattle, WA: Battelle Memorial Institute, Human Affairs Research Centers.

Nelkin, D. 1981. "Some Social and Political Dimensions of Nuclear Power: Examples from Three Mile Island," _American Political Science Review_, 75:132-142.

NRC, 1975. _Reactor Safety Study: An Assessment of Accident Risks in U.S. Commercial Nuclear Power Plants_. Washington, D.C.: U.S. Nuclear Regulatory Commission, October (NUREG 75/104 (WASH-1400).

NRC, 1978. _Planning Basis for the Development of State and Local Government Radiological Emergency Response Plans in Support of Light Water Nuclear Power Plants_. Washington, D.C.: U.S. Nuclear Regulatory Commission (NUREG - 3096, EPA 520/1 - 78 - 016).

NRC, 1980. "Emergency Planning: Final Regulations," _Federal Register_, 45:55402-55418.

Otway, H.J., 1977. "Risk Assessment and the Social Response to Nuclear Power," _Journal of British Nuclear Energy Society_, 16:327-333.

Perry, R.W. 1979. "Evacuation Decision-making in Natural Disasters," _Mass Emergencies_, 4:25-38.

Perry, R.W., and M.R. Green. 1982. "The Role of Ethnicity in the Emergency Decision-making Process," _Sociological Inquiry_, 52:306-334.

Rubin, D.M., et al. 1979. _Staff report to the President's Commission on the Accident at Three Mile Island_. Report of the Public's Right to Information Task Force. Washington, D.C.: U.S. Government Printing Office.

Social Data Analysts, Inc., 1982. _Attitudes Toward Evacuation: Reactions of Long Island Residents to a_

<u>Possible Accident at the Shoreham Nuclear Power Plant</u>. New York: Social Data Analysts, Inc.

Slovic, P., and B. Fischoff. 1983. "How Safe is Safe Enough?" In <u>Too Hot to Handle</u>? C.A. Walker, L.C. Gould, and E.J. Woodhouse, eds. New Haven and London: Yale University Press, 1985. pp. 112-150.

Slovic, P., G. Fischhoff, and S. Lichtenstein. 1979. "Rating the Risks," <u>Environment</u>, 21:14-39.

White, G.F., ed. 1974. <u>Natural Hazards: Local, National, Global</u>. New York: Oxford University Press.

White, G.F., and E. Haas. 1975. <u>Assessment of Research on Natural Hazards</u>. Cambridge, MA: M.I.T. Press.

Williams, H.B., 1964. "Human Factors in Warning-and-Response Systems." In <u>The Threat of Impending Disaster: Contributions to the Psychology of Stress</u>, G.H. Grosser, H. Wechsler, and M. Grenblatt eds. Cambridge, MA: The M.I.T. Press, pp. 79-104.

Withey, S. 1962. "Reactions to Uncertain Threat." In <u>Man and Society in Disaster</u>, G.W. Baker and D.W. Chapman, eds. New York: Basic Books, pp. 93-124.

Zeigler, D.J., S.D. Brunn, and J.H. Johnson, Jr. 1981. "Evacuation from a Nuclear Technological Disaster," <u>Geographical Review</u>, 71:1-16.

Introduction

John E. Seley

The last act of the 97th Congress was the Nuclear Waste Policy Act of 1982 (NWPA). This represents the latest efforts of the federal government to resolve the problem of high-level commercial radioactive waste disposal, a problem which has plagued the nuclear power industry for two decades. Indeed, by the projected opening of the first repository in January, 1998, there will be a backlog of some 48,000 tons (53,000 metric tons) of spent fuel from commercial nuclear power plants, and these plants will be generating an additional 3,500 tons (3,900 metric tons) a year of spent fuel.

The legislative mandate to bury our nuclear garbage does not mean of course that it will actually get buried on time; few believe that the rigid timetable established by the Nuclear Waste Policy Act (including selection of the first repository site by 1989) can actually be achieved. Irrespective of the timing of waste disposal, there is the much more significant problem of whether or not the agencies responsible for nuclear waste storage will be able to overcome the long legacy of mistakes, mistrust, and ill-will from past waste burial efforts. At the simplest level, it can be argued that as long as the same institutions largely responsible for past mistakes, like the U.S. Department of Energy, continue to be involved in future planning efforts, we cannot expect much change. This is simply the nature of bureaucratic inertia. A more generous interpretation would argue that new provisions in law and regulation require consideration of larger issues, even if in-place institutions show no particular inclination to change their ways.

The question becomes: to what degree do the latest efforts at resolving the nuclear waste storage problem address the major issues which have thwarted previous efforts? Indeed, over the past twenty years of failure, observers have developed a comprehensive understanding of many of the problems to which a waste management system must address itself. If these issues are addressed, we can then hope they will be resolved during the course of actual planning for waste sites. If these issues are not addressed, the prospects that the latest round of legislation will actually result in resolution of the waste storage problem are dim. What are the major issues which must be addressed in nuclear waste storage? Does the latest initiative (the NWPA) promise an answer to these issues? It is an interesting case study in policy formation.

DESIGN AND TECHNICAL CRITERIA

There are, of course, the so-called technical concerns with the design and operation of the storage system. These include the construction and operation of repositories, the design and functioning of a complex transport system for pickup and delivery of wastes, continued monitoring of the entire process, and some form of emergency response mechanism in the event of accidents either on-site at pickup or storage locales, or along transport corridors.

Of these concerns, the effort to design and find a site in an appropriate geologic medium has preoccupied planners, with comparatively little attention to any of the other issues. Even so, two respected journalists, among others, have questioned whether current programs are aimed at finding a safe disposal method, or simply at resolving the waste problem before it becomes more of an embarrassment. Luther Carter, a contributing writer to _Science_, cautions that the "cliche that 'radioactive waste disposal is a political but not a technical problem' reflects a misapprehension of the realities of geologic disposal" despite it's widespread acceptance by Congress and the nuclear industry (Carter, 1983). Indeed, the technical and political are not so easily separated. Carter cites the example of Hanford basalt, in which an oversight committee called the Department of Energy contractor to task for claiming that the site was being studied for its geological properties. Rather, the committee noted that the only reason for studying the site was the "sociopolitical fact" that it already existed as a federal nuclear reservation. Other concerns about the system of waste disposal promise to present severe challenges to our institutional imagination and organizational response.

INSTITUTIONAL COMPLEXITY AND
MANAGEMENT RESPONSE

One of the continuing institutional problems is the complexity of a waste disposal system which will disperse hundreds of trucks a year over the nation's highways. Recognizing this potential hazard, virtually every state, as well as over 200 counties and municipalities, has passed ordinances against transport of nuclear waste. The fact that the Department of Transportation may be given the legal right to override such local ordinances does not address the problem of designing a nearly error-free system. A recent report of the Steering Committee to Develop a National Strategy for the Transportation of Hazardous Materials and Wastes in the 1980s found "There is no agreed-on hazardous material transportation safety goal within the government-industry complex" (National Research Council, 1983).

In addition to a lack of clear goals, the potential for abuse is evidenced, for example, by safety violations of waste disposal due to the involvement of organized crime. Finally, there appears to be a gap in regulation due to interjurisdictional confusion over relative federal, state, and local roles.

A further institutional concern is the problem of longevity--assuming we can resolve the issue of clear and responsible authority, will the institutions established to oversee nuclear waste survive the normal vicissitudes of governmental change and support? On a more mundane level, a series of questions about bringing repositories "on line" remain. When will they begin accepting waste? (Cotton, 1982). Will certain wastes (those in storage longest, for instance) be given priority for long-term storage? Or will political power dictate storage priorities? What happens to existing waste if the repository schedule is not met, and will this present a serious problem at power plant sites where monitoring is bound to be less conscientious? Most importantly, under the severe timetable of the Nuclear Waste Policy Act, is there any incentive to be more responsive to accidents, errors, and misjudgements than has been true in the past, or are we condemning ourselves to a technological quick fix?

Other observers have emphasized the unique features of a radioactive waste management system. As LaPort notes (1979,p.2):

We face the extraordinary tasks of instituting nearly error-free operational systems for handling radioactive materials and wastes, as well as developing nearly escape-proof burial grounds for them. These are quite remarkable demands upon a society and an intellectual community which have deeply embedded within their institutions and workways a short-time perspective, confidence in incremental, pragmatic processes of policy and substantive improvement, and an aversion to comprehensive,

synoptic plans productive of constraining, in-
flexible programs.

The pragmatic or incremental approach to decision
making relies on relatively rapid discovery of errors
and an ability to correct mistakes. But leaks of
radioactivity are not easy to discover, and "the consequences of errors are so harmful and so nearly irreversible as to render absurd the notion of improvement
of the basis of them" (LaPorte, 1979, p. 15). Thus, a
comprehensive plan is necessary. This plan would have
to include all technical aspects as well as political,
social, and economic elements which might jeopardize
the integrity of the system.

ROLE OF THE STATES

At the least, a comprehensive waste management
system will include cooperation among waste generators
(utilities and, perhaps, the military), private industry (responsible for much research and development on
waste storage technologies and probably for transport),
all levels of government, and publics. The Nuclear
Waste Policy Act concentrates on states and Indian
tribes, to the virtual exclusion of the role of other
actors. Even then, the involvement of states and
Indian tribes is limited primarily to suggestions on
the mitigation of adverse impacts, and the ambiguous
permision to "consult and cooperate" with the federal
government. Their true power, however, lies in provisions for a "veto" over selection of a specific site,
subject to override by both houses of Congress. This
promises to generate much political bargaining as
states with more Congressional influence try to parlay
that power into concessions, while weaker states may be
forced to accept smaller payoffs.
Contrary to prior legislation on major federal construction programs such as low-income housing, requirements for analysis, recordkeeping, monitoring, or even
simply bookkeeping are not specified. The ambiguity of
State involvement has led several observers to conclude
that "We may. . .expect to see the best, and the worst,
examples of how we choose to govern ourselves before
the process [of state involvement in choosing a waste
site] has worked itself out" (Pierce, Hill, Haefele,
1983, p. 10).

PUBLIC CONCERNS

Of more significance is the fact that local governments and local publics are hardly mentioned in the new
legislation. Yet, many observers, like Kasperson,

Derr, and Kates in this section, have pointed to the importance of localities (host communities) in the acceptance or rejection of sites, and the inadequacy of relying on state governments to represent local concerns.

Whether or not the technology of waste disposal is safe, it is still necessary to deal with public attitudes about the likelihood and results of radiological releases. While it may be tempting to dismiss public concerns as unfounded, it has been reluctant publics who have been able to thwart major projects in the past and who continue to play an active role in questioning the intent and scope of government actions for a variety of locational decisions (Seley, 1983). Thus, from a purely practical standpoint, it will be necessary to involve the public along transport corridors and at potential repository sites. Unfortunately, government thinking has not changed in this regard from the early days of waste storage efforts.

Despite a history of complex decision making over controversial facilities and a backlog of techniques for involving publics, current waste management plans rely exclusively on the public hearing for input from non-elected officials. The current version of the public hearing, although somewhat more sophisticated than its early shouting matches, is nonetheless at one end of a continuum of actual public input into decision making.

Health and safety, which is a continuing concern of the public, presents another unresolved problem. As Kasperson, et. al. note, the issue may be much more worrisome for workers in the radioactive waste industry than for the public. Yet, if accidents do occur (and transport "incidents" are reported frequently), the lack of accident planning may create difficulties for local officials and citizens. The recent experience with trying to plan for evacuation around (stationary) nuclear power plants has not been promising. Controversy surrounding the plans at Shoreham on Long Island and Indian Point 3 in Westchester County, New York, has been particularly rancorous, leading the head of the NRC to ask Congress to allow power plant employees to be deputized so they can carry out evacuation plans in the absence of local cooperation. In short, the planning for accidents has not come very far from the confusion experienced during the Three Mile Island accident.

IMPACT ASSESSMENT

Another continuing thorn in the side of waste management planners is the difficulty of predicting the effect of a project like a waste repository on local

populations at sites or along transport corridors.
Although the Nuclear Waste Policy Act provides monies
for independent technical assessment and mitigation of
adverse effects, there is little recognition of the
complexity of the task involved in such procedures,
and, indeed, the Act provides a shortcut procedure for
impact evaluation. Again, if the role of the public
were more suitably defined, the effort to identify
effects would be enormously enhanced. In the absence
of either acceptable techniques for impact assessment
or methods of eliciting public opinion and feedback,
the effort to mitigate effects will be hampered
severly.

COSTS

Already under great financial pressures (Brody,
1983), utilities will be relieved of the burden of
waste disposal costs due to provisions in NWPA which
allow the Secretary of Energy to pass on the costs of
disposal to utility customers, to begin at 1.0 mil per
kilowatt-hour, presumably without the need for
regulatory approval. Even though projected costs are
small, this by-passing of local public service
commissions will not make disgruntled ratepayers any
happier. Congress will oversee the cost of waste
disposal, but if costs escalate in a fashion similar to
that of nuclear power plants, ratepayers may be
expected to take notice. Indeed, as Solomon notes
below, the costs of decommissioning and dismantling
power plants may be much higher than anticipated.

The greatest economic concern would seem to be not
how much is paid, but who is paying it. As Kasperson,
Derr, and Kates note, future generations will be asked
to pay for waste disposal although they may not be the
beneficiaries of the power which the wastes represent.
So, too, with the decommissioning and decontrol of
power plants, as Solomon points out.

THE FUTURE

The Nuclear Waste Policy Act must be viewed as the
latest initiative in a continuing effort to deal with
the nuclear waste disposal problem. It allows for a
number of promising directions, including the role of
states and Indian tribes in the planning process, provision of monies for independent technical review, a
research and development effort, and specification of
a mission plan for organizational management. Nonetheless, it also leaves unanswered a number of continuing problems; and allows little opportunity for them to
be addressed. These include nagging technical flaws,

institutional complexity, assurances that the public health and safety will be protected in the event of accident, the role of the public and, specifically, local communities in planning and impact identification, and a full consideration of costs (including those for decommissioning and decontamination of power plants).

The articles in this section explore many of these concerns in depth. Hare and Aikin review waste disposal technologies, emphasizing the front end of the fuel cycle and the problems of security of fissile materials. Kasperson, Derr, and Kates raise some larger issues in waste management and argue for the need to address equity concerns. By applying different equity principles, they conclude that lengthy interim storage is appropriate, as well as a more careful examination of the benefits and risks of waste disposal. Solomon highlights the problem of plant decommissioning. The dismantling of plants will add to concerns over transport, cost, and the full scope of solid nuclear waste management.

Each of these continuing concerns is exacerbated by a timetable which appears to be too fast, and will likely not be met. Nonetheless, the pressure to bury wastes "as soon as possible" creates an unnecessary burden for a political process which should be deliberative and open.

REFERENCES

Brody, M. 1983. "Nuclear Waste: Construction Delays Cast a Cloud Over Utilities," Barron's 31, pp. 13, 33, 36-37.

Carter, L.J. 1983. "The Radwaste Paradox." Science, 219:33-36

Cotton, T.A. 1982. "Statement Submitted to the Subcommittee on Energy and the Environment." U.S., Congress, House. Committee on Interior and Insular Affairs. 97th Congress, 1st Session.

LaPorte, T.R. 1979. Management Errors and Nuclear Wastes: Problems for Social Analysis. Berkeley, California: Institute of Governmental Studies.

Pierce, B.; D. Hill and E. Haefele. 1983. "High-level Radioactive Waste Management - A Means to Social Consensus." Presented at International Conference on Radioactive Waste Management, Seattle, Wash.

Seley, J.E. 1983. The Politics of Public-Facility Planning. Lexington, Mass: Lexington Books.

Steering Committee to Develop a National Strategy for the Transportation of Hazardous Materials and Wastes in the 1980s, Transportation of Hazardous

310

   <u>Materials: Toward a National Strategy</u>. Washington,
   D.C.: National Research Council, Transportation
   Research Board.
U.S. Congress. House. Nuclear Waste Policy Act of 1982.
   Public Law 97-425, 97th Congress, 2nd Sess., 1982,
   H.R. 3809.

13

Nuclear Waste Disposal: Technology and Environmental Hazards[1]

F. Kenneth Hare and A. M. Aikin

The waste from the nuclear industries require comprehensive treatment if man and nature are to be protected against radioactivity. Though nuclear wastes are not alone in being dangerous to life, they have acquired special visibility in the past few years. Their proper management has long been a matter of concern, and one in which considerable experience and skill has now been gained, after some initial errors. The final stage of such management, however--the ultimate disposal of long-lived radioactive wastes--has not yet been successfully achieved.

Radioactive substances are those in which the atomic nuclei are unstable. The nuclei decay at precise rates and emit radiation of several kinds, all of which may be harmful to man, plants and animals. Such substances are widespread in nature and also occur within the human body. Some human cancers and birth defects arise from this natural radiation. They are the price we pay for living in a world whose atoms are not all stable.

Much more strongly radioactive substances are generated, however, when the nuclei of certain metals are split, either naturally or by man. One abundant natural element is fissile uranium-235 (U-235), which is a small but universal fraction of natural uranium. The earth's crust contains about four grams of uranium per metric ton, and the sea about three grams of uranium per thousand metric tons. A constant 0.715 percent of this uranium is fissile U-235. Other fissile nuclei occur, but are extremely rare in nature. They include plutonium-239 (Pu-239) and uranium-233 (U-233).

Nuclear technology, whether for weapons production or power-generation, depends on the harnessing of fission. The reverse process of fusion, whereby light atomic nuclei combine (as in the sun), is at present used only in weapons. Power generation depends on the controlled fission of U-235 in reactors, and on Pu-239 formed in those reactors from U-238, the most abundant

311

type of uranium. It may later be possible to use thorium-232 (Th-232) as the source of energy after it has been changed to U-233. All these forms of fission create two product streams: (1) an enormous amount of heat, which is the source of energy for the electric power to be generated; and (2) fission products, which are lighter nuclei formed by the splitting of the original nuclei of U-235, Pu-239 or U-233. The first of these streams is the desired product from the reactor. The second is the price we pay. It could be a high price, since the fission products are mostly intensely radioactive, and hence dangerous. They constitute most of the nuclear waste that has to be dealt with, in terms of total radioactivity.

The operation of reactors also produces other wastes. Part of the uranium is converted into other heavy nuclides, called actinides, of which plutonium-239 is the prime example. Others include americium, curium, neptunium and other isotopes[2] of plutonium. These are also radioactive. Very few are useful aside from Pu-239, which is fissile. Still other radioactive wastes are formed from the materials of which the reactor is built, such as iron and cobalt, or from the air and water contained within the reactor.

Power-reactor operation thus creates many different radioactive substances, each with its characteristic (and unalterable) rate of decay. Some nuclei decay in seconds. Others linger for millions of years. Most are useless and are hence wastes of a dangerous and awkward kind. Reactors used to manufacture plutonium for weapons create similar wastes, which have been accumulating ever since World War II. Nuclear waste management and disposal must protect man and nature against the hazards presented by all these unwanted by-products.

Wastes also occur at other stages of the fuel cycle, to use the usual if paradoxical title. Large amounts of radium, radon and thorium are released, for example, in the mining, milling, and refining processes whereby uranium oxide fuel is manufactured for nuclear reactors. These heavy, radioactive substances occur in the tailings and waste heaps that accumulate around mines, mills, and refineries, from which they may be leached into water supplies, streams and lakes. In addition, radon gas, a radioactive decay daughter[3] of radium or thorium, occurs in the mine and diffuses out of tailings and waste heaps. The jargon calls all these products front-end waste, to distinguish them from the wastes from reactors, the back-end of the fuel cycle.

Back-end wastes accumulate in all countries that operate reactors for whatever purpose. Currently some twenty-six nations are known to have the problem. Front-end wastes, however, are accumulating only in the

smaller list of countries that mine and refine uranium.
Nevertheless these wastes are a problem for all mankind. Many are deadly and pose a threat to future
generations. Some are capable of world-wide dispersal,
so that they imperil all people.

The management of these wastes has had a long
history, going back to the first realization that they
were threats to human health. In more recent times,
the wastes from reactors have been reasonably well
managed in most countries. What has been neglected,
however, is the question of the disposal of the long-
lived wastes that will not simply decay in storage.
This problem now looms large in people's minds, and has
become a subject for heated argument. A solution will
have to be found--and the solution will have to satisfy
a skeptical and alarmed world. Disposal means the
planned, permanent placement of radioactive wastes in
some sort of repository, with no intention of subsequent recovery. All highly radioactive waste with a
long life expectancy should, if possible, be disposed
of in this fashion. The present report reviews the
options available for this process and examines the
possible environmental problems associated with these
options. It also looks at measures that may be taken
to offer long-term protection of man and biota against
the hazards involved. The report begins with an
analysis of the nature and origin of the wastes and a
review of disposal technology.

THE NATURE AND ORIGIN OF WASTES

The Fuel Cycles

The nuclear fuel-cycle begins with the extraction
of uranium ores from rock and ends with the disposal of
wastes. It is not, in this sense, a cycle, but a sequence with well-defined stages. Each stage produces
wastes, many long-lived, and thus candidates for disposal. In Table 13.1, we have identified the main
stages and the wastes associated with each. Before we
discuss these, we must first define the classes of fuel
cycle of which it refers. For practical purposes we
have to deal with:

The Once-Through Cycle. Uranium or uranium oxide
fuel is used in the reactors once and then removed.
The fuel rods are fabricated from mined uranium ore.
Once-through cycles are of two main families, those
using (1) heavy water moderated and cooled and graphite
moderated and gas-cooled, both using natural uranium,
with 0.715 percent U-235; and (2) light water, boiling
light water, pressurized light water, and advanced
gas-cooled reactors, all of which require enriched

TABLE 13.1.
Process States and Wastes in the Nuclear Fuel Cycle

Process	Radioactive Waste
Uranium Mining ⟶	Waste rock (low grade ore) Mine waters, (contain Ra-226 but are used in mill)
↓ ⟶	Ventilation air, (contains Rn-222)
Milling	Mill tailings Th-230, Ra-226, Rn-222, and sometimes Th-232
↓	Waste waters, mainly Ra-226
↓	Ventilation air, mainly Rn-222
Refining ⟶	Impurities from U, mainly Ra-226
UF₆ conversion ↓	
Enrichment ↓	Wastes are recycled
Conversion to oxide ↓	
Fuel Fabrication	
Reactor Operation ⟶	Reactor Wastes
Irradiated Fuel ⟶ ↓ (if no processing)	Irradiated Fuel for
Fuel Reprocessing ⟶ ↓	High level Waste Actinide wastes (for disposal)

uranium (usually about 3 percent U-235). Most power reactors now in use are of the light water kind (LWRs), and need enriched uranium.

   Advanced Fuel-Cycles. Now under intensive international study, and based on recycling fissile materials derived from the reprocessing of irradiated fuel. These materials include plutonium-239 and uranium-235, both of which are present in irradiated fuel extracted from reactors. These cycles may be based on the recycling of plutonium and uranium, or on thorium-232, which can be converted to fissile uranium-233.

Breeder Reactor. In a breeder reactor, natural uranium or depleted uranium from other reactors is mixed with plutonium-239 from fuel processing in such a way that more fissile material is produced (mainly plutonium) than is consumed by fission. In effect, such reactors breed their own fissile fuel from the irradiated fuel derived from non-breeder reactors, or from natural uranium.

The wastes that result from these cycles are all much the same in character, but there are significant differences in the details of waste management. We have laid major stress on the highly radioactive wastes (HLWs) derived from fuel reprocessing and on irradiated fuel that comes from once-through or advanced fuel cycle reactors.

Character of the Wastes

Radioactive wastes are produced at all stages of the nuclear fuel cycle, though the major quantities (in bulk and total radioactivity) occur at the front and back ends of the cycle. The front end includes the mining, milling and refining stages, and affects those countries and regions actively exploiting uranium ores. The back end refers to the materials that arise from reactor operation. The middle parts of the cycle produce less waste. To guide the reader, we have prepared Table 13.1, which lists the stages of the cycle and briefly indicates the nature of the wastes associated with each. They are of five main kinds:

1. naturally occurring radionuclides rejected during the mining, milling, and refining stages. These include isotopes of uranium, thorium, and their decay daughters, such as radium and radon gas;

2. heavy radionuclides, called actinides, formed from uranium and other heavy metals by neutron absorption in the fuel rods of the reactors;

3. fission products, i.e. lighter nuclides formed by the fission of uranium-235, plutonium-239 or uranium-233 in the fuel rods;

4. solid radionuclides formed in the reactor when neutrons are absorbed by the reactor structure, such as iron-59 and cobalt-60;

5. gaseous or liquid radionuclides formed by neutron absorption (e.g., argon-41; tritium, H-3; carbon-14) in and around the reactor.

Items 1 and 4 are mostly solid elements that can be contained and treated as wastes. Radon gas under 1, and the gaseous and liquid products under 5, are hard to contain, and most of them are deliberately dispersed to the hydrosphere or atmosphere. In this chapter we are concerned only with disposal, i.e., the permanent isolation from the biosphere, of dangerous, long-lived radionuclides that are retainable. These occur in 1 to 4 inclusive, though the major hazard for man is in 1, 2, and 3.

The Mining and Milling Operations: Waste Generation and Management

Uranium is mined in many countries, the largest current exporter being Canada. The ores contain uranium in oxides from UO_2 to U_3O_8, or in more complex mineral molecules such as carnotite or austenite, which contain other metals. Uranium content varies from very high (around 40 percent) in pitchblende to as little as 0.1 percent by mass. Much commercial mining involves ore containing less than 1 percent uranium, so that extraction produces large volumes of bulky wastes. Before the uranium can become nuclear fuel, it must be extracted and purified, which includes separating it from its own radioactive decay daughters. These natural readionuclides generally stay with the bulk of the ore and are discharged with the very large quantities of mine and mill waste, the principal nuclides of concern being thorium-230, radium-226 and radon-222. Uranium itself is of low radiotoxicity, but it has a chemical toxicity comparable to other heavy metals such as lead.

Uranium-bearing ores normally occur deep enough that they are not a hazard to life, but there are exceptions. In some areas of the world, radioactive minerals are close enough to the surface to raise the background radiation well above normal, both by direct radiation and by the release of radon gas. The radium content of some drinking waters is also above recommended levels. Statistically, it is difficult to determine whether there has been a decrease in life expectancy in these areas, but it is safe to assume that it would not be good to multiply these occurrences. Potentially, that is what uranium mining is doing. It brings into the biosphere large quantities of these naturally occurring but immobile radioactive materials and changes them into forms in which escape and dispersion are possible. This problem is well recognized, and measures have been taken to keep it under control. The International Atomic Energy Agency (IAEA) has issued a code of practice for the management of these mine and mill wastes. Many national agencies have also established regulations. The problem of long-term isolation of the radium-226, however, is still not adequately solved. Radium is rather easily dispersed by water and ingested by plants and animals.

In the production of nuclear fuels, the uranium is first mined, and then milled, usually at the mine site, to an impure yellow cake. It is then brought to nuclear purity in a separate refinery. Thereafter, since the radioactive daughter products have all been removed and grow back very slowly into the uranium, the subsequent enrichment and fuel fabrication stages are relatively non-hazardous. So it is wastes from mining, milling, and refining that must be managed with care. Wastes from the mining operation consist of (1) waste rock, or very low grade ore, which can be disposed of in the mine if it presents an exposure hazard; (2) liquid waste, a combination of mine drainage and water used in the mining operation, and which is generally all used in the mill; and (3) gaseous wastes, the exhaust ventilation, which contains radon gas as well as ore dust.

The mine wastes present little hazard, as they are generally at low radioactivity levels and management techniques are available and in general use.

The milling operation consists of grinding the ore to a very fine powder, which is leached with carbonate or sulfuric acid solution to dissolve the uranium. The uranium is then purified by a combination of ion-exchange, solvent extraction, and precipitation processes, the details depending on the nature of the original ore. The milling wastes, which are generally collected into one stream, consist of a slurry of the pulverized ore containing almost all the uranium daughter products, together with a small fraction of unrecovered uranium. Most of the chemicals used in the process are also present. Radium-226 is the most hazardous nuclide in the slurry, a typical concentration being 800 pCi^3 g^{-1} of dry tailings, or less than one gram in 1000 metric tons of tailings. (In secular equilibrium, production of 1000 metric tons of uranium would produce approximately 200g of radium-226 as waste, compared with about one metric ton of plutonium after irradiation in a nuclear power plant.) This slurry is discharged into a basin where the solids are retained behind a natural or man-made dam, and the liquid overflows to a waste treatment lagoon. Here the contained radium is coprecipitated with barium sulphate. Time is allowed for this precipitate to settle before the supernatant liquid is discharged into the natural water systems (rivers, lakes, sea or groundwater). This liquid often still contains some radium-226, but should be below the levels prescribed by IAEA and national regulations. As long as the mill is in operation, the processing of the liquid waste will continue, and there must be control over the discharge of effluents. It is known, however, that continued leaching of the finely-divided solid mill wastes by rain water will remove radium that will then appear in the seepage from the piles. Radium concentration in such seepage is likely to exceed the recommended levels. If such leaching occurs, the seepage will have to be continuously collected and processed to remove the radium, and this operation might be needed for centuries to come. The alternative is tolerance of higher radium levels in the surrounding waters. Neither of these alternatives is acceptable. Thus, methods of stabilizing these mill-tailings must be found. Many methods are under active study and testing, but because of the very large size of the tailings deposits, they are costly and difficult. All methods aim at physical stabilization of the wastes, and at a decrease of wind and water erosion to near zero. Some thought has been given to putting the solid wastes back into the mine from which they came, but only a little more than half could be handled in this way because of the increase in volume brought about by the pulverization. Revegetation of the tailings piles appears to be the optimum method of stabilization, but little is known yet about how effective this will be in decreasing radium leaching. Some consideration is also being given to removal of the radium and thorium from the ore during mill processes, and to their separate disposal. If removal were complete enough, the large volume of mill wastes would no longer present a radioactivity hazard. For the removal to be effective, however, over 99 percent of both these elements would need to be recovered. Given the nature of many of the ores, this would be very difficult and costly to achieve. Evaluation of this possibility is continuing, but it is unlikely to be the chosen waste management process.

The sheer bulk of the mining and milling wastes stands in the way of ready treatment. United States estimates, for example, give a waste-to-product ratio of 1,300 to 1. Tens of millions of metric tons of these low-grade but by no means harmless materials surround the mining and milling areas of producing countries. At present, these wastes are under control; but a long-term solution that will not require continual surveillance is not yet fully developed. The main hazard is the slow and continued leaching of radium into nearby water systems.

The Middle Stages of the Cycle: Refining, Hexafloride Conversion, Enrichment, Oxide Conversion, Fuel Fabrication

These middle stages comprise: (1) refining, to remove residual impurities (including radium-226 from the natural uranium); (2) hexafluoride conversion, in which the yellow-cake (U_3O_8) from milling is converted into uranium hexafluoride (UF_6), a readily volatilized compound; (3) enrichment of uranium-235 in the UF_6 from the original 0.715 percent of natural uranium to between 2.5 and 3.5 percent, for use in light-water and similar reactors; (4) conversion of UF_6 to UO_2, or oxide conversion; and (5) fuel fabrication, using the UO_2, to form ceramic pellets, rods to contain the pellets, and bundles to contain the rods. Heavy water moderated reactors, such as CANDU in Canada (and under development elsewhere) the Magnox type reactors, enable one to by-pass the hexafluoride conversion and enrichment stages, because such reactors use natural uranium. the middle stages produce wastes, but these are of low radioactivity and are generally recycled or handled on-site. Hexafluoride conversion produces calcium fluoride (CaF_2) ash, which is drummed and buried at a suitable facility. Enrichment plants have large volumes of liquid waste that are locally ponded, the radioactive sludges being buried on site. Fuel fabrication from UF_6 produces a further quantity of CaF_2, a waste that is avoided in heavy-water technology. The CaF_2 is usually uranium-contaminated, and is typically stored on-site. Under modern practice, these stages are quite tractable from the waste management standpoint. Local problems do occur, chiefly because of current public anxiety about all aspects of waste management. But by comparison with the front and back ends of the cycle, the middle stages present no difficult problems and do not raise questions of ultimate disposal on the same scale.

Irradiated Fuel from Reactor Operation

In a nuclear reactor, heat is produced from the fission process that occurs when a neutron is absorbed by the fissile nuclides uranium-235 and plutonium-239. The U-235 isotope occurs naturally in uranium, and the Pu-239 is formed in the reactor by neutron absorption in U-238 to form U-239, which decays radioactively to Pu-239. This fission produces a spectrum of fission product nuclides that accumulate in the fuel. Many of these are unstable and decay radioactively. Fission also produces neutrons about 2.3 per fission on average. One of these neutrons is absorbed by a fissile nuclide to keep the fission process going, and the others are absorbed by non-fissile nuclides in the fuel, the fission products, and the construction materials of the reactor core. As the concentration of fission product nuclides increases, they absorb more and more neutrons, so that ultimately a fuel bundle becomes a burden to the reactor and must be removed. Also in the fuel elements, there is a build-up of isotopes of heavy elements. These are formed through a series of neutron capture reactions and radioactive decays. They generally go under two names; the transuranic elements, or TRUs, i.e., elements with atomic numbers above that of uranium; or actinides, with atomic numbers higher than actinium. The important members of this group, as well as the longer-lived fission products of interest in waste management, are listed in Table 13.2.

TABLE 13.2
Significant Nuclides in Radioactive Waste Management

Element	Isotope	Half-life	Generation Mechanism
Hydrogen	H-3	12.3y	Fission and neutron capture
Carbon	C-14	5.7×10^3y	Neutron capture
Argon	Ar-41	1.8h	Neutron capture
Cobalt	Co-58	72d	Neutron capture
Cobalt	Co-60	5.3y	Neutron capture
Krypton	Kr-85	10.8y	Fission
Strontium	Sr-89	54d	Fission
Strontium	Sr-90	28y	Fission
Yttrium	Yt-91	61d	Fission and neutron capture
Zirconium	Zr-93	1.5×10^6y	Fission
Zirconium	Zr-95	66d	Fission and neutron capture
Niobium	Nb-95	35d	Fission and daughter of Zirconium-95
Technetium	Tc-99	2.1×10^5y	Fission
Ruthenium	Ru-106	1y	Fission
Iodine	I-129	1.7×10^7y	Fission
Iodine	I-131	8d	Fission
Xenon	Xe-133	5.2d	Fission
Cesium	Cs-134	2.1y	Fission
Cesium	Cs-135	3×10^6y	Fission
Cesium	Cs-137	30y	Fission
Cerium	Ce-141	33d	Fission
Cerium	Ce-144	285d	Fission
Promethium	Pm-147	2.6y	Fission
Samarium	Sm-151	93y	Fission
Europium	Eu-154	16y	Fission

Lead	Pb-210	21y	Daughter of Polonium-214
Radon	Rn-222	3.8d	Daughter of Radium-226
Radon	Ra-226	1.6×10^3y	Daughter of Thorium-230
Thorium	Th-229	7.3×10^3y	Daughter of Uranium-233
Thorium	Th-230	8×10^4y	Daughter of Uranium-234
Uranium	U-234	2.4×10^5y	Daughter of Protoactinium-234
Uranium	U-235	7.1×10^8y	Natural source, daughter of Plutonium-239
Uranium	U-238	4.5×10^9y	Natural source
Neptunium	Np-237	2.1×10^6y	Neutron capture and daughter of Americium-241
Plutonium	Pu-238	87y	Neutron capture and daughter of Curium-242
Plutonium	Pu-239	2.4×10^4y	Neutron capture
Plutonium	Pu-240	6.6×10^3y	Neutron capture
Plutonium	Pu-241	15y	Neutron capture
Plutonium	Pu-242	3.87×10^5y	Neutron capture
Americium	Am-241	433y	Neutron capture and daughter of Plutonium-241
Americium	Am-243	7.37×10^3y	Neutron capture
Curium	Cm-242	163d	Neutron capture
Curium	Cm-244	18y	Neutron capture

Source: mainly OECD (1977)

The actinides are generally long-lived alpha particle emitters, whereas the fission products are shorter-lived beta and gamma emitters. The radiological properties of the irradiated fuel are effectively determined for the first few hundred years by the fission products and thereafter by the actinides. Though there are differences in production rates of individual nuclides in different nuclear fuel systems, it matters little as regards waste management whether the principal fissile material is U-235, Pu-239, the thorium U-233 cycle, or even if fast fission is used. All systems produce radioactive fission products and actinides that require careful storage, handling, and ultimate disposal. Thus, the discharged fuel still contains useful quantities of fissile uranium-235 and plutonium-239, but to be used again they must be chemically separated from the neutron-absorbing fission products. This chemical reprocessing to recycle the fuel is expensive and must compete economically with the use of fresh fuel; but it would allow much more energy to be produced per unit mass of mined uranium.

Fuel Reprocessing: Waste Generation and Management

By 1985, there will be about 10,000 metric tons of irradiated fuel in storage at nuclear power plants around the world. Most of these storage systems were designed with the intent of storing fuel only for a few years before shipping it to a chemical reprocessing plant. There have been delays in building such commercial plants, so additional fuel storage capacity is needed and is being built. It is still assumed in most countries using nuclear power that fuel reprocessing will be profitable and acceptable; but it will be at least another decade, if not longer, before enough reprocessing plant capacity (and recycle fuel-fabrication plants) will be available to deal with the ever increasing supply of irradiated fuel. If nuclear power is to be a provider of large quantities of electrical energy during the next century, there is little doubt that fuel recycling will be needed to extend the usefulness of the available uranium. This is the basis for the assumption that, in the long term at least, all irradiated fuel will be reprocessed to recover the unused fissile and fertile material. Thus, the programs to develop long term safe methods for waste storage and isolation have generally concentrated on the waste streams from fuel reprocessing plants. It is only lately that any thought has been given (as we do later in the chapter) to methods for permanently disposing of irradiated fuel.

The common method of fuel reprocessing consists of cutting the fuel elements into short pieces (1-2 cm), dissolving the uranium oxide fuel in nitric acid, extracting the uranium and plutonium from this solution by solvent extraction, and recoverning these two products separately by selective backwash conditions. Two major waste streams result, the undissolved metal fuel cladding, usually zirconium alloy (Zircaloy) or stainless steel, and the acidic solutions containing the fission products, the actinides, and some unextracted uranium and plutonium. This acid solution is often referred to as high level waste (HLW) since it is indeed highly radioactive. As a result of military programs, large quantities of this solution have been produced. So far it has been concentrated to reduce the volume, sometimes neutralized, and stored in large tanks. Containment has not been perfect, and the few leaks have been widely publicized, though they have done little environmental damage.

More permanent methods of storage or disposal of this high level waste are now being developed, with most work concentrated on converting the waste to an insoluble solid. Glass has become the preferred form, and several countries are testing processes at the pilot-plant stage. A typical glass block will be a metal clad cylinder, 60 inches (1500 mm) long and 16 inches (400 mm) in diameter, containing the waste from 2.2-4.4 pounds (1-2 kg) of irradiated fuel. These glass canisters will be stored, with cooling and shielding, pending their placement in a permanent repository. Similarly, the fuel cladding wastes will be compacted and incorporated into an isoluble medium.

These processes have yet to be used on a commercial scale, and so there are still doubts concerning their ability to deal in a fully satisfactory manner with all the toxic wastes that arise in fuel reprocessing. These doubts have resulted in delays in authorization of the construction of fuel reprocessing plants. They will only be answered by further research and development, and ultimately by demonstration on a commercial scale. Sufficient international attention is being given to this waste management problem that an acceptable solution should be found.

Other Wastes from Reactor Operation (Reactor Wastes)

During the operation of a nuclear power plant, many of the neutrons from the fission process are absorbed by the materials in the reactor core. Thus, radioactive nuclides are produced in structural components, such as pressure vessels, pressure tubes, or control mechanisms. In addition, the coolant carries small quantities of corrosion products around the circuits, and as these pass through the reactor core, they become radioactive. They are eventually deposited throughout the piping system, in the boilers and pumps, thus causing radiation fields around this equipment. The principal radionuclides formed in this way are cobalt-60, iron-59 and manganese-54. Of these, cobalt-60 has the longest half-life - 5.26 years. In addition to the radioactivity due to these corrosion products, some fission products can, on occasion, be embodied in the coolant circuit through small holes in the fuel cladding. Such cladding defects are rare, and the fuel is removed from the reactor at the first opportunity; but the fission product contamination of the coolant circuit remains.

During operation, the radiation fields around the process equipment are not a hazard, since they are located behind shielding. Like all such systems, however, occasional maintenance is required, and station personnel must enter these areas to work on such items as pumps and valves. During these procedures gloves, rags, clothes and other such items become contaminated radioactively. The clothes are generally laundered at the plant, giving rise to potentially radioactive solutions. The gloves, rags and other disposable objects are bagged and become solid radioactive waste. Any replaced items of equipment, such as valve packings and pump seals, are also bagged or wrapped in plastic for disposal. To keep the radiation fields around the process equipment down to tolerable levels, the radioactive corrosion material and fission products are continuously removed from the coolant circuits by filtration and ion-exchange purification of a sidestream. These filters and ion exchange resins become another form of solid wastes, since they are generally used only once, and then removed when loaded.

Some radioactive gases are also formed in the reactor, and many are released directly to the atmosphere. The significant gaseous radionuclides are tritium (half life 12.3 years, H-3) carbon-14 (half life 5730 years) and krypton-85 (half life 10.8 years). they are formed by various nuclear reactions or as fission products, and because of their long half lives, are of environmental concern. Other radioactive gasses that are released, such as argon-41 or nitrogen-13, are shortlived and thus have little environmental impact. These gaseous wastes are usually released from stacks after various forms of treatment, such as filtration and delay systems, to allow the shortlived radioactivities to decay. More elaborate decay or collection systems, such as cooled charcoal traps or cryogenic distillation, have been developed, but they have not been widely used, since their small benefits have not so far been thought to warrant their cost.

The liquid wastes are treated in two ways. When the amount of contained radioactivity is low, they are often dispersed by blending with the condenser water discharge. Alternatively, they are purified by ion exchange, or on occasion concentrated by evaporation. The concentrate is then converted to a solid, either by incorporation in cement or, in some experimental cases, bitumen. The function of these processes is to convert the radioactive waste into solid form.

Various methods for the immobilization of the solid reactor wastes are in use, and others are being investigated. At some nuclear plants, the combustible wastes are put through a special type of incinerator to achieve a large volume reduction. At other plants, they are mechanically compacted. This reduces the magnitude of the waste problem, but does not provide immobilization for long-term disposal. In most countries, these solid wastes are stored in concrete bunkers. West Germany has taken the lead in setting up

a national disposal system. The evaporation concentrates from their power plants are incorporated in
cement, and the ion exchange materials in the thermo-
setting resins. These immobilized solids are then
placed underground in the Asse salt mine.

Most of the radioactivity in the reactor wastes
will decay to non-toxic levels in a few decades. For
this reason, there is still some controversy as to
whether it will be necessary to bury and/or dispose of
these wastes with the same requirements as for high
level waste. Engineered structures could contain them
for at least 100 years, which is probably long enough.
As the nuclear industry grows, however, the quantities
of these wastes will become large, and a central disposal site serving several nuclear plants would make
environmental control much easier. The situation today
is not dangerous, but is is also not tidy. Continued
and more extensive planning and regulation are needed
for the future.

Some consideration has also been given recently to
the decommissioning of nuclear facilities. Some research reactors have already been shut down and decontaminated, and the equipment sent to waste burial
sites. However, the dismantling of a nuclear power
plant will present problems on a much larger scale. As
few studies of this potential environmental problem had
been made, IAEA convened a Technical Committee to discuss this topic in October, 1975, and a report was
subsequently issued. Other national studies have also
been recently published, and estimates have been made
of the volumes of radioactive waste that will be produced. No problems of feasibility have been found, but
most plans will allow for the nuclear plant to sit idle
for a number of years before dismantling is started.
Much of the radioactivity in the structures is relatively shortlived, and a few years' decay will make the
job much easier. No detailed studies of the decommissioning of fuel processing plants have been published.
This disposal of nuclear facilities at the end of their
useful life is not an insurmountable problem, or a big
hazard to the environment if done carefully; but regulations and procedures have yet to be developed. It is
beyond the scope of the present report.

DISPOSAL TECHNIQUES

The Objectives of Disposal Systems

Growing uneasiness about the accumulation of wastes
in the nuclear industry has led in the past few years
to strong pressure for demonstrations of the feasibility of safe disposal. In several countries, there
has been a slowing-down of the industry's growth

because of this uneasiness. One of the most drastic reactions is that of the Government of Sweden, whose 1976 Conditions Act says that permission to commission a nuclear reactor... "may be granted only providing that the reactor owner

1. has produced an agreement which adequately satisfies the requirement for the reprocessing of spent nuclear fuel and has demonstrated how and where an absolutely safe final storage of the high-level waste obtained from the reprocessing can be effected, or

2. has demonstrated how and where an absolutely safe final storage of spent, unreprocessed nuclear fuel can be effected."

Similar pressure is being exerted on the industry in many other countries. Accordingly, there is under way a major study of the technology of disposal systems. Several technical studies have been published. Of special value to the present report are studies by the Swedish Nuclear Fuel Safety Project (Karn-Bransle-Sakerhet, KBS, 1978); by the American Physical Society (1977); by the U.K. Royal Commission on Environmental Pollution (1976); and by OECD (1977). Reference has also been made to a report from Canada's Department of Energy, Mines and Resources (Aikin, Harrison and Hare, 1977).

The distinction between ultimate disposal and long-term storage is important. Storage implies retrievability; one seeks temporary containment and isolation of the materials until one finds a further use for them (for example, the decision to reprocess irradiated fuel). Storage also implies monitoring and surveillance, the maintenance of security and safeguards, and vigilance for health and environmental protection. Disposal implies the reverse: the materials are put away with no intention of recovery and with the purpose of isolating them from the biosphere for an exceedingly long period - in effect, for all time. It implies the intention ultimately to relax monitoring and surveillance, since these are obligations we cannot put on future generations. The objective of permanent isolation from the biosphere arises from the very long lives of many of the radionuclides. It will be necessary to isolate the fission products for several hundred years and the actinides for some tens of thousands of years, even to reduce levels of radioactivity to those typical of high-grade uranium ore bodies. In practice, current planning aims at much longer isolation, on the scale of hundreds of thousands of years. Even the shortest of these periods is longer than that elapsing since the end of medieval times in Europe. The Neolithic revolution and the beginnings of urbanization are less far back than the 10,000 years needed to reduce even reprocessing wastes to something comparable to the radioactivity of rich uranium ore. It is this long isolation time and the difficulty of long-term guarantees of performance that make many critics skeptical of disposal technologies.

In this chapter we shall not challenge the need for such disposal. It may well be that present perception of the hazard is exaggerated and that public health and the natural environment could be protected with less drastic measures. Many calculations suggest that this is the case. But the level of public concern is high enough to justify defense in depth against the hazard. At some subsequent date, experience and public perception may call for less draconian measures. For the present, however, very elaborate, fail-safe technologies for disposal are necessary. All present plans for disposal are based on the multiple barrier principle, as in the case of reactor safety itself. In brief, the plans call for maximum possible immobilization of the wastes or irradiated fuel, especially against solution or suspension in water; encapsulation of the immobilized wastes in long-lasting, corrosion-proof canisters or capsules, often multiple; incorporation of the encapsulated wastes in buffering material designed to retard water penetration, while freely permitting heat release; choice of site for the engineered repository in deep rock, or in the ocean floor, so as to minimize water migration past the wastes, and to retard the migration of radionuclides in such migrating water; choice of site for the repository remote from earthquake or volcanic hazards, possible economic resources, and with good access to the required transportation links.

The major objective of the present report is to consider the potential environmental impact of such repositories. Clearly, if the above specific requirements are met, there will be no impact, apart from that associated with the preparation of the wastes for disposal, their transportation to the repository, and the operation of that repository until it is sealed off (as it will be when full). But it is vital to consider the possibility of miscalculations or accidents leading to failure. The ruling principle is to design the facility and the handling chain so as to minimize the likelihood of such mishaps. It is also vital to estimate their consequences, should they occur. In what follows, we shall concentrate on the disposal of high-level reprocessing wastes, or of irradiated fuel. It may also be necessary to dispose of waste from earlier stages of the fuel cycle, such as reactor waste containing transuranic elements, or materials derived from the decommissioning of reactors. But the big and urgent problem is to learn how to cope with the intensely radioactive fuel or reprocessing wastes that are being or will be produced in growing volumes every year. A solution to this problem can readily be adapted to look after other wastes.

Disposal of Irradiated Fuel

We have seen that fuel rods are removed from all kinds of reactors after one to five years of irradiation because of the accumulation within them of fission products that absorb too many neutrons. One metric ton of uranium fuel typically yields 25,000 to 35,000 MW(t) days of heat in light water reactors (much less in natural uranium reactors). This irradiated fuel contains wastes (which, if reprocessed, would yield about 150 litres of glass or 4 m^3 per 1,000 MW(e) y^{-1}). This fuel is intensely radioactive when removed from the reactor, but the radioactivity is mostly due to short-lived fission products that decay very rapidly. The fuel is then stored in water-filled cooling bays at the reactor site to shield workers and the public from the radiation, and to remove the great amount of waste heat generated. When removed from a light-water reactor it contains:

about 3 percent by weight fission products (mostly β and γ emitters);

about 95 percent unconsumed uranium, of which 1 percent is fissionable U-235;

about 1 percent plutonium (mostly fissionable Pu-239;

and, less than 1 percent other actinides (mostly emitters).

Fuel from heavy water reactors is poorer in fissionable U-235 and Pu-239.

After two years or more in the cooling bays at the reactor site, the fuel needs to be moved because the bays become full. Most nuclear engineers advocate storage at a central interim storage facility also under water -- though Atomic Energy of Canada Limited (AECL) is experimenting with dry concrete canister storage, which renders the constant filtration of cooling water unnecessary. In interim storage, the fuel continues to cool and to lose radioactivity as the radionuclides decay to stable nuclides. Throughout this period of interim storage, the fuel must be rigorously isolated from human beings and protected against accident, willful sabotage or theft. There is division of opinion as to how long the fuel should remain in interim storage. A recent Swedish study suggests a limit of 10 years, but longer periods have already been achieved.

A decision must then be taken as to reprocessing. If the plan is to reprocess the fuel to recycle its fissile contents (U-235 and Pu-239), it will be transported to a reprocessing plant -- which may well be nearby. We shall treat this option later. At present, we shall consider the course of action to be taken if it is decided to dispose directly of the fuel.

Figure 13.1 shows the hazard presented by irradiated light water reactor fuel (3.1 percent U-235) per

FIGURE 13.1
Hazard indices of various radionuclides, and total hazard, presented by one tonne of irradiated light-water reactor fuel, as a ratio of the hazard presented by uranium ore equivalent to one tonne of fuel. Burn-up 33,000 MWd(t) per tonne of uranium. Enrichment 3.1 percent U-235 (After KBS, 1978).

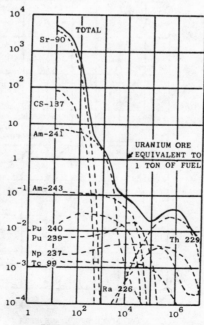

TIME AFTER DISCHARGE FROM
REACTOR, YEARS

330

metric ton, relative to that presented by the uranium ore that is equivalent to a metric ton of such fuel, together with the contributions made by the chief radionuclides (compare with Table 13.2 for identification of nuclides). Figure 13.2 shows the same data for high-level wastes from the reprocessing of one metric ton of fuel. In each case, the fuel is assumed to have undergone a burn-up in the reactor of 33,000 MW(t) days per metric ton of uranium (data from KBS, 1978). The hazard is assumed to be the radioactivity associated with each nuclide multiplied by that nuclide's relative toxicity to man. The details of such curves depend on the original fraction of uranium-235, the burn-up achieved, and the type of reactor. Canadian CANDU reactors, for example, have much lower burn-ups (about 7,5000 MWt days per metric ton) and the evolution of the waste is rather different. The following broad properties manifest themselves:

FIGURE 13.2
Similar to Figure 13.1, but is for the equivalent immobilized high-level waste. (After KBS, 1978).

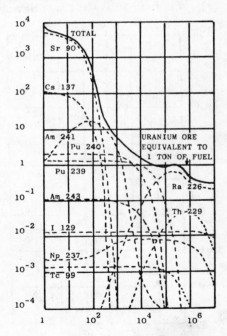

TIME AFTER DISCHARGE FROM
REACTOR, YEARS

In the first 600 years, it is the fission product nuclides that present the greatest hazards. The main activity is due to the very toxic nuclides strontium-90 and cesium-137, both having half-lives near 30 years, and both readily ingested by man. Other fission products are rather less hazardous. Most have short half-lives, but iodine-129 (half-life 17 x 10 years) is an exception that escapes to the atmosphere (being volatile). Other long-lived nuclides include technetium-99 and cesium-135. The fission products are mainly beta and gamma emitters.

After 600 years, the hazard is primarily due to the actinides and radium. For 50,000 years americium-241 and plutonium-239 and -240 dominate. These nuclides are alpha-emitters. After that, the main hazard is due to radium-226, which also emits hard γ-radiation.
Certain heavy radionuclides not present in the irradiated fuel at the time of its withdrawal grow into the fuel as decay daughters of other heavy nuclides. Examples include radium-226 and thorium-229.

Irradiated fuel's total hazard drops to that typical of uranium ore (i.e., to naturally occurring radiation hazards) in about 30,000 years, and then remains near that level for about another 150,000 years. The time taken to reach these levels depends, of course, on the burn-up of fuel achieved in the reactor.

It is generally assumed from analyses of this kind that isolation of spent fuel from the biosphere is required for about 600 years in the case of the dominant fission products and at least long enough for the longer-lived activity (chiefly due to the actinides) to decline to ore-body levels. Another figure often quoted is ten half-lives of plutonium-239 (244,000 y). Precise figures are not very logical. In practical terms, the presence in the fuel of long-lived actinides like plutonium-239 means that isolation must be planned to be permanent. Any escape from confinement while activity levels are still above those typical of nature itself must be avoided.
Though the direct threat to the biota (including man) is posed by the radioactivity of the wastes, it is the release of heat that poses the major problem from the point of view of their storage and subsequent disposal. This heat must be dispersed to avoid damage to the containment. Figure 13.3 shows the heat released per metric ton of uranium from irradiated fuel from pressurized water reactors as a function of time out of the reactor (with 3.15 percent uranium-235, and 33,000

FIGURE 13.3
Decay heat generation, in Watts per tonne of uranium, from irradiated fuel and immobilized high-level wastes (99.5 percent uranium removal) from a pressurized water reactor using 3.25 percent U-235, after a burn-up of 33,000 MWd(t) per tonne or uranium. (After KBS, 1978).

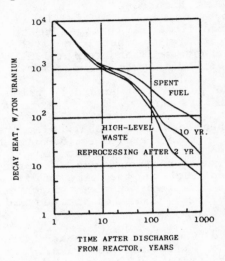

TIME AFTER DISCHARGE
FROM REACTOR, YEARS

MW(t) days per metric ton burn-up). Curves are shown for both irradiated fuel and reprocessing wastes. Figure 13.4 shows the equivalent cooling of Canadian CANDU fuel after 7,500 MWt days per metric ton with natural uranium. It follows from these and other data that much heat has to be dispersed from the irradiated fuel, especially in the first few years when the fuel is stored under water (whose high heat capacity helps in the dispersal process). After the fuel has been installed in a repository, the heat must disperse itself through the surrounding medium by conduction. The rate of heat generation decreases rapidly at first. In the first century, the fission products dominate the release, but the actinides take over thereafter. If we assume that irradiated fuel is disposed of in a repository after fifty years in surface storage, its heat output is likely to be about 600 W T^{-1} for LWRs and PWRs, and 100 W T^{-1} for CANDUs. We return to this subject later.

The immobilization of irradiated fuel should not pose severe problems. The fission products, actinides, and unused uranium remain within the ceramic pellets of the fuel rods, whose Zircaloy cladding will help to confine any that must be encapsulated in some form, to

FIGURE 13.4
Decay heat generation, in Watts per tonne (megagram), from irradiated fuel from a Canadian CANDU reactor, natural uranium, 7,500 MWd(t) per tonne of uranium burn-up.

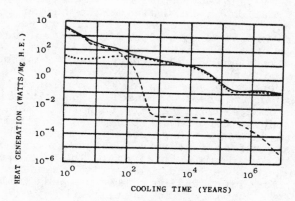

Dashed : Fission Products
Dotted : Actinides
Continuous : Total

minimize the migration of radionuclides. Little study was given to this problem until recently, since the direct disposal of irradiated fuel was considered undesirable for economic reasons. One recent published study (KBS, 1978) recommends encapsulation of the rods in pure copper sheaths to ensure prolonged containment. Within the next few years, it is to be expected that the major nuclear powers will publish designs and safety requirements for repositories for irradiated fuel.

Disposal of Reprocessing Wastes

As shown above, reprocessing wastes will be disposed of in the form of immobilized solid blocks, probably consisting of glass containing the fission products, small residual amounts of uranium and plutonium not removed in the reprocessing, and the remaining actinides. The glass is likely to be closed in a resistant metal canister. The wastes differ from irradiated fuel in lacking the radioactivity and heat release associated with the uranium and plutonium that have been removed. As Figure 13.2 shows, the effect of this removal is to permit a more rapid decay in radiological hazard than in the case of spent fuel. For the first three hundred years, strontium-90, and to a lesser extent cesium-137, dominate the hazard index. Thereafter, for a period of about 50,000 years, the actinides americium-241 and -243, and the residual plutonium-239 and -240 dominate. Thorium-229 (a decay daughter) builds in to dominate the hazard index from 50,000 years out to well over 10 million years. Over-all hazard drops to that of uranium ore equivalent to 1 metric ton of fuel after about 25,000 years. As in the case of irradiated fuel, these rough figures vary a little according to the type of fuel, the burn-up achieved, and the type of reactor. Figure 13.3 shows the accompanying heat release. The vitrified high-level wastes cool more rapidly than fuel, especially if reprocessing is carried out promptly after removal from short-term storage. Reprocessing wastes hence differ in waste-form from irradiated fuel and pose rather less long-term radiological hazards. The differences are not great, however, and are not in themselves a good argument for reprocessing. The demands made on the repository's capacity to contain the wastes and to disperse heat are not very different. In our judge-ment, repository design should permit the disposal of either irradiated fuel or vitrified high-level wastes. It is impossible to judge at this time how extensive future reprocessing will be.

SITING OF REPOSITORIES

The Options

The objective of disposal is the permanent isola-tion of the wastes from all contact with living organisms, especially from human beings. To do this, one must design an engineered site, called by conven-tion a repository. Options that have been considered for sites for repositories include (a) surface or shallow sub-surface permanent storage, involving permanent surveillance and monitoring; (b) burial in glacial ice, in Antarctica, or in Greenland; (c) dis-patch to space by rocket vehicle; (d) disposal in canisters on the sea-floor, probably in deep oceans, or burial of the canisters in holes in the sea floor, in each case in stable abyssal plains. A variate is disposal near subduction zones between tectonic plates; (e) geological containment on land, in sediments such as salt, clay, shale, or volcanic tuffs or in massive igneous rocks such as granite.

The first of these options is not, in our judge-ment, valid for disposal as normally defined. It has nevertheless been espoused by some leading authorities in Canada and elsewhere. Ocean disposal clearly has much merit, but technical and jurisdictional uncertainties are handicaps. At present, opinion usually favors on-land burial in an excavated rock repository within the home territories of the country. This is the method receiving most attention today.

Current Choices for Geological Containment

The sites being actively considered differ from country to country, depending on the geology, density of population, and the nuclear technology being employed. The kinds of rock being examined include clays, shales, salt beds and dunes, tuffs, and massive igneous rocks.

Salt beds or domes (diapirs) are being actively explored and tested in several countries, notably the United States, the Federal Republic of Germany (where storage in salt at Asse is a well-developed technique), and the Netherlands. Salt has many advantages. Its mere existence is proof that groundwater circulation is absent, since it is highly soluble. It may, however, contain entrapped water that will tend to migrate towards a heat source such as a repository, but this would not lead to the escape of radionuclides. Salt tends to flow under high pressure and hence self-heals any damage to excavated structures. It is an excellent heat conductor.

Igneous rocks, which are often massive and crystalline in form, are being examined in other countries, especially Canada, the United Kingdom, France and Sweden. Such rocks are often fractured and fissured, and water movement follows these lines of weakness. In Canada, the preference is for small quartz-poor, feldspar-rich intermediate or basic intrusions, which are abundant in the Laurentian Shield, the ancient core of the country. Similar rock-bodies are being explored in Sweden, where granites and granite-gneisses are the choice. These rocks contain quartz, but are also rich in felspars.

POTENTIAL ENVIRONMENTAL IMPACTS

General Considerations

Disposal aims to eliminate environmental impact due to the radiotoxicity or chemotoxicity of the wastes. In this section we examine the likelihood and magnitude of such impacts associated with the disposal technologies now available or under development. We also consider how the impacts can be minimized. Desirable levels of environmental and health protection have long

been established by the International Commission for
Radiological Protection (ICRP), whose standards form
the basis for most national codes. The emphasis has
naturally been upon human health, rather than on that
of other organisms. Apart from our desire to protect
ourselves, we appear to be more sensitive to ionizing
radiation than any other animal or plant species so far
tested. The United Nations Scientific Committee on
Atomic Radiation (UNSCEAR) has recently (1977) published detailed estimates of the radiological exposure
of human populations from all parts of the fuel cycle.
Hence, this subject will not be treated here. Environmentalists are likely to insist, however, that the
impact of disposal on other organisms be considered,
especially as regards the tendency of specific nuclides
to be concentrated by certain species. Effects on
overall ecosystem functions will also need to be specified, if these are detectable.

The potential environmental impacts of waste disposal can be considered in relation to specific states
of process: (1) preparation of the wastes for disposal; (2) transportation to the repository site; (3)
emplacement in the repository; and (4) after emplacement. In this report, we shall examine only the fourth
of these stages, since this is the vital concern. The
impacts of reprocessing and immobilization technology,
and of transportation, have been covered elsewhere --
for example in Flowers, 1976; UNSCEAR, 1977; OECD,
1977; KBS, 1978. The risks associated with emplacement
have not been examined in as much detail, but these are
likely to be confined to small occupational exposures,
similar to those experienced in earlier parts of the
fuel cycle.

Impacts After Emplacement in a Rock Repository

The irradiated fuel or vitrified high-level wastes
will be installed in excavated structures in salt,
igneous rock, shale, clay or tuff. The repository must
be at sufficient depth for the rock to absorb all the
radiation emitted and to disperse the heat so that
there is no significant rise of soil and surface
temperature. If there is to be impact on living organisms, there will hence have to be either (1) a breach-
ing of[4] the containment by earthquake, vulcanism,
erosion , meteorite impact, or faulting (catastrophic
effects); or (2) dispersion of radionuclides from the
repository by migrating groundwater, essentially a
long-term gradual process. Proper location of the
repository and its design will minimize each of these
possibilities. We shall consider them in turn.

Catastrophic Effects

Catastrophic processes are considered very unlikely to breach the containment of wastes. The following can be visualized:

Bedrock-Movement Due to Warping. Loading of the landsurface by glacial ice can downwarp the crust, in some areas by as much as 3280 feet (1000 m). Subsequent recovery can be quite rapid (of order 1 cm y-1). Vertical warping of this kind will not breach a repository, since it is uniform over large areas. The districts being explored by both Swedish and Canadian authorities have been repeatedly subjected to such movements in the past million years and may well be downwarped again. Renewed glaciation within 10,000 years is possible.

Bedrock-Movement Due to Faulting. In the choice of repository site, it must be a first requirement to avoid major fault-planes or fracture-zones that have a history of recent movement. In rigid Shield areas like the Laurentian Shield of Canada, the Angara Shield in the Soviet Union, and the Fenno-Scandian Shield in Norway, Sweden, Finland and parts of the Soviet Union, there are extensive areas in which little faulting has occurred in the past several hundred million years. The KBS study for Sweden, for example, estimates that the probability of a fault movement affecting a one km^2 repository is less than one in a billion per annum. Many other countries, however, are tectonically[5] more active. Japan, Italy, and New Zealand are good examples. In such countries, special care will be needed in locating areas with minimum likelihood of faulting. New faults are rare. Movement is nearly always along pre-existing fault-planes that can be readily detected, mapped and monitored.

Even in tectonically active areas, however, movement along a fault-plane rarely exceeds a few inches per movement, and this would be most unlikely to cause rupture of the respository, or damage to the encapsulated wastes. Open faulting, in which new, open fissures appear, is essentially a surface phenomenon.

Earthquake-Shocks. The seismic effects of faulting of the above kind include substantial shock-wave movement into the rock around the fault, extending out to several hundred miles. Damaging earthquakes occur in association with fault-movement along well-known axes (for example, the St. Lawrence Valley in Canada and the Pacific Coast of the United States). Minor earthquakes or tremors occur in many areas, occasionally with damage to surface structures. These shocks have maximum effect at the surface, and rarely damage mines, subways, and other engineered sub-surface structures. There is a long history of mine management in seismically active areas from which these effects can be confirmed and analyzed.

Meteorite Impact. Analysis of meteorite impacts over the past million years (KBS, 1978) suggests that the odds against a direct meteorite impact causing a 100 m crater in a given square kilometer are about 10 trillion (10^{13}) to one per annum. Canadian estimates are closer to 10^{11} to 1. Both are quite negligible hazards.

Stream Erosion. In hard rock areas, stream erosion does little work in periods of a hundred thousand years, except in mountainous terrain. Glacial erosion may overdeepen existing valleys. In the Laurentian Shield of Canada, successive glaciations in the past million years have locally excavated pre-existing valleys by 3280 feet (1000 m). In flat plateau areas, glacial erosion is minor, and consists chiefly of removal and redistribution of soil and subsoil. A total regional erosion of 10 to 100 meters per million years probably overestimates the effect.

Vulcanism. Active volcanic areas will be avoided in the choice of site. New vulcanism is usually in areas with a previous history of such activity. Impact of such new vulcanism on a well-chosen repository should be nil.

These and other possible catastrophic effects are best avoided, or at least minimized, by choice of a repository that is remote from active faults, and seismically or volcanically active areas; deep within the rock; or remote from deep valleys, especially those aligned on faults or fracture zones.

These conditions are readily met in most areas, but are harder to satisfy in countries like Italy, Japan, New Zealand and Turkey that have no seismically inactive areas. In such cases, self-sealing rock types such as clay and salt are probably better than rigid rocks -- but more research is needed to verify this. It is unnecessary to avoid areas potentially open to continental-scale glaciation (such as Canada and Scandinavia). The growth of a new ice-sheet would exclude man, plants and animals. Moreover, it would provide a thick, extra covering for the repository and would still further compress any existing fracture or fault planes. It would probably endure some tens of thousands of years.

Dispersion Processes

Dispersion of radionuclides by migrating ground-water is the chief hazard in waste disposal. Avoidance of such dispersion provides most of the site-selection and design criteria for the repository.

Groundwater originates from rain or melted snow at the land surface. In a few areas, such as the Sahara and the interior basins of Australia, the groundwater may be fossil, i.e., derived from previous epochs with a wetter climate. Elsewhere, the year's surplus rain tends to sink into permeable rocks in upland areas and to emerge from springs in valleys, having followed a curving path through the rock. Certain rocks, called aquifers, permit ready groundwater movement and store large quantities in pores and fissures. Others are highly impermeable. Groundwater may occur under high pressure below an impermeable cap layer -- for example in the sedimentary limestones of southern England, Northern France, the Low Countries, and in the Munster Basin of Germany. Groundwater movement is easy in porous and fissured rock and is hence most rapid near the surface. At greater depths, the pores and fissures are narrower or may be virtually absent. Motion is so slow that it may be unmeasureable. Calculations of flow at 500 and 1000 m depth in crystalline, fissured igneous rocks give velocities in the order of 0.1 to 0.2 liters per square meter per annuum, for areas of low surface relief, corresponding to permeabilities much below 10^{-9} m s^{-1}, possibly as low as 10^{-12} m s^{-1}.

Confirmation that groundwater movement is very slow at even moderate depths in level terrain can be derived from age determination of water sampled at those depths. This is difficult to do, since present-day water has to be injected into the hole during the drilling. Hence, determination should underestimate the true age. Using the carbon-14 technique, Swedish experiments (KBS, 1978 at one site have given ages of 4,400 years at 954 feet (291 m); of 4,275 at 1673 feet (510 m); of 11,055 at 1325 feet (407 m); and of 8,205 at 1617 feet (493 m), the pure experimental error being in each cast $\pm$ 100 years. Finnish measurements at site have given an age of 4,010 years at 446 feet (136 m). The rapid local variations are probably real, reflecting the fact that movement is local in fractured igneous rock with a hillocky surface. The rock has rapid variations of fissure-density, and hence of permeability. Age determinations of groundwater around a chosen repository will be valuable but not conclusive evidence of the slowness of movement (and hence desirability of the site).

The repository itself will be sited and designed so as to minimize the risk that such groundwater movement will leach radionuclides out of the wastes. The latter are in ceramic or vitreous form, encased within low-corrosion metal canisters. These are likely to be embedded in highly absorbent buffer material that entirely fills the excavated repository and backfilled shafts. Even rapid groundwater movement through such a medium will be greatly retarded. Great care will have to be taken, however, to minimize the risk that the actual construction of the repository does not open up new paths for more rapid circulation.

If, nevertheless, groundwater movement does penetrate the repository, it will eventually cause some leaching of the radionuclides in the irradiated fuel or vitrified wastes, insoluble though they are. It is necessary to attempt some estimate of how rapid this leaching and outward transport may be. If there is any danger that radionuclides may be transported to the surface before their radioactivity has decayed, it will be necessary to compute the degree of dilution that will occur and to estimate their subsequent progress through the drainage basin and its ecosystems. A low rate of leaching will depend upon the following succession of barriers: (a) a low rate of groundwater migration; (b) a high integrity of the encapsulation; (c) a low rate of solution or suspension of radionuclides from the glass or fuel; (d) a high capacity of the buffer material to retard the passage of the radionuclides; and (e) a high capacity of the bedrock to sorb the radionuclides.

As we have just seen, the rate of groundwater migration decreases with depth. Hence, a deep repository is less likely to suffer leaching. It is also more remote from hazards due to seismic activity, erosion, meteorite impact, or human interference. Swedish plans call for a repository at 1640 feet (500 m) in igneous rock, where the present groundwater is at least of order 4,000 to 11,000 years of age. It would be reasonable to assume an equivalent period for the groundwater around the repository to return to the surface. In fact, their plans assume the improbably fast return time of 400 years (KBS, 1978, Vols. II and IV). A Canadian study calls for a repository at 2624-3280 feet (800-1,000 m) (Aikin, Harrison and Hare, 1977), where the age of groundwater is probably considerably more and the rate of flow even slower. The greater the depth, the greater the security against leaching and the other hazards mentioned above -- this is the ruling assumption. But the assumption needs to be qualified with the warning that the structural integrity of the repository may be less easily maintained at these greater depths. The optimum depth for a given site should be experimentally determined from local field measurement, coupled with modeling of groundwater flow.

The metal canisters in which the wastes will be encapsulated can be made corrosion-resistant by a suitable choice of metal -- for example titanium or copper -- with known long-term resistance properties. If, as in some plans, a cladding of lead is included within the canisters, resistance will be even higher. It is assumed in several studies that such canisters are likely to be breached in about 1,000 years and effectively removed within the next few thousand years, gradually exposing the wastes to circulating waters. This is probably conservative. In calculating the effects of canister failure, allowance must also be made for accidental damage, either during installation or as a result of rockfalls or earth movements.

The rate of solution or suspension of the radionuclides within the glass or fuel has been studied both theoretically and empirically. It is not certain that this leaching is simply equivalent to the solution rate for the glass or ceramic, since some radionuclides may migrate towards the interface with the groundwater. Initial leaching rates at 68°F, using borosilicate glass, are reported to be about 10^{-4} g cm^{-2} y^{-1}. The nepheline syenite glass developed by the Chalk River Nuclear Laboratories of Atomic Energy of Canada Limited started at 10^{-5} and decreased after ten years of continuous leaching to 2×10^{-8} g cm^{-2} y^{-1}. Thus, under prolonged solution, leaching rates decrease with time because of the accumulation of a coating of corroded material, which decreases diffusion rates. All authorities agree that the rate of leaching increases rapidly with rising temperature -- perhaps thirty or forty fold by 212°F (100°C). Hence it is important to avoid excessive rises of temperature due to the heat release. Equivalent leaching rates for UO$_2$ fuel have not been as well determined, but they, too, are likely to be very low.

The retardation of movement of dangerous materials away from the wastes towards the surface depends on the capacity of the buffering material and the rock along the migration path to sorb the radionuclides that have entered solution. Even though groundwater may move quite freely through fissures or porous rock, it is common for the movement of dissolved or suspended substances to be retarded by ion exchange or adsorption, mechanical filtration, precipitation, and mineralization; this is how, in the remote past, many ore-bodies were formed. The combined process is called sorption.

Research in several countries, most notably France (de Marsily, Ledoux, et al., 1977), suggests that high sorption capacity is the key to optimum performance by deep rock repositories. In rock with good sorption capacity, the radionuclides mostly move much more slowly than the groundwater itself. The same should be true of well-chosen buffer materials (especially along

the boreholes, shafts, and drifts, which might otherwise act as migration pathways). The radionuclides
differ as regards retardation:

> Technetium-99 and iodine-129, both with very
> long half-lives, are poorly sorbed, or not at
> all. It can be assumed that they will move
> with the same velocity as the groundwater,
> and hence ultimately escape. Strontium is
> also poorly sorbed, moving at 103 percent of
> groundwater velocity; its shorter half-life,
> however, means that it will decay long before
> it has travelled far.

> The actinides, especially americium, are
> moderately or strongly sorbed, and will move
> much more slowly than the groundwater. Nep-
> tunium is less uniformly sorbed than the
> other actinides.

Clearly, then, a series of barriers exist between
the wastes and the surface ecosystems. Their probable
overall effectiveness has been calculated by parameters
that govern the transfer processes (e.g. ERDA, 1976;
KBS, 1978). These models predict the length of time
required for the main radionuclides to reach the
earth's surface and to have impacts on critical population groups. They may also be used to compute the
radiation doses received by these groups as functions
of time. The use of models is essential, since no
empirical test can be devised on so long a time-scale.
The models are simplified and crude, but one can have
some confidence in their results, since fail-safe
values of the parameters are used.

KBS in Sweden has applied such a model for an
installed capacity of 10,000 MWe (KBS, 1978) to a
critical group consisting of a population drawing its
main water supply from a well near a repository that
undergoes leaching of its wastes after encapsulation is
breached (1,000 years from emplacement). Figure 13.5
shows the calculated impact in radiation doses to
individuals within the critical group, in rems per
thirty-years (i.e. per generation). The diagram com-
pares the doses likely to be experienced with other
levels of exposure typical of a Swedish population -
natural background, and the dose due to radium-226 in
drinking water. The maximum impact will be deferred
until between 100,000 and 1,000,000 years hence. The
peak dose is likely to be of the order of 0.1 rem (100
millirem) in 30 years, compared with typical background
doses of 100 mrem per annum, or 3 rem in 30 years.

The same model was used to compute the collective
dose to the world's population due to the operation of
the proposed Swedish repository. In the most unfavour-

FIGURE 13.5
Calculated upper limiting and expected radiation doses
to persons living near a well close to a deep
geological repository in Sweden, from KBS, 1978. Dose
loads from natural sources, and established Swedish
dose limits, are also plotted. (After KBS, 1978).

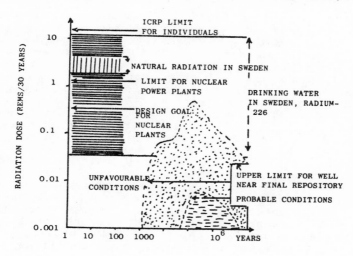

able 400 years, the dose is expected to average .007
manrem[6] per MWe per annum, which is well below the
recommended limit for the entire nuclear fuel cycle
(one manrem per MWe y[-1]). The assumptions and numerical values used in these calculations were highly
conservative, and the actual impact is expected to be
substantially less than those given above. It, therefore, appears that the effect of the multiple barriers
introduced into the disposal system should (1) defer
the impact of the leaching of the wastes until a very
remote period of the future, when much of the radioactivity will have decayed; (2) reduce the doses to
individuals in critical groups to extremely low values;
and (3) similarly reduce lifetime impacts to the entire
world population to very low values. It is conceivable
that catastrophic processes of the sort described
before could accelerate and magnify these impacts, but
the probability of such catastrophes is exceedingly
low.
     All the calculations described above were carried
out as part of exercises to demonstrate the feasibility, in principle, of the concept of disposal in deep
rock repositories, in salt or igneous rock. In the

case of an actual repository, it will be necessary to combine field measurement with a far more exhaustive modeling exercise.

Thermal Effects

Rocks vary significantly in their thermal characteristics, both as regards conductivity and physical properties. The waste heat from the repository will have to be dispersed in such a way that undue rises of temperature occur at no level, from the waste itself to the ground surface. Rising temperatures will affect corrosion rate, glass or ceramic leaching rate, gas formation and water migration. A rock with high conductivity will, if possible, be chosen, after thermal testing *in situ*. Tests are already in progress in mines in several countries, and the methodology is being well developed. The rate of heat release from irradiated fuel or vitrified high-level wastes can be accurately calculated (Figs. 13.3 and 13.4). In the case of vitrified wastes, the rate of release per unit volume of waste can be controlled by varying the fission product loading of the glass. French practice has favored 10 percent fission product content, which gives substantial heating in the first few centuries. Some Swedish opinion favors a 9 percent load, which will enable them to keep core temperatures in the glass below 194°F (90°C) and surface temperatures of each canister below about 140°F (60°C). These figures depend critically on the duration of pre-disposal storage, on the reprocessing date in relation to withdrawal from the reactor, and on the geometry of the actual emplacement of waste in the repository; one can space out the waste to optimize heat disposal, subject to considerations of cost and physical integrity. The same applies to fuel. The heated area around the repository will gradually enlarge. Maximum temperatures in the waste or fuel will be reached within a century, and the repository itself should then cool. The heated area will continue to expand for many centuries and will ultimately affect surface temperature slightly - possibly by as much as +1.8°F (+1°C). These effects will need to be accurately calculated for the specific site chosen.

Future Security

We shall not treat the problem of present-day security in the nuclear cycle, since this does not seem to us to be an environmental issue. We must, however, at least raise the problem of future security. It is highly unlikely that future generations will deliberately seek to re-enter a sealed-off repository. But it is possible that they may do so inadvertently. The possibility exists of a future Dark Ages, in which there is a loss of technological skills, and perhaps a rupture of information memory.

The only precautions that we can advocate to guard against such an event are (1) to leave as many and as varied records in other forms as we know how; and (2) to avoid using sites which may in future be of value in mineral or geophysical exploration. It is the second of these measures that seems to us to argue against the use of salt formations. Salt has a high economic value, and is often associated with other valuable materials, such as oil and gas.

Environmental Survey, Monitoring and Modelling

Following the choice of an actual repository, it will be necessary to conduct exhaustive field surveys and testing, combined with hydrological, geochemical, ecological and ecotoxicological surveys, and modelling exercises. The environmental impact analysis will have to predict, as far as is humanly possible, the consequences of potential failure of containment by the repository at various future dates. A first requirement will be the establishment, through the entire drainage basin in which the repository is to be located, of background levels of the chief radionuclides and of radiation dose. This includes waterbodies, soils, selected species of the biota, rocks and atmosphere. The background will include contributions from nuclear weapons testing fallout (which is still in progress), as well as from naturally occurring radionuclides. The establishment of these baseline data is necessary if future increases due to migration of radionuclides from the repository are to be detected. Such migration is not likely within the first millenium, so that a major purpose of the exercise is to bequeath to posterity an adequate record of preexisting levels of radioactivity. Another purpose, however, is to provide for the unlikely contingencies of failure in handling at the repository itself, of sabotage, of catastrophic disturbance -- and of the calculations made by the designers.

A second, and equally vital, requirement must be exhaustive hydrogical, geochemical and ecological surveys of the drainage basins. The most accurate possible picture of water movement through the whole hydrologic cycle should be sought, especially as regards groundwater movement, soil moisture relationships, and the role of the chief reservoirs (for example, the great northern bogs). This should be

346

coupled with an adequate review of fluid geochemistry, and of fluviatile, lake, and bog ecosystem function. The latter should include, if possible, the identification of potential bio-accumulators or food-web multipliers within the biota. There is already much experience of heavy metal behaviour in aquatic ecosystems, including the vital questions of metal speciation, complexing and accumulation within species.

The design of adequate experiments of this kind will be a vital exercise for the various national licensing agencies. It is a new kind of environmental impact analysis in three main senses. First, it aims at the identification of potential impacts that are very improbable, and may be literally inconceivable with present experience. Second, it aims at very distant epochs. And thirdly, it aims at the design of monitoring systems not so far visualized. Ecotoxicological methods of today, as exemplified in the brilliant modelling exercises of UNSCEAR, aim much more specifically at impacts on contemporary man.

Finally, one must stress that all such efforts should aim at satisfying widely-felt public anxieties that have been aroused by alarming statements about the impacts of nuclear wastes on future generations. Whether these statements are justified or not, it is indeed a heavy responsibility to pass on to posterity large amounts of very toxic material. This alone seems to us to justify the large effort and considerable expenditures represented by the above suggestions, at least to the extent needed to allay these concerns -- and to protect future generations.

CONCLUSIONS

We have arrived finally at what appears to be a comforting conclusion--that the ultimate disposal of nuclear wastes should be technically feasible and very safe. We find that the environment and health impacts will be negligible in the short-term, being due to the steps that precede the emplacement of the wastes in the repository. Disposal itself, once achieved, offers no short-term threat--unless an unforeseen catastrophe of very low probability occurs. The risks appear negligible by comparison with those associated with earlier stages of the fuel cycle.

Ultimately -- millinnia hence -- a slow leaching of radionuclides to the surface might begin. But it would be so slow that great dilution of each nuclide will occur. This phase is likely to be reached somewhere in the period 100,000 to 1,000,000 years hence. We cannot visualize what the world will be like at that time. Past experience suggests that it will pass through a sequence of ice ages and warmer interglacials. We do not believe that these will of themselves disrupt a repository. But they will certainly disrupt human society. Man will by that time probably have evolved dramatically. We cannot guess what he will like, or what his capacities and needs will be. In the end--millions of years hence, when most of the radionuclides will have decayed into harmless daughters--the remains of the nuclear age will accumulate in two main reservoirs. These will be the ocean and the continental rock bodies.

We are dissatisfied with the situation at the front end of the nuclear cycle. No satisfactory technique has yet been reached for the disposal of the heavy radionuclides that accumulate in the tailings from the milling stage. Isotopes of radium and thorium are present in these tailings and are leached outwards by percolating rainwater or snowmelt. These heavy nuclides pose the same sort of threat to mankind as the actinides contained in the high-level wastes from reactors. It does not make sense to contain the one without the other.

We are convinced that the environmental and health effects of waste disposal are small beside those posed by the accumulation of fissile materials in storage (mostly plutonium-239 for military purposes at present, but soon to include Pu-239 and other fissile nuclides from fuel reprocessing). They are also small by comparison with the maintenance of adequate security and safeguards for such materials.

NOTES

1. This chapter is a slightly revised version of a paper published originally in <u>Nuclear Energy and the Environment</u>, Essam, E. El-Hinnawi, ed. Oxford: Pergamon Press, pp. 168-199. Reprinted with permission.
2. All types of atomic nuclei are collectively called <u>nuclides</u>. A radionuclide is any species of nuclide that emits α, β or γ radiations, or neutrons. All the different nuclides of a single element, which differ only in the number of neutrons present, are called isotopes of the element.
3. Radionuclides decay into other nuclides, called daughter products. Four decay chains are know, only three of which occur in nature.
4. Erosion is normally slow and hence not catastrophic. But it is important only if accelerated by some drastic change in environmental conditions.
5. Tectonic refers to the tendency for the rocks of the crust to be folded or faulted due to internal compressive stresses, or to tension.

348

6. i.e., one rem impacting one man.

REFERENCES

Aikin, A.M., J.M. Harrison, and F.K. Hare. 1977. The Management of Canada's Nuclear Wastes, Ottawa: Energy, Mines and Resources Canada, Report EP77-6.

American Physical Society. 1977. "Nuclear Fuel Cycles and Waste Management," Reviews of Modern Physics.

Energy Research and Development Agency (ERDA). 1977. Alternatives for Managing Wastes from Reactors and Post-Fission Operations in the LWR Fuel Cycle. Springfield, Virginia: National Technical Information Service, 5 Vols.

Flowers, B. 1976. Nuclear Power and the Environment. Sixth Report of the Royal Commission on Environmental Pollution, London. H.M.S.O.

Federal Republic of Germany. 1976. Radioactive abfalle in der Bundesrepublik Deutschland, Bonn: Bundesministeriums fur Forschung und Technologie, 5 Vols.

International Atomic Energy Agency (IAEA). 1976. Management of Radioactive Wastes from the Nuclear Fuel Cycle. Vienna, 1 Vol.

IAEA. 1976. Effects of Ionizing Radiation of Aquatic Organisms and Ecosystems. Vienna, Technical Report 172.

Keeny, S.M. 1977. Nuclear Power, Issues and Choices. Report of the Nuclear Energy Policy Group, Ford Foundation and Mitre Corporation, Cambridge, Mass.: Ballinger.

Karn-Bransle-Sakerhet (KBS). 1978. Handling of Spent Nuclear Fuel and Final Storage of Vitrified High Level Reprocessing Waste. Stockholm, 5 Vols.

Marsily, G. de, E. Ledoux, A.Barbeau and J. Margot. 1977. "Can the Geologist Guarantee Isolation?" Science, 1970. 519-528.

Organization for Economic Cooperation and Development (OECD). 1977. Report on Objectives, Concepts and Strategies for the Management of Radioactive Waste. Paris, 273 pp.

Pigford, T.H. 1974. "Environmental Aspects of Nuclear Energy Production," Annual Review of Nuclear Science, 24, pp. 515-559.

Pochin, E. 1976. Estimated Population Exposures from Nuclear Power Production and Other Radiation Sources. Paris: Nuclear Energy Agency, (OECD).

United Nations Scientific Committee on the Effects of Atomic Radiation (UNSCEAR). 1977. Nuclear Power Production. UN General Assembly Document A/AC.82/R.243.

14

Confronting Equity in Radioactive Waste Management: Modest Proposals for a Socially Just and Acceptable Program[1]

Roger E. Kasperson, Patrick G. Derr,
and Robert W. Kates

The United States has entered the 1980s with one of
its major potential energy sources--nuclear power--
immobilized by a myriad of institutional, economic, and
social obstacles. Preeminent among these obstacles is
the waste management problem, a concern which heads the
list of public worries about the use of nuclear power
to generate electricity.

Recognition that the long-standing lack of an
effective and acceptable waste management program can
be traced to past neglect of difficult social and
institutional--rather than purely technical--problems
has come very late. But it has come, and in its 1979
report to the President, the Interagency Review Group
on Nuclear Waste Management forcefully pointed out that
the resolution of these social and institutional issues
was likely to be much more difficult than overcoming
the remaining technical problems (U.S. Interagency
Review Group on Nuclear Waste Management, 1979.

Unfortunately, despite the passage of the Nuclear
Waste Policy Act of 1982, federal radioactive waste
management continues its preoccupation with technical
problems and technical solutions. The Department of
Energy has yet to initiate a broad-based research
program on social and institutional problems, and gave
scant attention to them in its final environmental
impact statement on commercial wastes (U.S. Department
of Energy, 1981a), its <u>Statement of Position</u> (U.S.
Department of Energy, 1980) on the nuclear waste confidence rulemaking procedure, and the latest draft of
its <u>National Plan</u> (U.S. Department of Energy, 1980b).
While the National Research Council's Committee on
Radioactive Waste Management (1980) and others have
lamented the past neglect, the social and institutional
obstacles continue to make it difficult for DOE even to
gain access to the various states where it must seek
prospective waste-management sites.

This chapter will suggest one way in which these
difficult social and institutional problems might be

fruitfully addressed: namely, by way of an explicit consideration of the equity issues raised by nuclear waste management. In particular, this chapter will show how one might approach the waste-management problem with the goal of defining an equitable and socially acceptable waste-management system. Whether the resulting system will resemble the system dictated by a predominantly technical analysis remains to be seen. Because it gives explicit consideration to equity, the analysis which follows here will consider two different kinds of information. These are:

1. A statement of the distribution of beneficial and harmful impacts over some specified population which would result from a given decision or policy. This requires an empirical analysis which includes

    (a) A specification of those 'things'--whether social goods, opportunities, harms, or experiences--whose distribution is being investigated;

    (b) An explicit delineation of the population and relevant subpopulations--possibly including past or future populations--to be considered in the analysis; and

    (c) A statement of the actual impact distributions--as defined by (a) and (b)--which would result from alternate proposed solutions to the radioactive waste problem.

2. A set of standards or principles by which the equity or "fairness" of particular distributions may be judged and by which the social preferability of one distribution over another may be judged.

The standards or principles included in (2) are, of course, principles of equity. We begin below by surveying several such possible principles. Three are selected for actual application to the locus, legacy, and labor/laity aspects of the radioactive waste management problem, and it is this application--the actual equity analysis--which the rest of the chapter addresses.

It must be noted that, although we find the three selected principles to be plausible (and endorsed by substantial public opinion), the analytic approach which we suggest does not depend upon the inherent justice or social acceptability of precisely these three principles. What is to be demonstrated is how moral analysis using a set of equity principles can be applied to technological choice. The determination of the most proper equity principles to be used in such analysis is itself one of the neglected social questions which so urgently requires more thorough and explicit investigation.

SOME CANDIDATE EQUITY PRINCIPLES

Utilitarianism

Utilitarianism assumes as morally ultimate the obligation to promote human happiness by maximizing aggregrate utility for the relevant community. Although both Bentham and Mill defined utility as nothing other than pleasure together with the absence of pain, Mill explicitly extended his concept of pleasure to include those kinds of aesthetic and intellectual pleasure which (so he argued) distinguished human beings from pigs.

Whether as broad as Mill's or as narrowly hedonistic as Bentham's, the concept of utility has proven versatile and underlies large parts of modern economics. Utilitarianism's general imperative to maximize utility for the relevant community, however, can be understood in two interestingly different ways. In one interpretation, the maximization of utility is taken explicitly to omit any consideration of the distribution of the produced utilities; questions of 'how much happiness to whom' are per se irrelevant. Of course, it is possible that distribution may somehow influence the total aggregated utility (for example, through diminishing returns of pleasure for certain goods); but it is only the aggregated utility which is itself morally relevant. Ronald Green (1977), points out that, in this interpretation, a utilitarian principle of equity would counsel indefinite population increase so long as the aggregate gains in happiness outweighed the corresponding losses in per-capita well-being.

In a different interpretation, Mill's "greatest good for the greatest number" is taken to require the distribution of goods to more, rather than fewer, persons. In addition to counseling the maximization of utility, this variety of utilitarianism would require the distribution of those utilities in as wide a manner as feasible.

Whether these two interpretations of utilitarianism have genuinely different social implications may be an empirical rather than a philosophical question. As Ben-David, Kneese, and Schulze (1979) have shown, the first interpretation becomes identical to the second if one assumes (1) diminishing marginal utility, and (2) the same utility function for all individuals regardless of their actual present wealth.

Because it so emphatically focuses on the consequences of proposed policies or courses of action, the utilitarian approach underscores the crucial role which a definition of the relevant population (1-b above) will play in any equity analysis. What should be the intergenerational scope of the population considered in the analysis? Should it include all future generations which may be harmed or benefitted by radioactive waste-management decisions? If so, it would seem that considerations of aggregated utility over future generations would overwhelm concern for the well-being of the present generation. Should future harms and benefits be discounted, as indeed they typically are in economic calculations of future utility? If so, even a small discount rate will effectively ignore the future in only a few generations: at a five percent discount rate, one death today will count as much as 1,738 deaths in 200 years and 3,000,000,000 deaths in 450 years. But even for the fission product portion of radioactive waste, the period of risk is 700 years.

Pareto Optimality

The concept of Pareto optimalization may be construed as a principle of equity which addresses distributional procedures rather than distributional results. Specifically, it would endorse those distributional changes which increase utility for some or all members of the relevant community without making anyone worse off; if a proposed change makes some members of the community worse off, it would not be counseled.

When it is thus transformed into a systematic principle of equity, however, Pareto Optimality suffers from severe conceptual difficulties and is unlikely to be generally applicable to the radioactive waste management problem. These include the likelihood that there are no alternatives that do not actually (as opposed to theoretically) make someone worse off in noncompensable ways.

The Equality Principle

In its classic formulation by Aristotle, the equality principle--the claim that equals must be treated equally--requires that "the same equality [must] exist between the persons and between the things concerned; for as the latter--the things [distributed] are related, so must be the former; if the persons are not equal, they will not [receive] what is equal"

(McKeon, 1941). Accordingly, cases which are similar in certain specified respects must be treated similarly, and those which differ in the specified respects must be treated differently. In addition, some equality principles include minima below which no person is allowed to fall or maxima above which no person is allowed to rise. Different formulations of the equality principle identify different sorts of equality as morally relevant. We consider three such specifications below, giving to the resulting versions of this principle the labels: (1) burden/ benefit concordance, (2) burden/benefit equalization, and (3) proportionality of benefits to need and burdens to ability.

Burden/benefit concordance. This entails a simple principle: those who reap the benefits of some activity should bear the burdens. Put slightly differently, it is justifiable to impose a harm only if there is some corresponding proportional good. The distribution of burdens and benefits of energy systems, and particularly of radioactive wastes, is particularly troublesome, for what is beneficial for some people often is harmful or burdensome to others. High degrees of uncertainty characterize both benefits and burden, and beneficiaries and those burdened are separated by great distances or long expanses of time.

For the legacy problem of radioactive waste, we will argue that benefits are largely concentrated in this and the next generation. This principle, then, would obligate this generation to internalize harm and avoid exporting it to the future. One issue which must be addressed, however, is whether harms and benefits for the distant future can be calculated with sufficient accuracy and whether uncertainties can be narrowed enough to make meaningful the application of this principle.

Burden (or benefit) equalization. This mandates an even distribution of harms (or benefits) regardless of the corresponding side of the ledger. Conceivable arguments for such a principle are several. Where the risks of an activity are potentially large (and perhaps catastrophic), highly uncertain, and feared by the public, then justice is best had by distributing risks equally among all members of the society. Or, alternatively, if associated benefits cannot be calculated with sufficient exactitude to implement the principle of benefits/burden concordance, then the fairest allocation of harm is one of equalization.

The 1979 closing of low-level commercial waste facilities was provoked, in part, by the fact that the burden was not a shared one. It is difficult (and perhaps inappropriate), however, to separate the

application of this principle to nuclear waste from its
application to the entire nuclear technology. For
example, both the Nevada test grounds and the Hanford
reservation in Washington are currently candidate sites
for a high-level waste repository. It can be argued
that these areas have already borne more than their
fair share of the nuclear burden from weapons testing
and military uses and that other peoples and places
should absorb future additional risk.

   Proportionality of benefits to need and burdens to
ability.   A very different equality principle is
involved here. It is predicated on inequalities in
existing distributions and allocates burdens and benefits to achieve more equal end states. The notion that
burdens should be distributed in proportion to ability
to bear them has been a central tenet in Marxist
thought ("from each according to his abilities, to each
according to his need") and also underlies, for
example, progressive tax structures. Applying this
principle, Hanford might logically become a preferred
site for a high-level waste repository because its
accumulated knowledge and experience would better equip
it to absorb nuclear risks than communities lacking
such experience. Similarly, experienced workers would
more easily absorb risks than publics unfamiliar with
radiation hazards.
   Such a principle, it is apparent, threatens to
concentrate risk, thereby enlarging the discrepancy
between the enjoyment of benefits and the experience of
burden. As with the equalization of burden principle,
it may also lead to enlarging rather than reducing risk
burdens. This concern has arisen in the controversy
over the genetic screening of workers: the screening
out of workers who are particularly vulnerable to
certain hazards, it is argued, will encourage the per-
petuation of higher exposure levels, will deny employment opportunities to some groups, and will encourage
neglect of risk-reduction strategies.
   In regard to the legacy issue, it could be argued
that the future is more likely to be able to bear
burdens than the present. With time, for example, it
seems likely that more developed technology, increased
scientific knowledge, and even greater societal wealth
will be available. There may even be a cure for
cancer, or at least an enhanced medical capability.
Such a linear view of progress, however, particularly
for distant future generations, would appear to be an
imprudent assumption for this generation. It is possible, for example, that continued population growth,
rapid exhaustion of natural resources, or cataclysmic
events could result in a diminished future capacity to
absorb burdens.

The National Council of Churches' _Energy and Ethics_ (1979:17) argues that because energy is as essential for survival as food, housing, or clean air, "the needs of those who are below the minimum standard take preference over the wants of those above the average." (1979). This is clearly an application of the principle of apportioning benefits to needs.

Stewardship

The roots of the stewardship principle are ancient. In both Jewish and Christian thought, the principle derives from the frequently cited passages of Genesis: "God blessed them, saying to them, 'Be fruitful, multiply, fill the earth and conquer it. Be masters of the fish of the sea, the birds of heaven and all living animals on the earth,' "and "Yahweh God took the man and settled him in the garden of Eden to cultivate and take care of it."

The principle has other sources in post-Platonic philosophy (Passmore, 1974) and need not be grounded in any religious belief. The belief that human beings are responsible to those who will come after them might be defended by an appeal to the biological connection among generations or to the human community's attempt to preserve and develop what it loves (Passmore, 1974).

Whatever the source of the principle, balancing the claims of the present generation against the stewardly obligation to future generations is a very difficult ethical problem. Precisely what directives does the principle entail? In one view, it suggests an obligation to provide our descendants with a nature made more fruitful by this generation's efforts. In a different view, it suggests an obligation to refrain from any irreversible destruction of such natural goods as scenic beauty, abundant species, and so forth--this is Green's suggested minimal requirement (Green, 1977). Thus it has been invoked by both pro- and anti-nuclear forces; the former wishing to leave a legacy of improved technology, the latter a legacy of unimpaired nature.

If the principle's implications for nuclear waste management are not clear across generations, they are simply invisible within generations. Accordingly, it is not likely to provide a standard for evaluating the institutional and technical options in waste management.

Merit (Moral Desert)

A wide variety of equity principles mandates the distribution of benefits in proportion to civic merit.

Whether the relevant merit is defined as virtue, race, or perceived market value, the principle provides little guidance in the present context: the merit of future generations cannot be determined in advance. Within the current generation, a tacit acceptance of this principle may underlie the traditional practice of placing noxious facilities in poorer, rural communities; since they have failed to attract more desired enterprises, they have less merit.

Contractual Principles

One accepted and established procedure for allocating burdens and benefits is entrance into a binding contract. The process requires that agreement be voluntary, that relevant information not be withheld, that all parties abide by the agreed terms, and that the contract be enforceable (usually by law).

The contractual process was used to establish the Western New York Nuclear Services Center. Explicit parties to the process included the Atomic Energy Commission, the State of New York, and the W.R. Grace Company. While the host region was not a formal signatory, it was clearly a willing participant in the process. To defend the agreement on the basis of the contractual principle of equity, one need only point out that the various parties entered into the contract under conditions which fulfill the requirements for valid contractual agreements. That contracts should be honored is taken for granted in this approach; it is difficult to imagine any society which could survive if members assumed no duty to honor agreements and commitments.

The contractual principle of equity, however, has obvious limitations. Contracts between DOE and several states with respect to the search for high-level waste repository sites have not deterred subsequent charges of unfairness. It is inherently difficult to meet the requirement for adequate information when dealing with first-of-a-kind facilities. And, of course, it is simply not possible to contract with future generations.

Freedom of Choice

The principle of free choice is closely related to the contractual principle just discussed, but it is more general in application and less often explicitly formulated. The principle underlies the institution and processes of Western democracies, and assumes some sorts of equality--particularly equality of opportunity and of basic legal and political rights.

In contexts involving the imposition of risks or other burdens, the principle has been codified in practices of informed consent. Generally, the requirement of informed consent involves two distinct considerations: (1) the provision of sufficient information and understanding to enable the prospective consentee to comprehend the alternative choices and the consequences likely to be associated with each, and (2) the voluntary assumption of the specified risks and burdens without the influence of coercion or duress.

In practice, these conditions are difficult to fulfill. Recent research on medical consent forms, for example, indicates that relatively few individuals who provided their written consent actually had adequate understanding at the time of choice (Brody, 1980). Baruch Fischhoff (1983) indicates a variety of reasons why individuals have difficulty judging probabilities and ranges of consequences, especially low probability/high consequence events, even for hazards much simpler than nuclear power or radioactive wastes. Seley and Wolpert (1983) note that individuals in communities hosting large industrial facilities tend to exaggerate benefits while underestimating the adverse socio-economic impacts. Given the long time spans involved, the intensity of the nuclear controversy, and the residual uncertainties in isolating radioactive wastes from the biosphere, achieving adequate information for meaningful consent by those who will experience the risks is a challenging task at best.

But voluntarism in choice, with the absence of coercion or duress, is also difficult to achieve. Individuals desperate for work are not in a position to refuse jobs, and communities with inadequate or declining tax bases and high unemployment rates may find it difficult to refuse a waste repository, whatever the present or future uncertainties as to public health risks. Fischhoff (1983), shows that institutional factors can also erode genuine consent. Compensation without equality may constitute bribery, and the particular environmental conditions which prevail in the consent situation may produce subtle forms of coercion.

Rawlsian Procedures

In his influential **A Theory of Justice**, Rawls argues that procedures for developing equitable allocations stem from the first of his two principles of justice, "each person is to have an equal right to the most extensive basic liberty compatible with similar liberties for others" (Rawls, 1971).

To assure just allocational procedures, Rawls proposes a veil of ignorance. Under this veil, the

deliberating parties are assumed to be ignorant of a
variety of facts concerning their particular status.
No one knows, for example, his place in society, class
position, social status, or natural abilities.
Further, the deliberating individual does not even know
the particular circumstances of his own society or even
to which generation he belongs. The veil of ignorance,
in short, obliges an evaluation based solely on general
considerations.

As an analytic procedure, Rawlsian principles would
appear to have potential application to radioactive
waste management issues. A thoughtful application of
Rawls theory to the future generations problem has
recently been undertaken by Barbour (forthcoming), in
which he argues for the sustainability of future
resources and burden/benefit concordance as ethical
principles.

THE SELECTED PRINCIPLES

Our purpose in this chapter is to demonstrate how
an explicit consideration of equity changes the manner
in which alternative radioactive waste managenent proposals are analyzed and evaluated. We do not propose
to identify here an absolute set of ethical imperatives
but to select some plausible candidate principles for
use in the analysis which follows. Accordingly, our
selection of appropriate principles is partly pragmatic. We note, first, that both distributional (or
'end-state') and procedural principles of equity are
widely perceived as relevant to 'fairness' or equity.
Second, we note that there is substantial consensus for
certain principles in a variety of policy statements by
a range of groups interested in the radioactive waste
management problem. We suggest the following three
principles as plausible and as likely to find support
across diverse interests and positions.

1. The beneficiaries of an activity should bear
    associated burdens proportional to the benefits
    enjoyed, and, conversely, the imposition of a
    harm or burden should be accompanied by a
    proportional benefit.

2. The experience of risk should be shared rather
    than concentrated within a sub-population of
    beneficiaries.

3. The imposition of a harm or burden should be
    made as voluntary as reasonably achievable
    through observation of practices of informed
    consent.

The first principle, benefit/burden concordance, is frequently recognized in discussions of equity. In fact, the first stipulation--that beneficiaries should bear the burden--is sometimes assumed to be <u>the</u> distributional test of equity. It is widely recognized in American legal precedents, as Harold Green (1983) demonstrates. The principle is recommended in the discussions of such diverse parties as the DOE, ex-President Carter, the National Council of Churches, and the Sierra Club. The second stipulation of the principle--that benefits should accompany the imposition of harm--mandates that harm may not justly be imposed without a concomitant effort to restore the previous (equitable) balance.

The second principle--that actual risks should be shared broadly among the beneficiaries--imposes a particular constraint upon Principle 1. It, too, is a distributional ethic. Bearing the burden of waste isolation from the biosphere and even compensation of risk-bearers may not necessarily involve the individual assumption of bodily or social risk. Risks will of necessity be imposed upon certain populations for the larger good (it is impossible to make risks and benefits wholly congruent). Principle 1 requires only that the burden for harms created, not the harm itself, be borne by the beneficiary. Principle 2 takes account of the facts that (1) future harms can be anticipated with only considerable uncertainty, particularly in accidents or for the long time periods involved; (2) there is substantial and possibly noncompensable public fear over the risks involved; and (3) institutions may not suffice, especially for distant generations, places, and workers, to ensure full compensation for harm even if it it were provided by those that benefit.

Principle 3 recognizes procedural requirements for the permissible means of imposing harms. The wording is "as voluntary as reasonably achievable" because purely voluntary means are not possible in locating noxious facilities or in allocating feared risks if larger social goods are to be obtained. Thus, full consent will not be realizable. But if consent is to be overridden for a larger societal good, reasonable means should have been exhausted for informing the risk-bearers, for their full participation in public proceedings, and for obtaining the maximum degree of achievable consent. Such a social imperative will certainly exceed current program practices and existing institutional mechanisms but does not, in our view, present insuperable difficulties.

We now turn to the application of these principles to waste management, considering in turn the locus, legacy, and labor/laity problems. For each, we describe the issues of burden and benefit distribution and then suggest, management options for addressing each of these principles.

THE LOCUS PROBLEM

Undoubtedly the most visible and volatile equity problem currently is the geographical separation of beneficiaries and those who will bear burdens within this generation. This we have termed the "locus" or "backyards" problem. The sources of inequity are several, but the effects are unmistakable. Nearly every state in the United States has adopted or considered legislation restricting the transport of radioactive materials and/or the search for a high-level waste repository site. By 1981, for example, over one-half of the states had adopted bans or moratoria on the siting of such a repository. An equal number of states had restrictions on radioactive waste transport through their territories. In 1979, for both technical and political reasons, the three remaining commercial low-level waste sites threatened to suspend receiving wastes, which prompted Congress to enact the Low Level Radioactive Waste Policy Act of 1980 assigning responsibility to the states for the disposal of such wastes. Federal-state conflicts have spilled over into Congress and constitute a prominent focus in the Nuclear Waste Policy Act of 1982.

Distributional patterns differ with type of waste. Current high-level waste inventories, for example, are dominated by the defense wastes generated over the past 35 years as by-products of the production of plutonium for atomic weapons. Much of this accumulation of waste is now two or three decades old, with reduced levels of radioactivity, and new defense wastes are being generated at much lower rates (although the Reagan administration plans a major increase in the production of nuclear weapons). The arguable benefit of this 350-year history is the national security provided to the populace as a whole, although some minor site-specific benefits of employment and local business at government installations are also entailed. But defense waste results from national decisions taken in the national interest; the nation as a whole, in short, must be considered the beneficiary. Current policy recognizes this in designating public funds as the appropriate means for dealing with defense wastes.

Much the same can be said for the radioactive wastes from medical and research facilities. This waste, nearly all of which is low-level, comprises an insignificant amount of the total waste burden. Nevertheless, the lack of low-level waste storage capacity is a pressing problem because the radio-pharmaceutical firms which supply these nuclear materials and the hospitals and research laboratories which generate the waste lack storage space. Current plans to deregulate biomedical wastes will ameliorate this problem somewhat (Marshall, 1980). The importance of the benefits was highlighted by the 1979 suspension of waste acceptance at low-level waste burial sites, an act which nearly forced medical treatment and research to be discontinued at widely separated facilities. That this waste is viewed differently from waste from electricity generation is suggested by the exclusion of medical waste from the restrictions (subsequently preempted) enacted on out-of-state waste at Hanford by the State of Washington. As with defense waste, the beneficiary pattern is very widespread but concentrated in major urban centers, a fact which creates some disparity of interest on a local, state, or regional scale.

In comparison with the benefits of defense and medical/research activities, the benefits of commercial power reactors are more regionally concentrated. The 78 reactors currently licensed and the 50 or so others likely to be completed are strongly concentrated in the Northeast, Midwest, and Far West. Chicago, for example, currently draws 50 percent of the electricity from nuclear facilities. Comparable figures are: New England--30 percent; New York and New Jersey--20 percent; Michigan--20 percent (Gilinsky, 1981). Whatever the concentration, an initial question must be whether the benefits are real or only imagined. Benefits exist only if nuclear power is actually a more efficient and reliable source of electricity than other production alternatives, an issue which is hotly debated. If the comparative advantage of nuclear power depends upon national subsidies (such as the Price-Anderson Act), then regional benefits actually reflect a form of transfer payments from the nation as a whole.

Considered opinion suggests that the coal/nuclear cost comparison is too uncertain to compute (Stobaugh and Yergin, 1979). Let us assume for this analysis the accuracy of utility calculations of nuclear power's relative regional advantage. The direct beneficiaries of nuclear plants are the users of electricity, including industry, commercial users, and residential users; the benefits are the savings in electric cost as opposed to fossil fuel generation or doing without electricity. These benefits are localized in utility districts with heavy nuclear development for, although there is an integrated grid system with some long distance electricity transport, over 90 percent of electricity is consumed within its utility of origin. An estimate of the magnitude of such benefits in a high fossil-fuel-cost state, although complicated by various hidden subsidies, may be as high as the $140 million which Governor Joseph Brennan of Maine argued that the closing of the Wiscasset nuclear plant would cost Maine consumers, or as low as the $43 million conceded by its opponents.

The degree of regionalization of benefits, however, is likely overstated by designation of nuclear electri-

city consumers, for it ignores the distribution of
secondary benefits. Presumably, the manufacture of
products at lower prices provides net savings to consumers elsewhere in the nation, to an enlarged gross
national product, and to a more favorable trade
balance. Nonetheless, if the Maine voters in the 1980
referendum had decided to close the Wiscasset plant
(and had the plant actually been closed), then surely
Maine residents should have been absolved of that
portion of the waste burden.

The mixture of national and regional benefits from
nuclear power contrasts with likely geographical
patterns of costs. Current nuclear waste storage and
disposal sites are relatively few--comprising some 12
commercial and defense low-level waste sites, three
high-level defense waste sites, and an additional two
sites (Morris, Illinois and West Valley, New York) and
some 78 reactors where spent fuel is currently stored.
Because of its relatively small volume, all of the
nation's high-level waste, defense and commercial, will
probably end up at somewhere between two and six repositories. Similarly, the mining and milling of uranium
have produced geographically concentrated tailings
which are too voluminous to be moved long distances.
These are located primarily in the West, remote from
most commercial nuclear power reactors. This pattern
of inequity is quite apparent for low-level wastes
where the major waste generating states are quite
remote, particularly from the two western disposal
sites.

An inventory of costs has yet to be calculated for
a fully operational radioactive waste system serving
some 150-300 GWE of nuclear electricity at a mature
state of development (say in year 2020). But the
system would include the full network of waste-
producing, interim storage, reprocessing (if it
happens), and long-term storage facilities, as well as
a transport system for moving the wastes. A full cost
accounting would include not only the obvious logistical requirements, but also the risks to public health
and safety, the long-term management costs, the burdens
of regulatory infrastructure, the threat of nuclear
proliferation, and such less apparent social costs as
public fear at sites and along routes, community dis-
ruption, and long-term institutional demands.

Locus Management Options

In assessing management options, each of our three
equity principles will be covered sequentially.

Benefit/burden concordance. National policy on the
locus problem has been clear since the Carter adminis-

tration: "all costs of storage...will be recovered through fees paid by utilities and other users of the services and will ultimately be borne by those who benefit from the activities generating the wastes" (U.S. President, 1980).

With the caveats noted above concerning the complexities in impacts distributions, this provision, as recognized in the Nuclear Waste Policy Act of 1982, would achieve the equity principle adequately were all relevant costs internalized. Unfortunately, this may not be the case. It is unlikely, despite the provisions for liability and social impact mitigation in the Nuclear Waste Policy Act of 1982, that beneficiaries will in fact bear the full associated costs. Missing will be the public fear of exposure and its effects and a variety of ill-understood social costs. Clearly, current policy needs to be broadened to embrace the full range of adverse impacts at facility sites and at nodes along corridors in the waste system.

Turning to our requirement that benefits accompany the imposition of harm, we note that the provision of benefits raises a number of vexing questions. Benefits, to be sure, in the form of employment, increase business, and increased tax revenues accompany the development of waste storage site facilities. There are also various government programs designed to mitigate the impacts of large-scale facility developments in small, rural communities. Suffice it to note here that the benefits accruing at the site will be distributed highly unevenly over the host-region population, will not occur in timely fashion, and will tend to be overestimated by experts.

By comparison, social costs are poorly understood, both because of the underdeveloped theoretical state of social impact assessment and because these are first-of-a-kind facilities. Meanwhile, government programs designed to redress adverse impacts are beset by fragmentation, restrictions, and lack of timeliness. The Nuclear Waste Policy Act of 1982 addresses this problem through its provision for a contract between the federal government and host state for a program of impact mitigation. There is no assurance, however, that these funds will be dispersed of in equitable fashion to the host locality.

How should the size of such benefits be calculated, how should they be distributed, and what mechanisms would be most effective? First, there should be adequate liability and insurance provision to compensate for public health impacts occurring at waste facilities and along routes. If the Price-Anderson Act may be construed as applying to waste transport and storage, then damage from minor accidents should be recoverable. But since it is catastrophic releases that underlie much of public concern, it is especially important that liability provision be extended to cover the full range of accidents. The extent to which accident costs can be displaced to local victims is suggested by a recent Federal Insurance Administration sensitivity analysis of the Three Mile Island accident which found that, even with no medical or personal injury expenses included, a more severe accident involving evacuation would have resulted in an average Harrisburg family loss of $67,500 with only $2,247 recoverable from the $560 million Price-Anderson pool (Federal Insurance Administration, 1979; Kehoe, 1980). Means should be found so that the burden of liability is not placed upon the victims and that government does not end up in an adversarial position of resisting liability payments to the local residents.

Second, a variety of means is available for providing benefits to risk-takers. Stipulations may be placed on employment pools which both minimize social costs (by reducing in-migrants) and maximize benefits (by enlarging host-region employment). Compensation for adverse impacts may be, as in the Nuclear Waste Policy Act of 1982, provided through impact aid or in lieu of taxes. There is precedent for a system of incentives--several countries (including France) have instituted programs to reduce electricity rates or local taxes for residents around nuclear power plants. A similar system of incentives has been suggested recently for U.S. nuclear power plants (Starr, 1983) and such plants entail more benefits than waste storage facilities. An effective compensation program would need to address such issues as eligibility criteria, valuation criteria, mode of compensation, financing, and distributional mechanisms (Cole and Smith, 1979; O'Hare and Sanderson, 1977).

Finally, we propose consideration of an alternative means of site compensation, one that may by-pass some of the difficulties in defining the pool of compensation according to the burden of uncertain risks or the good will of the state in dispersing funds to host localities. Incentives may be offered beyond compensation for anticipated damages, recognizing that projected damages are essentially incalculable and may easily be underestimated and that the service provided (furnishing a waste storage or disposal site) is itself extremely valuable for the beneficiaries. Appropriate sites for storing toxic and radioactive wastes are rapidly becoming among the more valuable pieces of real estate in the United States. To that end, we propose that the waste management program consider a geology rental fee to be paid to the host region. Such a fee could be calculated by reference to the comparative cost of engineered safeguards required to provide protection equivalent to the afforded by the geology. Sufficient information is becoming available to make such a comparison, at least crudely, and the pool of funds is likely to be substantial enough to serve as a means of compensation and to provide an escrow fund to ameliorate the consequences of future accidents.

Sharing the risks. There is an understandable reluctance to become the waste storage facility or the waste repository. The 1979 closing of low-level waste repositories was intended, in part, to force a broader sharing of the waste burden. Beneficiaries, it may be argued, should not be allowed easily to buy out of the burden of such uncertain risks (as one, for example, might hire a tree surgeon or a roofer to climb where one fears to climb). In the face of risks never experienced before with both catastrophic potential and particularly dreaded consequences (cancer), beneficiaries should ideally share in the experience of hazard.

Increasing the multiplicity of sites is a major means of achieving risk sharing as well as simplifying the transport system. Other considerations, however, might conflict with this equity objective. For example, a highly decentralized system for high-level waste disposal could possibly enlarge the aggregate risks of waste storage or increase social conflict and expand costs. Relocating uranium mill tailings to beneficiary areas makes no public health or economic sense.

Within appropriate technical and regulatory constraints, however, there are opportunities for enlarging the degree of risk sharing associated with both low-level and high-level wastes. Mill tailings, by contrast, present a more formidable problem because of their volume.

It is apparent that the national capacity for low-level waste storage requires major expansion, and current government policy calls for state responsibility in adding more disposal facilities. Given the relative ubiquitous generation of medical and research and waste from geographically concentrated electricity-producing facilities, the intent is for regional siting. Under the Low Level Radioactive Waste Policy Act, individual states may locate disposal facilities within their borders or contract with neighboring states which might well host such a repository. A consortium of regional groups had been formed by June, 1981 in six regional areas and the State Planning Council has created a model interstate compact to facilitate such arrangements. While this approach moves in the direction of a risksharing objective, its ultimate success is unclear for most early progress has been in the regions already possessing facilities.

For high-level waste, the picture is quite different. Here, current governmental efforts appear to be headed away from this goal. The network design of interim storage of spent fuel in a mature waste system provides an opportunity for reducing considerably the overall burden of radioactivity to be handled and transported and at the same time a wider sharing of the responsibility. But, as in other aspects of the waste disposal system, little has actually been done as debate continues as to whether at-reactor or away-from-reactor facilities are the best interim storage solution, whether reprocessing will become a reality, and whether the West Valley, New York, Barnwell, South Carolina, and Morris, Illinois facilities could be converted into the limited governmentally-sponsored interim storage provided in the Nuclear Waste Policy Act of 1982.

As for the repositories, current plans have the danger of a relatively centralized system (i.e., few sites), with the leading candidates (Hanford, the Nevada Test Site, and the Waste Isolation Pilot Plant [WIPP]) in places which are remote from nuclear power beneficiary areas but which have already assumed a significant portion of the societal risk from nuclear technology. The Nuclear Waste Policy Act of 1982 calls for site selection of the first repository from three (after five) candidate sites. The second site could well be in granite, an option that would enhance the prospects for regional siting. Meanwhile, the Act also calls for the development of long-term monitoring at-ground storage.

We propose three alternative innovations to help ensure a greater degree of risk sharing in the high level waste disposal system. First, a visible institutional means is needed to ensure "fairness" in actual site selection. Without such a means, concern will remain that the selection process is stacked against the politically weak. The need for this mechanism has been enhanced by public relations efforts designed to show that suitable geologic depositories are very widespread. Thus a site, even if chosen only for initial exploration has, and will surely evoke a cry, "Why here?" from residents of the surrounding area. It is clear that no state with candidate geological formations should be able to exclude itself from consideration, as Louisiana, for example, appears to have done. Second, we would suggest at an intermediate stage in the site identification process, a lottery among candidate sites designating those which will be exempted from further consideration. A requisite number of technically qualified candidates would be needed to ensure success, of course, but the exemption of many areas by random process would reduce unnecessary tensions over siting. Lottery systems are characteristically employed when fairness is desired in allocating unwanted, often dangerous assignments. The drawing of straws for dangerous missions and the military draft are examples.

Third, the host site will also need assurance that risk sharing will actually occur, for policy reversals have frequently characterized the waste program to date. There is considerable concern in New Mexico, for example, that if the WIPP demonstration project is developed expeditiously but other repositories are not, it may well at some future time be converted to a permanent repository for domestic and/or foreign wastes. One simple means of providing such insurance is a legal storage limit to the amount of waste to be accepted at any site.

Finally, the mill tailings problem appears to be quite resistant to risk sharing procedures. Most of the tailings are in low population density areas, and the volumes are very large. Stabilization in situ appears to be the only viable course.

To summarize, we see a strategic option for waste management that is technically sound and which realizes a high degree of risk sharing. The key elements in the strategy are:

- State responsibility for low-level wastes as currently envisioned in multiple sites.

- Continued reliance upon existing facilities for defense waste storage for the near term but with prompt immobilization of the high level waste.

- Expanded reliance upon at-reactor storage for commercial waste storage, possibly augmented by one or several regional AFRs, for 40-50 years of interim storage (as in the Swedish solution) of commercial high-level waste.

- Delay of current timetables for one or two high-level waste repositories in order to create a regional system of repositories (with co-location with AFR's a possibility) in which multiple sites would begin accepting waste simultaneously.

- Two institutional mechanisms to ensure fairness in risk sharing--a lottery of site exemption among qualified candidate sites and a legal storage limited to the amount of waste to be accepted at any given site. Such a strategy recognizes that there is, in fact, no pressing need to put high-level radioactive waste in the ground and that there is substantial advantage in regionalization of the waste handling and disposal network.

More voluntary risk assumption. Our third equity principle calls for the assumption of risk to be as voluntary as reasonably achievable. We use the term "as voluntary as reasonably achievable" in much the same way that it is used in regulatory language for risk reduction. There is, in current parlance, no such thing as zero risk. Similarly, there is no such thing as completely informed consent for risk assumption. But, beyond this, a purely voluntary system will likely deliver no sites for radioactive wastes, hazardous chemical wastes, or other noxious facilities. This is a case where sacrifices must be made for the common good. The argument here is that the burden is on the developer to inform and achieve consent by meeting the objections and concerns of those who will bear the risks.

The current waste management program has performed poorly on this criterion, and it begins with a major, perhaps insuperable, debit. The Department of Energy, as an institution, bears the legacy for the past failures and inadequacies in the waste-management program. At a time of general institutional distrust, particularly in regard to the management of nuclear energy, the Department of Energy suffers from a serious lack of credibility. The lead agency undoubtedly faces difficulties in recovering the public trust, as does the Nuclear Regulatory Commission for its regulatory responsibilities.

Many mechanisms are available for achieving more voluntary assumption of risk in radioactive waste locational decisions, but their suitability depends on one's view of the extent to which the public is rational or irrational about risk issues generally, and nuclear risk issues in particular. There is a deep-seated belief among the community of technical experts that opposition to nuclear power is anchored in the public's ignorance of nuclear technology, its inability to make risk comparisons, and its irrational response to radiation risks. The attitude is important, because one does not design processes for achieving more voluntary risk assumption if one is convinced that enlightened and rational choice is unachievable. Despite some ingrained folklore, the relevant evidence to date does not suggest that the public is either more poorly informed on nuclear questions than other difficult technological or social policy issues or that it is irrational in its perception of technological risks. For example, research conducted by Paul Slovic and his colleagues (1979) at Decision Research indicates that laypersons generally-rank order risks quite well, although they tend to overestimate some rare but well publicized risks (e.g., botulism) and underestimate more chronic common killers (e.g., alcoholism, smoking). It is also apparent that the public assigns more weight to consequences, particularly those potentially catastrophic, than to the probabilities of accidents

(Slovic, et al., 1979). However, the departure between expert and lay assessment on nuclear power is unusually large, a fact which should underscore rather than deny the need for extensive public participation in decision processes. Specifically, we recommend a number of specific steps to improve the provision of information to inhabitants of prospective host regions:

- The development of objective information and improved means for public participation should be placed under the auspices of a competent group outside the DOE and be amply funded.

- Such a group would develop a public educational research and demonstration program, adequately funded and subjected to careful professional and lay person review.

- Technical and financial resources should be committed to prospective host regions and states in order that local capabilities for technical and management review be created (a useful precedent exists in the WIPP context).

- Aid to groups that would take part in the process (interviewers).

Full voluntary consent to risk bearing is not possible, in our view, if a solution to the radioactive waste problem is to be achieved. While we are hopeful that the more equitable management program we suggest would alleviate some of the conflict which has characterized efforts to date, we are not so naive as to assume that full voluntarism can be achieved, any more than it is possible in locating prisons, town dumps, or methadone centers. Nor would voluntary consent prove a viable means for establishing the hundreds of toxic waste-disposal sites that EPA says are required over the next decade. Similarly, states and localities should not have veto power over the transport of wastes through their territories.

Having said that, we do wish to make it difficult to override the concerns of those who will bear risks that are presently, for all the technological optimism, still quite uncertain and that, we know, provide deep public anxieties. So while veto is not feasible nor mandatory for equity, the host regions of AFRs (if instituted) and waste repositories should certainly have the right and the capability for independent competent review, for negotiating improved safety assurance and social impact reduction, and for effective appeal. Such a right to negotiation was incorporated in the DOT's final rule on route selection for transporting radioactive wastes. It is not, however, a feature of the Nuclear Waste Policy Act of 1982. Since the capability of independent host region review is so essential in an atmosphere of institutional distrust, we do not find the notion of a public defender for the site desirable. In regard to appeal, the host region, as well as the host state, should have the right of appeal (or disapproval), for review by the state legislature in the former and for appeal to Congress in the case of the latter. Inadequately assessed or adverse social impacts should be specifically recognized as appropriate grounds for appeal. In such an appeal proceeding, the burden of proof should be on the developer for demonstrating procedural and substantive safety adequacy (a balance which is unclear in the Nuclear Waste Policy Act of 1982).

Finally, intervenors may be expected to provide a key role in testing the validity of developer assumptions and analysis. They provide systematic means for doubt identification that the developer should have to overcome. The pluralism that they provide may also lend greater credibility for the decision process as a whole. In fact, the site developer may wish to consider funding parallel studies by advocates and skeptics as a means of bracketing areas of dispute. For these reasons, substantial funding and technical resources should be provided to intervenors as well as the host state and region.

THE LEGACY PROBLEM

If the locus problem is the equity issue provoking the most intense and visible conflict--the legacy problem, harming future generations for current gain-- appears as the more pervasive and troublesome. Because radioactive wastes remain dangerous for thousands of years, choices must be made as to the nature of our obligation (if any) to the future and the ways of resolving conflicting obligations between present generations and those of the future.

It could be argued, of course, that obligations exist only where individuals have rights and that only individuals who exist actually have rights. The future, as only possible persons, can then have no claim on the present. Or, put another way, the future lies outside of our morally relevant community. We reject this line of argument that considers only the actualized person at the expense of those who we have every reason to believe will come into existence. And while we cannot anticipate what the desires and values of distant future persons will be, we can reasonably inquire as to the kind of legacy we will leave for them.

Distributional Issues

The distribution of benefits and burdens over the many generations likely to be affected by radioactive wastes is not known nor is it knowable, despite discussions of risk, such as that of the Final Environmental Impact Statement (U.S. Department of Energy: 1981a) or the CONAES report (National Research Council, Committee on Nuclear and Alternative Energy Systems: 1979), that assume the contrary. Therefore, the discussion to follow notes only gross characteristics which are pertinent to equity considerations.

The benefits of radioactive wastes are essentially the benefits of nuclear energy used for defense purposes or the production of electricity. As noted earlier, the use of nuclear energy for weapons production is an arguable benefit for the current generation of U.S. citizens. But the massive build-up of nuclear weapons, both in terms of their legacy of weapons and proliferation threat as well as the radioactive waste generated, must be seen as a major harm currently being exported to the future. The benefits of nuclear energy to the future, then, must be found primarily in the value of electricity production and related capital and technology, not in its contribution to the growing world arsenal of destructive weapons.

The use of nuclear energy for electricity production has a number of potential long-term contributions. First, to the extent to which nuclear energy is the cheapest means of meeting actual energy needs, the greater societal accumulation of wealth which results and which may be passed on to future generations in the form of capital stock is a net benefit. It is, however, at least for the current generation of light-water reactors, a marginal advantage largely confined to the current and next generations during the transition from fossil-based to renewable energy-based systems.

Another possible long-term impact involves the effects of nuclear power development upon the inventory of nonrenewable resources. To the extent that nuclear power releases pressure upon hydrocarbons and permits their greater transfer to future generations, then this may be a tangible benefit for the future. Yet the advantage appears quite dependent upon fuel recycle and the deployment of the breeder, for the once-through fuel cycle is also consuming easily accessible uranium supplies, a nonrenewable resource in short supply. Moreover, within the time scales of radioactive wastes, this generation of nuclear reactors, limited by uranium supplies, appears to have benefits largely concentrated within the current and perhaps next several generations. By then (i.e. about 2050) the global transition to renewable energy sources should be complete.

Still another favorable legacy may be the technology of power generation by relatively renewable sources --the advanced converter, the breeder, and fusion. To the extent that the technology is linearly developed, one might argue that the current light water reactor and its legacy of waste are necessary concomitants to a legacy of proven and sustainable technology.

Finally, nuclear power does offer some possibility for reducing a possible global catastrophe associated with major increases in coal-burning--namely the carbon dioxide hazard and related climatic change. It is notable that Global 2000 designated this as the potentially most severe of the long-term global impacts of energy systems (U.S. Council on Environmental Quality and U.S. Department of State, 1980).

Turning to the risks and burdens imposed by radioactive waste inventories, there are three major clusters of risk over time. For this and the next generation, the decision on reprocessing of nuclear wastes, with the attendant if uncertain risk of proliferation, dominates risk considerations. An active debate exists over the degree to which reprocessing adds to the proliferation threat, and the degree to which technical fixes (e.g., denatured fuel cycles) alleviate the problem. The CONAES report concluded that the magnitude of proliferation risks cannot be assessed in terms of probabilities and consequences (National Research Council, (CONAES, 1979) but the risks were viewed serious enough for then President Carter in 1977 to defer the reprocessing of waste. Though that decision has been reversed by the Reagan administration, it is unlikely that reprocessing will actually take place without major government subsidy. In any event, reprocessing would enlarge benefits to this and the next several generations by extending uranium and fossil-fuel resources, but at an unknown increment in global proliferation risks.

The other significant component of near term risk-- one which has received scant attention to date--lies in the above ground activities involved in the deployment of the waste disposal system and the operation of repositories prior to closure. Once high-level wastes are sequestered in a multibarrier system within repositories 2,000 feet (1829 m) underground, public health risks would appear quite minimal. But the system of interim storage, handling, and transport of wastes to repositories will undoubtedly have its failures and accidents. While catastrophic risks appear highly unlikely, smaller risks, both radiological and nonradiological, are entailed.

A second cluster of risk will come into play after repository closure and extending over the next 600-700 years, sometimes known as the period of fission product hazard. By 1000 years after implacement, both the radiation and heat generated by the decay of the wastes have diminished by about three orders of magnitude (U.S. Nuclear Regulatory Commission: 1981a). This is a period of significant potential risk to future genera- tions, but is amenable to control through the multiple barriers to human exposure currently required by NRC regulations. The near-term risk is also the time of concern for low-level wastes, a several- hundred-year legacy problem in which failures in sound management practice appear to be the source of most of the hazard.

Finally, there is the period of very long-term risk to distant generations, extending from 1,000 to 1,000,000 years from the present. Although predominant attention has been given to high-level wastes, mill tailings are, in fact, the more significant problem during this period, due to their large volumes, their storage on the surface of the earth, and the fact that 85 percent of the radioactivity in the original uranium ore remains in the tailings. Over long periods of time, these tailings in the form of fine, pulverized rock, will likely be exposed by wind and water erosion or human intrusion. Currently some 140 million tons of such tailings exist at 22 inactive and 21 active sites, and some 10-15 million more tons are being produced annually.

Assuming a 1978 U.S. population, the mill tailings would, if left uncovered but not dispersed, result in about three premature cancer deaths per year averaged over the long term (according to NRC estimates) from random exposure alone. Although this number is not large, especially compared with the estimated 1,594 annual premature cancer deaths from radon exposure in buildings, the absolute numbers of deaths in the future--when calculated over the period of hazard-- cumulate to the hundreds of thousands or millions. Current NRC approaches would reduce fatalities by a factor of 100 but still leave a significant residual risk over time, even without major failures or human intrusion. If populations increased markedly in the mill tailings areas over the very long term, the aggre- gate risk to the future would increase correspondingly.

Compared with mill tailings, high-level wastes from both commercial and defense sectors appear to represent a more tractable risk over the long term. Although such wastes also remain hazardous over very long periods, the risks decline below those of mill tailings after 5,000 years. The emerging standard for high- level waste of the Environmental Protection Agency will, it is estimated, result in 1,000 fatal cancers for the first 10,000 years. Thereafter, the risk is calculated at approximately that of the natural ore body from which the uranium occurred, a risk apparently seen as "acceptable" by the Agency.

The comparison with the natural ore body is one frequently made by regulators and other technical experts in discussions of "risk acceptability" for protecting future generations from radioactive waste hazards. At first glance, such a comparison appears appropriate and helpful. If, after all, the emplaced waste adds no risk, then surely it should define the limit of responsibility. This reasoning is very much behind EPA's current thinking on its general HLW standard and the NRC technical criteria for high-level wastes which "end" the future at 10,000 years (U.S. Nuclear Regulatory Commission: 1981a). But, for both technical and social reasons, the analogy may be greatly misleading:

- The comparison generally fails to take into account that other releases occur during the fuel cycle or that most of the radioactivity remains in the mill tailings.

- The hazards of nature inherited by humans provide no guidance for the acceptability of hazards created by human actions; rather, the use of uranium poses choices which must find their justifications elsewhere.

- The disposal of radioactive wastes exposes different people, usually at concentrated locations, from those exposed at uranium ore body locations.

Finally, it is also important to note that, beyond public health risks, there is the danger that the burden of dealing with risks may be passed on to the future. The high-level defense waste problem has been passed on to the future for the past three decades and, given the formidable price tag associated with such wastes, this export of burden to the future could well continue. Several options for the current waste program--long-term monitored surface facilities and current approaches to mill tailings--involve a large potential displacement of burden onto the future.

While this discussion has indicated that the distributions of risks and benefits over very long time periods can be described in only a gross way, the major time discrepancy is unmistakable. Whereas benefits are largely concentrated in this and the next generation, there is substantial export of risk, and possibly managerial burdens, to distant future generations. The legacy problem is, in short, at least as fundamental an equity concern as the locus problem.

Legacy Management Options

To assess legacy management options, we refer again to our three equity principles:

Benefit/burden concordance. Since the benefits of nuclear power are concentrated strongly in this and the next generation, responsible management suggests that the obligation to reduce risks extends beyond that which is acceptable to this generation. For various waste classes, this would suggest time-phased regulatory objectives--with greater stringency for the protection of distant future generations who do not share in benefits. Specifically, the objective should be for near zero risk after 100 years (or roughly near the end of the period of retrievability as now planned for high-level wastes), and approaches should emphasize best available technology rather than "feasibility" or "reasonable achievability."

Current approaches to mill tailings badly miss this objective. In the past, mill tailings have been neglected; tailings have not been properly stabilized or controlled and have been removed for building purposes. In fact, as recently as 1976, the Nuclear Regulatory Commission had only two people working part-time on this issue. A remedial program is now under way under the Mill Tailings Radiation Control Act of 1978 to clean up the inactive sites. A program is also beginning for the long-term immobilization of tailings. But it is strikingly less ambitious and protective of the future than efforts on high-level waste, despite the fact that mill tailings pose the greater long-term problem.

The final NRC rules (1980b) on uranium mill licensing recognize the long-term hazards and the need to control the hazards without active care and maintenance by the future. But the NRC chose a disposal technology which offers only minimal long-term protection despite the low cost involved in more adequate disposal technologies. The envisioned disposal involves earth covering of tailings, means to reduce seepage (e.g., liners), and below-grade burial. While providing for a token $250,000 pool to cover long-term surveillance, the Commission decided against an insurance fund to guard against unforeseen events on the grounds that the likelihood of such occurrences was "small" and design against them was "impractical" (U.S. Nuclear Regulatory Commission, 1980b). Despite the fact that increased protection could be achieved for the future through advanced treatment (as through fixation or nitric acid leaching) for less than half a mill per Kw hour of electricity, the Commission decided in favor of the less adequate technology. The envisioned program for mill tailings is, in our view, the most inadequate component of radioactive waste management in regard to the legacy problem and should be substantially upgraded to include, at minimum, advanced treatment of tailings as well as continued attention to means of extracting the long-term actinides. Further, as described below, adequate insurance funds should be included as part of responsible long-term protection.

Turning to high-level waste from commercial power reactors, the current program for continental geological disposal appears consistent with the long-run protection of the future. In particular, it is much preferable to the indefinite storage of such wastes in engineered surface or near-surface facilities (still, however, a possible disposal option). Continued attention should be given to seabed disposal and to the strategy of many widely dispersed deep holes up to four kilometers with resistant waste forms as proposed by Ringwood (1980), for both may prove preferable over the long-term.

Several changes in the high level waste management program would further reduce potential inequities. First, as noted, there is considerable advantage to a program of lengthy (40-50 years) interim storage followed by a decision whether or not to reprocess spent fuel or dispose of it immediately. Spent fuel can be safely stored for such a period, the storage simplifies the disposal problem, and reprocessing, if it can be achieved without adding to the proliferation problem (which may be resolved by then), would enlarge benefits while reducing the very long-term hazards. Second, high-level defense wastes should be subjected to the same licensing requirements as commercial high-level waste, and a program for immobilization and disposal of such wastes, despite the financial temptations, should not be continually deferred.

Low-level wastes do not constitute a major legacy problem, but poor past management has passed on significant financial if not public health burdens. Improved disposal, ensuring safe isolation of such wastes over several hundred years, is clearly needed; the recently proposed NRC criteria appear adequate to that end (U.S. Nuclear Regulatory Commission, 1981b). One potential problem in the greater locus equity achieved through state responsibility for such waste is an increased risk for the future through potentially poorer or delayed state management.

Finally, an institutional mechanism is needed to meet our equity principle that "the imposition of a harm or burden should be accompanied by a proportional benefit." Whatever the success in reducing risks to the future, accidents will occur, assumptions will be contraverted, and harm and burdens will be imposed. An equitable management program would recognize this and provide for means of compensation, perhaps as a financial pool, similar to the superfund for hazardous waste, held in perpetuity.

Sharing the risks. Since the beneficiaries of nuclear power, as noted above, are largely concentrated in this and the next generation, this principle is moot for the distant future. The distribution of risk will likely be spread over the next 100 years, but benefits also will occur during this period.

More voluntary risk assumption. Inherently, the future cannot provide its consent to actions taken by this generation so that the transfer of risk necessarily is involuntary. This is added reason for the obligation to reduce risk to the maximum extent possible.

Several opportunities exist, however, for ameliorating this problem. First, it is important to preserve options for the future. Above, we have proposed lengthy interim storage followed by a decision on reprocessing. This retains flexibility on a central issue in waste management. Beyond this, the retrievability of the waste once stored in a repository maintains the capability of the future to respond to errors and current gaps in knowledge. In its technical criteria, the U.S. Nuclear Regulatory Commission, (1981a) proposed a 110-year retrievability capability, divided into three phases. When added to lengthy interim storage, some 150 years of choice would be built into the disposal of high-level waste, a period sufficiently long to provide options only to the first several generations. The commission in its final rule, however, relaxed the retrievability requirement.

Finally, a substantial problem exists because the future cannot participate in its own behalf in decisions occurring now on waste management. In the previous discussion of the locus problems, the importance of developing the capability of the host community to assess impacts and participate in siting decisions was noted. The decision process is inherently flawed because most of those who will bear the risk cannot participate in the process. In such situations where individuals cannot represent their own interests, it may be necessary to create institutional mechanisms, however artificial, to ensure that those concerns are inserted into the decision process. One such mechanism would be a _public defender for the future_, equipped with a technical staff and possessing the ability to challenge proposed regulation and developmental plans.

THE LABOR/LAITY PROBLEM

Of the three equity problems treated in this chapter, the differential treatment of workers and publics is undoubtedly the least visible and debated in deliberations over radioactive waste management. In fact, Melville (1983) illustrated that the use of temporary workers has emerged as a widespread industrial practice in nuclear power and has been allowed to grow without evident public concern. In 1981, the Reagan administration began a concerted campaign to relax workplace health standards, with the likely outcome that workers would be forced to bear burdens for the benefit of society more generally. And, as Shakow (1983) indicated, there is strong reason to suspect that the market will not suffice to compensate workers. Since the processing, handling, and storage of radioactive wastes necessarily will involve exposure to workers, it is likely that workers will disproportionately bear risks in this generation.

Distributional Issues

Presently there is not a good understanding of the allocation of risk during the next 50-100 years of worker involvement in radioactive waste management. But it is likely that workers, and not the public, will bear the major radiation exposure burden. The DOE's final environmental impact statement on commercially generatd waste contains only a fragmented and cursory account of occupational exposure, but suggests that both the predisposal (i.e., above ground) activities and the operation of the repository carry greater worker than public risk. The DOE estimates (U.S. Department of Energy, 1981a) that routine radiological releases from the normal operation of geologic repositories would produce negligible impacts (one-person rem) upon the regional public of two million persons, but an estimated zero to 130 health effects in a workforce of about 8000 (with individual worker doses averaging about 1 rem per year). Similarly, it estimates operational accidents as producing less than 6000 person-rems for 20 years of waste emplacement but potential fatalities (in a canister drop) among workers. Similarly, patterns of differential exposure are expected to occur in the operation of an AFR and in waste transport.

The exposures are very small, however, when compared with either natural background or the risks presented in more dangerous occupational environments. On the other hand, the accuracy of such estimates will not be validated until the waste system is actually deployed. At Three Mile Island, the accident and associated waste clean-up involves greater worker exposure (10-20,000 person-rems) than public exposure. Other such accidents will surely occur and must be anticipated and included as part of the overall systems cost of radioactive waste management. Decommissioning and decontamination of reactors, at at-reactor or away-from-reactor storage facilities, and repositories will all have a worker radiation burden. In the meantime, standards permit ten times as much exposure to workers as to members of the public and workers apparently do not receive additional salary for working in radiation environments.

Labor/Laity Management Options

To assess responses to these inequities, we turn to our equity principles:

Benefit/burden concordance. Several means are available for producing greater concordance among benefits and burdens. First, to encourage the maximum use of remote control handling of the waste, the occupational exposure standard could, despite recent action to the contrary, be lowered by a factor of ten for most workers. This would recognize that a "safe" level of worker exposure is the same as that of the public. Beyond that, strict ALARA (as low as reasonably achievable) principles should be observed in the design and operation of the waste handling and disposal system.
Second, risky work in nuclear waste facilities should be compensated through higher wages. For those for whom the occupational exposure standard is not lowered, wage compensation would allow for a less satisfactory means for greater concordance of benefits and burdens.
Third, it is presently impossible to discern trade-offs between worker and public exposure in waste system technical options. The various environmental impact statements and other documents supporting policy decisions should make such tradeoffs explicit so that technical choices could reflect such difference.
Finally, the industry practice of using temporary workers as a way of meeting the formal requirements of radiation standards should not be extended to the waste system. An extension of this practice would almost surely increase the total occupational exposure burden, decrease the use of remote control equipment, and make more difficult the voluntary assumption of risk.

Sharing the risk. The proposed reduction of the occupational health standard restrictions on the use of temporary workers would decrease the probability that waste management risks would be concentrated in

380

workers. The costs of increased use of remote control
equipment and compensation of workers for risky jobs
should be fully recovered, as noted above, by the
charge upon electricity users and from taxes for
defense wastes.

   More voluntary risk assumption. Finally, although
nuclear power is in the forefront of industries in
regard to the provision of accurate information
concerning workplace hazards and in monitoring actual
exposure, more can be done to inform workers as to
risks they will bear. In particular, the Nuclear
Regulatory Commission proposed (but failed to enact) a
new information dissemination procedure (U.S. Nuclear
Regulatory Commission, 1980a), which promised fuller
and more helpful information on radiation risks to
workers, particularly in providing the content needed
to assess the danger involved. This proposal should,
in our view, be reintroduced, adopted and vigorously
implemented.

TOWARD A JUST AND SOCIALLY
ACCEPTABLE MANAGEMENT SYSTEM

   A satisfactory resolution of the radioactive waste
problem is inextricably linked with the fate of nuclear
energy as a technology. It is doubtful that just or
socially acceptable solutions for radioactive wastes
can be found as long as the conflict over nuclear power
continues. But the proposals offered here head in the
right direction. Although these proposals depart
significantly from the government's emerging waste
management system, they are, in our view, technically
sound and offer potential for winning the needed social
acceptance and public confidence.

   Key features of the proposed management system are:

- Interim storage of spent fuel at reactors,
  possibly augmented by one or several regional
  AFRs, for 40-50 years, with the decision of
  whether to reprocess made at a later time.

- Delay of current timetables for high-level
  waste repositories in order to create a
  regional system of technically qualified
  repositories in which multiple sites will
  begin accepting waste simultaneously.

- A lottery of site exemption among qualified
  candidate sites for high-level waste in the
  site selection process and a legal storage
  limit to the amount of waste to be accepted at
  any given site.

- State responsibility for low-level waste, as currently envisioned, but with substantial upgrading in disposal technology as indicated in the recently proposed NRC licensing requirements.

- Defense wastes to continue to be stored at existing sites, but with prompt immobilization and waste subject to commercial licensing criteria for eventual disposal.

- Substantial upgrading in program plans for mill tailings, to include at minimum advanced treatment but with continued attention to the feasibility of separating out long-lived actinides.

- Creation of a legacy fund, to be funded from the mill rate on nuclear electricity use and from general taxes, to be used for site mitigation and compensation for future impacts at repository and tailing sites as well as at key nodes in waste transport corridors. At sites, the size of the fund will be determined by references to a geology rental fee.

- The creation of an independent technical and financial capability in host localities and host states so that those bearing risks can participate effectively in their own behalf. Funding for these new institutions and intervenors would be provided by the beneficiaries of nuclear power.

- The right of localities to appeal siting decisions to the state legislature and states as currently contemplated, to appeal DOE decisions to the Congress.

- A public defender for the future, equipped with an independent technical staff and possessing the ability to challenge proposed regulations and developmental plans.

- Lowering of the standard of occupational radiation exposure by a factor of ten for most markets, strict adherence to ALARA principles for such exposure, and/or compensation for risky work.

- Restrictions on the use of temporary workers in nuclear waste facilities.

NOTES

1 This chapter is, with some minor revisions and updating, taken from Chapter 15 of Roger E. Kasperson, ed, Equity Issues in Radioactive Waste Management. (Cambridge, MA: Oelgeschlager, Gunn & Hain, Publishers, 1982, pp. 331-368. Reprinted with permission.

REFERENCES

Barbour, I. G., forthcoming. "Nuclear Energy and Future Generations," In Nuclear Power: Ethics and Public Policy. Warner Klugman ed.

Bell, W. 1974. "A Conceptual Analysis of Equality and Equity in Evolutionary Perspective," American Behavioral Scientist, 18, No. 1 (Sept./Oct.), 8-34.

Bell, W. and R.V. Robinson. 1978. "An Index of Evaluated Equality: Measuring Conceptions of Social Justice in England and the United States," Comparative Studies in Sociology, 1, 253-270.

Ben-David, S., A.V. Kneese, and W.D. Schulze. 1979. "A Study of the Ethical Foundations of Benefit-Cost Analysis Techniques," University of New Mexico, Department of Economics, Program in Resource Economics, Working Paper Series (August).

Brody, J.E., 1980. "Consent Form: A Patient's Chance to Challenge the Doctor's Orders," New York Times, July 30, p. 12.

Callahan, D. 1978. "Ethical Uses of Risk/Benefit Analysis," In Symposium on Risk/Benefit Decisions and the Public Health, J.A. Staffa ed. Rockville, MD.: Food and Drug Administration, pp. 71-74.

Cole, R.J. and T.A. Smith. 1979. Compensation for the Adverse Impacts of Nuclear Waste Management Facilities: Application of an Analytical Framework to Consideration of Eleven Potential Impacts, B-HARC-311-022, Seattle: Battelle Human Affairs Research Centers.

Federal Insurance Administration, 1979. "Sensitivity Analysis of the Three Mile Island Incident," Washington, D.C.: The Administration.

Fischoff, B. 1983. "Informed Consent for Nuclear Workers." In Equity Issues in Radioactive Waste Management, Roger E. Kasperson, Ed. Cambridge, MA: Oelgeschlager, Gunn & Hain, Publishers, Inc., pp. 301-328.

Gilinsky, V. 1981. "Remarks before the College of Natural Science Alumni Association, Michigan State University, April 9, 1981."

Green, H.P. 1983. "Legal Aspects of Intergenerational Equity Issues," In Equity Issues in Radioactive Waste Management, Roger E. Kasperson, ed. Cambridge, MA: Oelgeschlager, Gunn & Hain, Publishers, Inc., 1983, pp. 189-202.

Green, R.M. 1977. "Intergenerational Distributive Justice and Environmental Responsibility," Bioscience, 27, No. 4 (April), 260-265.

Kehoe, K. 1980. Unavailable at Any Price, Washington, D.C.:Environmental Policy Center.

Lipset, S.M. 1968. "Social Class," In International Encyclopedia of the Social Sciences, D.L. Sills, ed., Vol. 15 New York: MacMillan and Free Press, pp. 296-316.

Marshall, E. 1980. "NRC Plans to Deregulate Biomedical Waste," Science, 210 (12 December), 1228-1229.

Martin, K.A. 1980. "Biblical Mandates and the Human Condition," Journal of the American Scientific Affiliation, 32, No. 2 (June), 74-77.

McKeon, R. (ed.). 1941. The Nicomachean Ethics: The Basic Works of Aristotle New York: Random House, Book 5, Chapter 3.

Melville, M. H. 1983. "Temporary Workers in the Nuclear Power Industry: Implications for the Waste Management Program." In Equity Issues in Radioactive Waste Management, Roger E. Kasperson, ed. (Cambridge, MA: Oelgeschlager, Gunn & Hain Publishers), pp. 229-258.

National Council of Churches. 1979. Energy and Ethics New York.

National Low-Level Waste Management Program, The. 1980. Managing Low-Level Radioactive Wastes: A Proposed Approach Idaho Falls, Idaho: EG&G Idaho, Inc.

National Research Council, Committee on the Biological Effects of Ionizing Radiation. 1980. The Effects on Populations of Exposure to Low Levels of Ionizing Radiation, BEIR III Report, Washington, D.C.: National Academy Press.

National Research Council, Committee on Nuclear and Alternative Energy Systems. 1979. Energy in Transition, 1985-2010, San Francisco: W.H. Freeman and Co.

National Research Council, Committee on Radioactive Waste Management. 1980. Letter to Dr. Colin A. Heath, U.S. Department of Energy, April 18.

O'Hare, M. 1977. "Not on My Block You Don't--Facilities Siting and the Strategic Importance of Compensation," Public Policy, 25 (Fall), 407-458.

O'Hare, M. and D.R. Sanderson. 1977. "Fair Compensation and the Boomtown Problem," Urban Law Annual, 14, 101-133.

Passmore, J. 1974. Man's Responsibility for Nature N.Y.: Scribner's.

Pitkin, Hannah, 1965. "Obligation and Consent--I," American Political Science Review, 59 (December, 1965), 990-999.

Rawls, J. 1971. A Theory of Justice (Cambridge, Mass.: Harvard University Press).

Ringwood, T. 1980. "Safety in Depth for Nuclear Waste Disposal," New Scientist, 88, No. 1229 (27 November), 574-575.

Seley, J. and J. Wolpert. 1983. "Equity and Location," Equity Issues in Radioactive Waste Management, Roger E. Kasperson, ed., Cambridge, MA: Oelgeschlager, Gunn & Hain, Publishers, pp. 69-93.

Shakow, D. 1983. "Market Mechanisms for Compensating Hazardous Work: a Critical Analysis." Chapter 12 In Equity Issues in Radioactive Waste Management, Roger E. Kasperson (ed.), Cambridge, Publishers, pp. 277-290.

Slovic, P., B. Fischhoff, and S. Lichtenstein. 1979. "Rating the Risks," Environment, 21 (April), 14-20, 36-39.

Starr, C. 1983. "Coping with Nuclear Power Risks: A National Strategy." In The Analysis of Actual vs. Perceived Risks, V.T. Covello, W.G. Flamm, J.V. Rodricks and R.G. Tardiff, eds., New York: Plenum Press, pp. 251-257.

Stobaugh, R. and D. Yergin, ed. 1979. Energy Future, N.Y.: Random House.

U.S. Council on Environmental Quality and U.S. Department of State. 1980. The Global 2000 Report to the President. 3 vols. Washington.

U.S. Department of Energy, 1980. Statement of Position of the United States Department of Energy in the Matter of Proposed Rulemaking on the Storage and Disposal of Nuclear Waste, DOE/NE-007. Washington, D.C.: The Department, 15 April.

U.S. Department of Energy, 1981a. Final Environmental Impact Statement: Management of Commercially Generated Radioactive Waste, DOE/EIS-0046F. Washington, D.C.: U.S. Department of Energy, 3 vols.

U.S. Department of Energy, 1981b. The National Plan for Radioactive Waste Management, Working Draft 4 Washington, D.C.: The Department, January.

U.S. Interagency Review Group on Nuclear Waste Management. 1979. Report to the President, TID-29442. Washington, D.C.: U.S. Department of Energy.

U.S. Nuclear Regulatory Commission, 1980a. "Draft Regulatory Guide and Value/Impact Statement: Instructions Concerning Risk from Occupational Radiation Exposure." Washington. May.

U.S. Nuclear Regulatory Commission, 1980b. "Uranium Mill Licensing Requirements: Final Rules," Federal Register, 45, No. 194 (October 3), 65521-65538.

U.S. Nuclear Regulatory Commission, 1981a. "Disposal of High-Level Radioactive Waste in Geologic Repositories," Federal Register, 46, No. 130 (July 8), 35280-35298.

U.S. Nuclear Regulatory Commission, 1981b. "Licensing Requirements for Land Disposal of Radioactive

Wastes," _Federal Register_, 46, No. 142, (July 29), 38089-38105.

U.S. President. 1980. _Message from the President of the United States Transmitting a Report on his Proposals for a Comprehensive Radioactive Waste Management Program_, 96th Congress, 2nd Session, House Document 92-266. Washington.

15
Decommissioning Nuclear Power Plants

Barry D. Solomon

A few years ago, a feature article in a leading electric utility industry trade journal referred to commercial nuclear power plant decommissioning as a "nagging back-end problem", albeit one for which "it is wise to ponder the problem and its ramifications now" (Friedlander, 1978). But while the nuclear power industry ponders this problem, plant obsolescence, reactor embrittlement, and other unresolved technological problems are steadily increasing the number of plant sites where early retirement could prudently be considered. The back-end phases remain the least addressed aspect of the nuclear fuel cycle. This lack of attention is a reflection of both the industry's preoccupation with a variety of other problems and the still-limited experience with decommissioning of licensed nuclear power plants. To date, the largest decommissioned reactor was rated at about 250 MWt, less than a tenth the size of the most recent commercial reactor units (U.S. Department of Energy, 1982b).

Decommissioning can be defined as the ultimate disposition of a nuclear power facility after it is permanently taken out of service (Manion and LaGuardia, 1980). The retirement of a nuclear power plant can be considered more important than the decommissioning of a coal-fired power plant due to the more complex and extensive ensemble of reactor, equipment at a nuclear plant, as well as its high level of residual radionuclides. In addition to nuclear power plants, all uranium mills, uranium hexafluoride conversion plants, fuel fabrication plants, and fuel reprocessing plants must also be decommissioned. Of this latter group of facilities, fuel reprocessing plants appear to pose by far the most significant costs and problems (Schneider and Jenkins, 1977; Comptroller General of the U.S., 1977).[1]

Previous discussion of nuclear power plant decommissioning has been almost totally restricted to decommissioning technology, cost and cost recovery (see

Solomon, 1982, for an overview). Topics of specific interest to energy geographers on this issue, however, include spatial problems at individual decommissioned plant sites, environmental impacts, nuclear waste transport to waste repositories, and post-decommissioning land use. All of these issues are directly influenced by the amount and type of radioactivity at a retired nuclear power facility, although some generalizations appear to be possible (U.S. Nuclear Regulatory Commission, 1981).

The paper begins with a discussion of the alternative decommissioning methods and associated issues. We will follow with a review of actual decommissioning experience in the U.S. and abroad, environmental impacts, the transport of radioactive nuclear waste, and post-decommissioning land use options. Our discussion will conclude with future decommissioning problems and potential mitigation strategies.

DECOMMISSIONING METHODS AND ISSUES

Six methods for decommissioning nuclear power facilities are recognized by the U.S. Nuclear Regulatory Commission (NRC): mothballing, entombment, prompt dismantlement, a combination of either mothballing or entombment with deferred dismantlement, and site conversion.

The first two options do not require any change in land use at the radioactive nuclear power plant site and therefore cannot be considered as viable, long-term decommissioning options. Under these options, a significant periodic inspection and structural testing program are required to verify the plant's integrity for an indefinitely long time period (Manion and LaGuardia, 1976). Mothballing a nuclear facility requires its placement in protective storage following the removal of all spent fuel and source material and the disposal of all liquid and solid waste. Entombment requries similar treatment, except that the radioactive materials and components must be encased in concrete or steel.[2]

The power plant dismantlement option requires the disposal of all radioactive structures, components, and systems so that the site can be restored and reused. Dismantlement requires about 5-8 years to complete, depending on the reactor size. However, the high level of induced radiation embedded in a retired nuclear power plant requires an in-depth assessment of the biological hazards that would be presented to the decommissioning workforce. To prevent these workers from receiving excessive doses of radiation, much of the cutting up of the nuclear reactor would be performed by remote controlled equipment under water. Nevertheless, one study has argued that 70-110 years of deferral would be required before radioactivity at a commercial nuclear facility would decay to levels safe enough for dismantling (Comptroller General of the U.S., 1977). This conclusion is based primarily on the expected residual levels of cobalt-60, which would emit gamma-ray energy from the stainless steel cladding of the pressure vessel and reactor core components.

The site conversion decommissioning option utilizes the exising turbine-generator system with a new steam supply system, either fossil fuel, or nuclear. While the original nuclear-generated steam supply system would be discarded, most of the electrical generating equipment could be adjusted for use in the new power system. This option, then, is actually a form of dismantlement, but it has only recently been considered by the nuclear industry as a possible alternative use for a nuclear reactor site.

The NRC currently stipulates that any decommissioning method is acceptable, but that a nuclear power plant licensee must address plans for decommissioning the nuclear facility before an operating license is issued. An actual decommissioning plan and environmental impact statement (EIS) is not required until the operating license is terminated, and only a few such EIS's have thus far been issued.[3] In addition, since January, 1978, the U.S. Security and Exchange Commission has required that nuclear power utilities must disclose the estimated cost of plant dismantlement or decontamination, as well as the cost recovery method. However, the NRC has proposed a rule to eliminate any financial assurance criterion for these utilities (Fed. Reg. 46 FR 41786, 1981).

Few electric utility companies expect nuclear power plant decommissioning costs to be very large, although full decommissioning experience is limited to only three small commercial nuclear power reactors as of June, 1983. Such optimism regarding cost is based primarily on the conclusions of the major government studies, which have estimated immediate dismantlement costs for a 1100-1200 MWe nuclear reactor or fuel reprocessing plant to be about $40-60 million in current dollars (e.g. Smith et al., 1978; Oak, et al., 1980; Schneider and Jenkins 1977). These costs might be equivalent to 5-10 percent of the capital investment costs of most existing nuclear power plants; if the utility company recovers these costs over the operating lifetime of the power plant, they may amount to 1-1$\frac{1}{2}$ percent of the plant's life-cycle busbar costs (Solomon, 1982). Nondismantlement methods would cost relatively less, although the costs of mothballing or entombment would accumulate over an indefinitely lengthy time period (Manion and LaGuardia, 1976).

The general dismissal of nuclear power plant decommissioning costs as insignificant by most electric utility companies is questionable for several reasons. First, decommissioning costs of small-scale nuclear reactors already taken out of service have sometimes been 25-100% or more of the facility's investment costs (Iwler, 1978). Second, although significant economies of size are expected in decommissioning costs of larger nuclear reactors, considerable experience in the nuclear power industry suggests that rapid cost escalation may be more likely (Bupp and Derian, 1978). The latter possibility is underscored by Swedish and several U.S. utility company cost estimates of $100 million or more for immediate dismantlement (Mandahl et al., 1979; Schwent, 1978). Thus, the only certainty about nuclear power plant decommissioning costs appears to be their uncertainty. Until the industry gains more experience in decommissioning commercial-scale nuclear reactors, reliable estimates of decommissioning costs will be unavailable.

Customers of nuclear utilities will inevitably pay the cost of decommissioning their nuclear power facilities, and state regulatory bodies have the responsibility of determining the method of cost recovery from rate payers. At least seventeen state regulatory commissions allow the utility companies to treat decommissioning costs as a negative salvage value and to recover them through depreciation accounts (Lovins and Lovins 1981)[4] and a few utilities in Pennsylvania and Connecticut have been required to collect these funds through interest-bearing revenue or corporate surety bonds (Ferguson, 1980). It is generally agreed that no matter which decommissioning cost recovery method is chosen, electric utilities should allocate these costs over the productive life of a nuclear facility (Fako and Dickson, 1981). Unfortunately, several utility companies with nuclear power plants have totally neglected these costs until recently (Comptroller General of the U.S., 1977). This de facto policy is inequitable, since future rate payers will be forced to cross-subsidize the earlier consumers of nuclear power. One result might be rate payers and consumer advocates arguing against the inclusion of these costs in a utility company's rate base during rate increase hearings in such cases.

The principal environmental hazard of decommissioning nuclear power plants is the occupational radiation dose, estimated at 1200 man-rems for immediate dismantlement of a large-scale nuclear reactor (Smith et al., 1978). However, the deferment of dismantlement up to 30 years could reduce the total dosage by about 60 percent, although most of the radiation exposure to workers would accumulate during the preparation for safe storage period (Smith et al., 1978). The maximum

expected radiation dose is estimated to result in one
or less fatal cancer or serious genetic effects on the
decommissioning labor force over the duration of the
decommissioning project (National Academy of Sciences,
1980), although such estimates are highly uncertain.
Another little explored problem is the cumulative
environmental impacts on radioactive waste repositories
and surrounding communities from the decommissioning of
several nuclear power plants.

Actual environmental impacts of decommissioning
will greatly depend on the engineering success of a
plant retirement at a specific site. The radiation
dose to the workforce, as well as to the general
public, will increase in the event of an accident
during decommissioning or during solid waste transport
to a radioactive waste repository. Positive or less
significant impacts on terrestrial ecology, aquatic
systems, aesthetics, and air quality are expected,
although significant but temporary noise pollution will
result during decommissioning from exterior blasting,
demolition equipment, and operation of trucks (U.S.
Department of Energy, 1982a).

Socioeconomic impacts of decommissioning may not be
large, since the required workforce will be about at
the same level as the power plant operation and maintenance workforce. However, after decommissioning is
completed, the loss of perhaps 200 predecommissioning
jobs and the resulting income and spending power could
be significant to a local economy, particularly in a
very rural area. Similarly, the absence of the nuclear
power plant could remove the major source of property
tax receipts for a rural local government.

No matter how decommissioning is accomplished,
appreciable amounts of solid radioactive material will
have to be transported to a radioactive waste burial
ground. Such material includes neutron-activated
material (primarily the steam pressure vessel and its
internals), contaminated piping and equipment, and
other radioactive waste (solidified liquid concentrates, combustibles, and other dry waste). Most
commercial radioactive waste in the U.S. will be
shipped in either shielded containers or 55-gallon
drums by truck to one of three major commercial nuclear
waste repositories, located at Hanford, Washington;
Beatty, Nevada; and Savannah River, South Carolina
(Barnwell). The Beatty facility accepts only low level
nuclear waste, which includes some decommissioning
waste.

The dismantlement of a nuclear reactor also
requires the removal of the steam pressure vessel, to
be transported to waste burial grounds via barge or
railroad. The pressure vessel would either be seg-
mented, or removed in one piece after the removal and
separate shipment of the internal equipment and

residual material. The trade off between waste shipment costs and the radiation releases from possible transport accidents needs to be considered for each radioactive waste disposal option available at a nuclear power plant site. A conflict may arise between choosing the shortest trip distances to minimize transport costs, and avoiding major population centers where the impact of a potential nuclear transport accident would be greatest. In addition, actual transport routes for nuclear waste shipments will be constrained by U.S. Department of Transportation regulations and local government ordinances banning or restricting nuclear waste traffic. These contraints will be elaborated on below.

The major decommissioning options will result in varying impacts on land resources, depending on the volume and the destination of nuclear waste. All shipments of solid radioactive waste will require the commitment of land to radioactive waste disposal. Although the three current commercial repositories in the U.S. will continue to receive nuclear waste in the near future, a more permanent disposal site will clearly be required before the end of the century (Hill et al., 1982).

If a nuclear power plant is mothballed or entombed, a commitment of the site for 70-110 years or more may be required. This commitment would have to be justified on optimum long run cost, and minimization of environmental and health or safety impacts to the workforce and the general public. Most of the non-research-scale reactors decommissioned to date have been mothballed. Only the dismantlement options result in unrestricted subsequent use of the affected land; a variety of alternative land uses (including a new energy facility) can be considered at that point. Important factors influencing future land use are the socioeconomic and fiscal impacts on the local community. Large industrial facilities, for example, are typically desirable because they result in huge property tax revenues for local governments.

DECOMMISSIONING EXPERIENCE

The greatest concentration of experience with nuclear power plant decommissioning is in the United States. Several categories of nuclear power plants have been decommissioned in the U.S., from experimental research reactors to licensed power reactors. Between 1968 and 1983, sixteen nonresearch scale reactors have been decommissioned: five licensed power reactors, four demonstration power plants, six licensed test reactors, and one "power production" facility (Lear and Erickson, 1975). In addition, well over 50 research reactors have been taken out of service, usually by dismantlement (Table 15.1).

The first nuclear reactor ever to be dismantled completely was a 1 MWt Oak Ridge National Laboratory research reactor in 1954 (Lear and Erickson, 1975). The first plant of any significant size to be decommissioned was the 65 MWt Carolina-Virginia Tube Reactor (CVTR) in Parr, South Carolina, which was mothballed in 1968. This action was followed by the entombment of the Hallam demonstration plant in Hallam, Nebraska in 1969. Rated at 256 MWt, the Hallam plant is the largest nuclear facility to be decommissioned as of June, 1983.

Only three decommissioned nuclear power plants ever generated commercial electricity: the former Rural Cooperative Power Association's Elk River plant, Detroit Edison's Fermi I, and Philadelphia Electric Company's Peach Bottom 1. The Elk River plant was dismantled in 1974, while the latter two facilities were mothballed in 1975. Operations at the latter two plants were unexpectedly cut short--Fermi I suffered a partial core melt-down in 1966 (Fuller, 1976), while Peach Bottom 1 was taken out of service in 1974 due to the high cost of retrofitting modifications required to meet more stringent Atomic Energy Commission safety criteria (Smith et al., 1978). Nonetheless, the world's first commercial nuclear power plant to generate electricity - Duquesne Light's plant in Shippingport, Pennsylvania - only recently was approved for partial dismantlement (U.S. Department of Energy, 1982a). The Shippingport reactor, operating intermittently since December, 1957, was rated at 72 MWe.

Nuclear power plant decommissioning activity is also accelerating in several other countries around the world. Data on non-U.S. decommissioning experience are limited; the only information that exists is for Japan and member nations of the Commission for European Communities (CEC). In addition, a few of the oldest nuclear reactors in the USSR may have been mothballed, although this fact has not been reported in the open literature. A recent report of the Organization for Economic Cooperation and Development (OECD), Nuclear Energy Agency lists examples of decommissioned nuclear facilities in Japan and Western Europe, yet the list is neither comprehensive nor as detailed as the analogous U.S. data (Radioactive Waste Management Committee, 1982) (Table 15.2). The most significant non-U.S. experience appears to have been gained in France. The most common decommissioning method outside the U.S. appears to be plant dismantlement, although this merely reflect the bias of the non-U.S. data toward smaller nuclear reactors, which are easier to dismantle.

In the next five years, nuclear power plant decommissioning experience is expected to increase appreci-

TABLE 15.1
Nonresearch Scale Nuclear Power Plants Decommissioned in the United States

Facility Name and Location	Reactor Type	Reactor Category	Power Rating, MW_t	Decommissioning Method	Year
CVTR Parr, SC	Pressure Tube, Heavy Water	Power Reactor	65	Mothballed	1968
Hallam Hallam, NB	Sodium-Cooled, Graphite-Moderated	Demonstration Plant	256	Entombed	1969
Piqua Piqua, OH	Organic-Cooled and Moderated	Demonstration Plant	45.5	Entombed	1969
BONUS Ricon, PR	Boiling Water Reactor	Demonstration Plant	50	Entombed	1970
Pathfinder Sioux Falls, SD	Boiling Water Reactor	Power Reactor	190	Mothballed and Partially Converted	1972
Saxton Saxton, PA	Pressurized Water Reactor	Test Reactor	23.5	Mothballed	1973
SEFOR Strickler, AR	Sodium-Cooled, Fast	Test Reactor	20	Mothballed	1973
Elk River Elk River, MN	Boiling Water Reactor	Demonstration Plant	58.2	Dismantled and Partially Converted	1974
Fermi 1 Monroe Co., MI	Sodium-Cooled, Fast	Power Reactor	200	Mothballed	1975

Facility	Reactor Type	Category	MW	Status	Year
Peach Bottom 1 York Co., PA	Gas-Cooled, Graphite-Moderated	Power Reactor	115	Mothballed	1975
B & W Lynchburg, VA	Pool	Test Reactor	6	Partially Dismantled	1975
GE EVESR Alameda Co., CA	Boiling Water Reactor	Test Reactor	17	Mothballed	1975
NASA Plumbrook Sandusky, OH	Light Water	Test Reactor	0.1	Mothballed	1975
VBWR Alameda Co., CA	Boiling Water Reactor	Power Reactor	50	Mothballed and Partially Converted	1975
Westinghouse Test Reactor Waltz Mill, PA	Tank	Test Reactor	60	Mothballed	1975
Sodium Reactor Experiment Santa Susana, CA	Sodium-Cooled, Graphite-Moderated	Power Production	30	Dismantled	1982

Sources: Lear and Erickson (1975) and J. D. White, U.S. Department of Energy, in personal correspondence, December 2, 1982.

TABLE 15.2
Nuclear Power Plants Decommissioned in Japan
and Western Europe (Including Research Reactors)

Country	Facility Name	Decommissioning Method
France	CESAR Cadarache	Dismantled
France	CHINON	Mothballed
France	EL2	Entombed
France	G1 Marcoule	Entombed
France	Minerve Fontenay Aux Roses	Dismantled
France	PEGASE Cadarache	Dismantled
France	PEGGY Cadarache	Dismantled
France	ZOE Fontenay Aux Roses	Entombed
Japan	Aqueous Homogeneous Critical Facility	Dismantled
Japan	Component Test Loop	Dismantled
Japan	Fission Gas Release Loop	Dismantled
Japan	Hitach Training Reactor	Entombed
Japan	Japan Research Reactor	Entombed
Japan	Mitsubishi Critical Facility	Dismantled
Japan	Ozenji Critical Facility	Dismantled
Japan	Sumitomo Critical Facility	Dismantled
Great Britain	BEPO Graphite Moderated Reactor	Entombed
Great Britain	Jason Research Reactor	Dismantled
Great Britain	Lido 300 kw Pool Reactor	Dismantled
Great Britain	Merlin Research Reactor	Dismantled
Great Britain	Various Zero Energy Reactors	Dismantled
West Germany	Nieder Aischbach Nuclear Power Plant	Mothballed
West Germany	Nuclear Ship NS "Otto Hahn"	Dismantled
Sweden	Agesta Nuclear Power Station	Mothballed
Sweden	RI Research Reactor	Mothballed
Norway	Jeep Reactor	Entombed
Norway	NORA Reactor	Dismantled
Netherlands	Schiphol Demonstration Reactor	Dismantled
Netherlands	Technical University Education Reactor	Dismantled
Switzerland	Experimental Nuclear Power Station	Mothballed & Entombed
Commission of European Communities	ISPRA-1	Mothballed

Source: Radioactive Waste Management Committee (1982, Appendix A).

ably, as several large commercial facilities are scheduled for retirement. U.S. reactors should again dominate, although medium-scale plutonium production reactors are expected to be decommissioned in Great Britain (Windscale) and France (Marcoule G2 and G3) in the near future; three large nuclear power plants in West Germany are also planned for decommissioning during this period (Radioactive Waste Management Committee 1982, Appendix A). Table 15.3 lists the commercial scale nuclear power plants in the U.S. that will be eligible for decommissioning during 1983-1988, based on reactor vintage and or a variety of operating problems. Another nuclear power plant which could be added to this list is Metropolitan Edison's Three-Mile Island II, site of the 1979 accident near Middletown, Pennsylvania (Kasperson et al., 1979; Zeigler et al. 1981).

The history of the nuclear power industry tells us that the future may be very different from the past, particularly when a change in scale of facilities is involved (Bupp and Derian 1978). An assumption that experience with commercial scale nuclear power plant decommissioning will approximate that of the small-scale facilities already decommissioned would be premature. Nonetheless, the discussion in the remaining sections of this chapter is primarily based on past decommissioning experience, with caveats where necessary.

ENVIRONMENTAL IMPACTS

Many of the environmental impacts of nuclear power plant decommissioning will be positive, partially offsetting negative impacts of plant construction and operation. Site specific impacts will greatly vary with the decommissioning method chosen, yet eventual plant dismantlement may inevitably be required in all cases. The most important environmental impacts appear to be the occupational radiation dose and the added land requirments for solid waste repositories.

Substantial controversy exists surrounding estimates of radiation exposure at nuclear power plants and the resultant health effects on plant workers and the public (Goffman, 1981). This controversy has included plant decommissioning. For example, although one influential federal study has estimated the work force radiation dose at 1200 man-rems during immediate dismantlement of a reference 1175 MWe PWR (Smith et al., 1978), the environmental impact statement for the 72 MWe Shippingport nuclear plant assumes 1275 man-rems. The expected high dismantlement radiation level at Shippingport can be explained in part by its early construction date and lower safety standards.[5] Actual

TABLE 15.3
Commercial Nuclear Power Plants in the United States
Eligible for Decommissioning, 1983-1988

Facility Name and Location	Operating Utility	Reactor Type	Power Rating MW_e	Power Generated to date (10^6 MWH)
Shippingport Shippingport, PA	Duquesne Light	PWR	72	6.6
Indian Point 1 Westchester Co., NY	Consolidated Edison	PWR	265	30
Humbolt Bay Humbolt Co., CA	Pacific Gas & Electric	BWR	65	5
Dresden 1 Grundy Co., IL	Commonwealth Edison	BWR	207	17
Yankee Rowe Franklin Co., MA	Yankee Atomic Electric	PWR	175	24
Big Rock Point Charlevoix Co., MI	Consumers Power	BWR	64	7
Haddam Neck Middlesex Co., CT	Connecticut Yankee Atomic Power	PWR	582	59
Monticello Wright Co., MN	Northern States Power	BWR	545	39
Oyster Creek 1 Ocean Co., NJ	Jersey Central Power & Light	BWR	650	44
San Onofre 1 San Diego Co., CA	Southern California Edison	PWR	436	37
Nine Mile Point 1 Oswego Co., NY	Niagara Mohawk Power	BWR	620	43
Vermont Yankee Windham Co., VT	Vermont Yankee Nuclear Power	BWR	514	31

Source: Estimated from the U.S. Department of Energy (1982b).

radiation levels may even vary from plant site to plant site for reactors of similar vintage, due to variations in reactor design, construction practices, and operations management. Such estimates of radiation releases during dismantlement should therefore be interpreted with extreme caution. Public radiation doses, however, can generally be assumed to be significantly lower than those on the decommissioning workforce.

Health effects of radiation exposure are subject to even more controversy than the estimates of radiation releases. The leading government and industry source of information on health effects due to radiation exposure at nuclear power plant sites is probably the BEIR III study of the National Academy of Sciences (1980). For a large-scale nuclear power plant, occupational health effects, such as fatal cancers and serious genetic effects, are assumed to total no more than about 1.02, with virtually zero health effects on the public. Other scientists have argued that the expected health effects due to low-level radiation will be much greater (Goffman, 1981; Sternglass, 1981), especially during an accident. The BEIR III estimates are summarized in Table 15.4 for the Shippingport plant.

The nuclear waste disposal problem in the U.S. will become more salient as several commercial scale nuclear power plants are decommissioned in the near future at a variety of locations. By June, 1983, the U.S. government still had not chosen a permanent radioactive waste disposal site or sites. Nuclear waste disposal can be seen as a classical test of the federalist system, an issue which will assuredly cause increasing state/federal and interregional conflict (Hill et al., 1982). Indeed, many environmental groups argue that any nuclear waste disposal site will leak and will therefore require some local sacrifice in the national interest. But until a permanent waste repository is established, the three existing radioactive waste burial sites may be hard-pressed to meet the cumulative space needs of decommissioned commercial nuclear power plants. The Hanford, Beatty, and Savannah River facilities have already been shut down on several occasions due to violations of safety standards during packaging and shipment, a problem which could easily increase in frequency as nuclear waste continues to be produced.

Since nuclear power plant decommissioning results, at worst, in no net change in land use requirements, no new negative impacts will occur on terrestrial ecology due to decommissioning alone.[6] Similarly, although radioactive liquid distillates will normally be discharged into local watercourses during decommissioning, only minor additional waterborne radiation will be produced due to decommissioning (U.S. Department of

TABLE 15.4
Expected Health Effects From Decommissioning the
Shippingport Nuclear Power Plant

Decommissioning Option	Occupation Radiation Dose (man-rems)	Occupation Health Effects	Public Radiation Dose (man-rems)	Public Health Effects
No Action/ Continued Operation	37/year	3.7×10^{-3}/year to 29.6×10^{-3}/year	1/year	1×10^{-4}/year to 8×10^{-4}/year
No Action/ Continued Surveillance	25/year	2.5×10^{-3}/year to 20×10^{-3}/year	0	0
No Action/ No Surveillance	0	0	N.A.	N.A.
Immediate Dismantlement	1275	0.13 - 1.02	28	2.8×10^{-3} to 22.4×10^{-3}
Mothballed/ Deferred Dismantlement	505	0.05 - 0.40	21	2.1×10^{-3} to 16.8×10^{-3}
Entombment	617	0.06 - 0.49	16	1.6×10^{-3} to 12.8×10^{-3}

Source: U.S. Department of Energy (1982a, Table 4.1-1).

Energy, 1982a). Air and noise pollution impacts could be significant during nuclear power plant decommissioning, particularly with dismantlement, although these effects will be short-lived. Such impacts will be due to the operation of fuel-driven equipment during decommissioning, particularly during exterior blasting and concrete demolition operations. However, during most of the decommissioning period, such impacts will be comparable or less than those occurring during nuclear power plant construction or operation phases (U.S. Department of Energy, 1982a).

The impact of nuclear power plant decommissioning on aesthetics is subjective and will depend upon the decommissioning method employed. All nondismantlement options will result in the retention of an unproductive nuclear power plant on the land with no discernible public benefit. In such cases, the decommissioned facility could be seen as an eyesore or public nuisance, with strong support for its removal. Conversely, plant dismantlement would probably improve aesthetics, depending on the subsequent land use. But no matter how the community feels about nuclear power production, few people would approve of the continued existence of an unproductive nuclear power plant in their own backyard.

Socioeconomic impacts of plant decommissioning are highly dependent upon post-decommissioning land use. On the one hand, when any industrial facility (including a nuclear power plant) goes out of business, loss of jobs and income will result at the applicable facility and economically-linked industries. The level of these losses will depend upon the labor intensiveness and the job skill levels in these industries. Although nuclear power plant operation is clearly not labor intensive, the direct loss of 200 or more plant operation jobs may be problematic in a given community. However, these losses could be easily offset by the development of a more labor intensive industry on the same site or general area following decommissioning, especially when the required labor skills can be locally supplied.

A final important impact of nuclear power plant decommissioning is the effect on local, usually county, property tax receipts. Until the plant is totally dismantled, the local government tax assessor could tax the facility at the value of unimproved land, at the very large value of an operating power plant, or more likely, at a rate between these extremes.[7] The resulting revenue loss could be highly detrimental on the local tax base, but in some places it could be offset by any future industrial development in the area. But to the extent that such an important revenue loss is not replaced, local government services and facilities, such as schools and utilities, will inevitably suffer in quality.

NUCLEAR WASTE TRANSPORT

The decommissioning of nuclear power plants results in a large volume of solid and liquid nuclear waste for disposal. In addition to radioactive solid waste, mostly spent fuel rods, decommissioning also requires the disposal of a radioactive steam pressure vessel, other solid waste, and liquid waste. But due to the high concentration of radioactivity in spent fuel rods, such waste has been stored in most cases in spent fuel pools at existing power plant sites since the shutdown of the West Valley reprocessing facility in 1972 and the 1976 presidential ban on commercial reprocessing. Nuclear spent fuel rods will continue to be stored in this manner in the near future, until the federal government resolves the high level waste disposal problem. In the interim, many nuclear power plants are running out of spent fuel storage capacity. All other radioactive solid waste at nuclear power facilities will continue to be transported to one of the three major commercial nuclear waste repositories in the U.S. The remaining solid waste at these sites must be disposed of either by selling it for scrap value, shipping it to a sanitary landfill or hazardous waste disposal site, or using it as backfill material at the decommissioned site. Purely liquid radioactive waste, however, will typically be disposed of in a river or other permanent waterbody at the power plant site following decontamination procedures. This disposal contrasts with the fate of radioactive liquid concentrate waste, which is solidified before disposition off the power plant site (U.S. Department of Energy, 1982a).

The transport of radioactive waste and steam pressure vessels from nuclear power plants to waste repositories involves trade-offs between safety and dollar cost (Robbins, 1981). The U.S. Department of Transportation, which regulates these shipments, prefers electric utility companies to utilize the waste repository nearest the power plant site, while traveling the most direct and safest routes possible to avoid major population centers. In practice, this policy usually amounts to truck transport on interstate highways, excluding segments in downtown areas of large cities (Figure 15.1). Precise transport patterns of these wastes are partly obscured due to lack of full DOT and NRC public disclosure for security reasons.

Ironically, the safest routes for transporting radioactive wastes are also often the heaviest traveled by other motor vehicles. Thus, while these highways may have a lower probability of accidents involving nuclear waste shipments, the radiation releases and public health consequences in the event of an accident would be more serious than on less traveled routes.

FIGURE 15.1
Transport Routes for Nuclear Wastes.

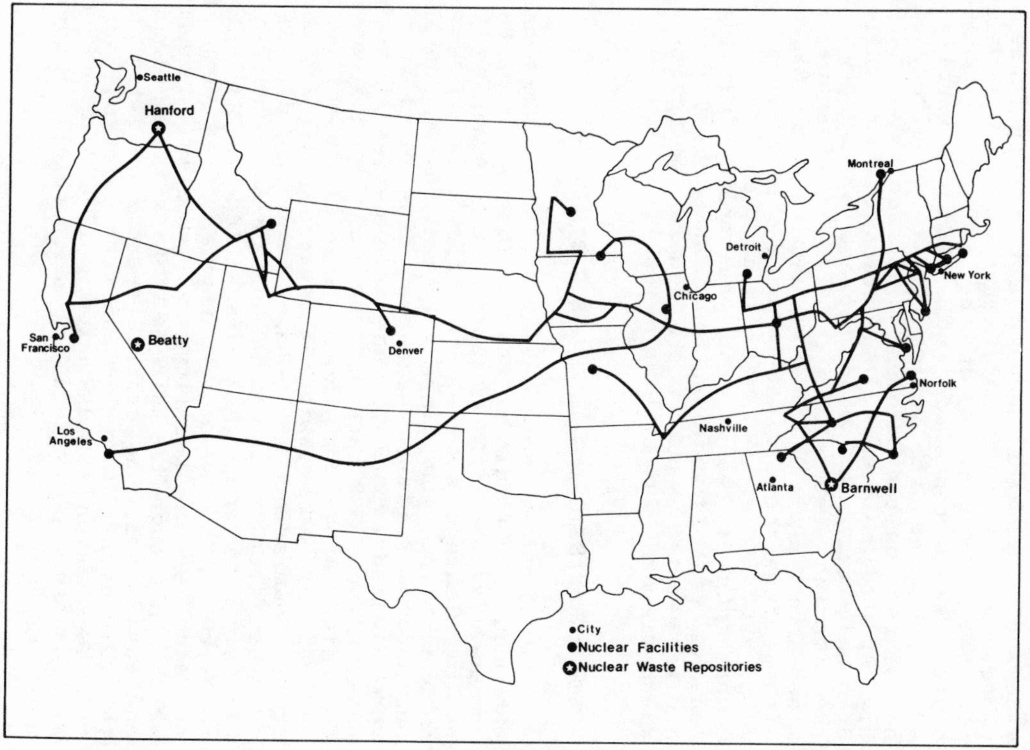

Although this risk may be worth taking, many local government officials are ill-prepared to deal with such an accident and are often oblivious to the existence of nuclear waste shipments through their communities (Cameron, 1981). In light of these health and safety concerns, dozens of local governments have passed ordinances restricting or preventing nuclear waste shipments through their communities (Haber, 1982). The DOT promulgated a rule to take effect in February, 1982 preempting these ordinances, although the rule is currently being challenged in the federal courts.

Decommissioned steam pressure vessels must be transported by barge or railroad, due to their large size and weight. Although the transport routes of the vessels will thus be more restricted and usually more circuitous than those of the radioactive waste, the nuclear waste repository utilized will usually be the same. However, the nearest waste repository will not necessarily be the cheapest; railroad rates are usually highest for shipments in western states, while waste disposal costs are also highest at the two western repositories (Table 15.5). Thus, nuclear waste shipments from nuclear power plants in the Midwest and sites farther west might be more cheaply sent to Barnwell, S.C. This problem is further complicated by land constraints at the waste repositories, possible state bans or restrictions on the transport or disposal of nuclear waste,[8] and future federal policy on nuclear waste disposal (Hill, et al. 1982).

POST-DECOMMISSIONING LAND USE OPTIONS

A retired nuclear power plant that is mothballed or entombed will remain in that condition for up to 70-110 or more years. In contrast, a dismantled facility could release a site to unrestricted land use within a decade or so. But since only three non-research-scale nuclear power plants have been dismantled in the U.S, very limited experience has been gained in this area. Moreover, the NRC has not even formally addressed post-decommissioning land use, because it is not considered part of the decommissioning process (U.S. Nuclear Regulatory Commission, 1981; Manion and LaGardia, 1980).

The first of the non-research-scale nuclear power plants to be completely dismantled was the Elk River facility in Elk River, Minnesota. This demonstration nuclear reactor facility was totally dismantled in 1974. Although the site was backfilled, landscaped, and paved over, the turbine systems were isolated from the reactor building for use with existing fossil fuel boilers (Manion and LaGuardia, 1976). Thus, the Elk River nuclear power facility was partially converted to a fossil fuel power plant.

TABLE 15.5
Radioactive Waste Disposal Costs at Three Sites

Cost Component	Cost Factor
Burial (at Hanford, WA and Beatty, NV)	
General Bulk Burial Rate (0-0.2 R/hr at container surface)	$4.75/ft^3
Weight Surcharge (10,000 lb and greater)	$50 + $0.01/lb (for weights over 10,000 lb)
Curie Surcharge (for 100 or more curie)	$375 + $0.05/curie (over 300 curie)
Cask Handling (min)	$250/cask
Burial (at Barnwell, SC)	
General Bulk Burial Rate (0-0.2 R/hr at container surface)	$3.60/ft^3
Weight Surcharge (5000 lb and greater)	$115/lb (for weights over 5000 lb; avg)
Curie Surcharge (0-500 curie)	none
Cask Handling (min)	$150/cask

Source: Manion and La Guardia (1980, Table 11.5).

The conversion of a nuclear power plant to another fossil fuel or nuclear power facility, or the ecological restoration of the site are the only known options considered by the nuclear power industry for post-decommissioning land use. Electric power companies are seemingly reluctant to discuss long term planning for existing power plant sites. The site-conversion option would normally follow the procedure at Elk River of utilizing the existing turbine system with a new steam supply system, following dismantlement. Whenever a decommissioned power plant site is owned by an electric utility company, any subsequent land use will normally be associated with the electric power industry (U.S. Department of Energy, 1982a). However, after the dismantlement of the Shippingport nuclear power plant, a new electric power facility will not follow; due to limited acreage, the power plant site will be restored by Duquesne Light to the approximate original, pre-nuclear plant ecology (Peters 1983).

The site conversion option and permanent dedication of a facility to nuclear application have also been considered at non-reactor sites. For example, refurbishment is being planned at the Eurochemic fuel processing plant in Belgium. Greater attention is expected to be given to this option, as acceptable nuclear plant sites become more difficult to find (Manion and LaGuardia, 1980).

Non-energy-related post-decommissioning land use options might also be considered in the future. After a nuclear power plant is decontaminated and dismantled, the land may possibly be reused without radiation hazards. But considering the large property tax levies that a nuclear power plant usually provides to local governments, often several million dollars annually, another large industrial facility might be most acceptable. This change could be accomplished if the electric utility company sells the land, an action which would allow local government economic development planning and land use controls to be more effective. However, many communities may prefer commercial development, agriculture, or open space on the recovered land, any of which could be viable in a given area. A problem with the nonindustrial, and thus non-energy, land use options would be the compatibility with surrounding land uses, particularly in a highly industrialized region. Such a constraint might be embodied in a local zoning ordinance or comprehensive plan, although amending these controls is also possible.

All of the above considerations will have an effect on the future of nuclear power in the United States. Decommissioning and nuclear waste disposal options will present some difficult choices for government and industry officials. Some predictions and recommendations for decommissioning are made in the next section.

FUTURE PROBLEMS AND MITIGATION STRATEGIES

Nuclear power plant decommissioning is in its infancy. Since a large number of "first generation" nuclear reactors are expected to be decommissioned in the mid-to-late 1980s (U.S. Department of Energy 1982b), much more will be known about the problems of decommissioning at a later date. Nevertheless, the following interrelated problems will probably predominate:

The lack of commercial experience. In a general sense, this will result in uncertainty about the optimal timing, costs, and resultant health and safety problems of decommissioning large-scale nuclear power plants.

Unavailability of decommissioning funds. As more immediate problems in the nuclear fuel cycle continue to take precedence over decommissioning, the funding of plant retirements will continue to be perceived by electric utilities as a minor issue. However, there are sound reasons to require nuclear power utilities to set aside liquid assets for decommissioning such as through revenue bonding, requirements which are currently the exception rather than the rule (Solomon, 1982).

Occupational radiation dosage and health effects of decommissioning. Since current estimates of radiation exposure and health problems from decommissioning are based on experience with small-scale nuclear power plants, commercial experience will again be critical. Deferring plant dismantlement reduces the radiation exposure during decommissioning. Thus, the accumulation of near-term commercial plant-dismantlement experience will conflict with environmental health objectives. Another approach to minimizing health risks may be to use a larger decommissioning workforce, subjecting each individual to smaller doses of radiation. Either strategy could be effective in reducing occupational radiation exposure and the resultant health problems. Inevitably, a trade-off will have to be made between the benefits and risks of a program of near-term nuclear power plant dismantlement.

Intensification of the nuclear waste disposal problem. Decommissioning will add reactor steam pressure vessels and their internals, as well as solid radioactive waste, to the existing solid nuclear waste stream. How much, and in what time frame, decommissioning will exacerbate the national nuclear waste problem will depend upon the method of decommissioning chosen for a significant number of nuclear power plants. The safe and permanent disposal of nuclear waste is one of the major unresolved problems of national energy policy.

Clearly, progress can be made on resolving the
nuclear waste disposal problem, which will require a
thoughtful program of federal-state-local communication
and cooperation. Although nuclear waste is a national
problem, regional and local interests must be considered for repository siting to truly succeed (Hill, et
al. 1982). Nuclear waste from decommissioning is
simply another symptom of the overall nuclear waste
disposal problem; more work will be required on both
the geologic implications of nuclear waste disposal, as
well as the optimal political system for its management
(e.g., autocratic, federalist, democratic, etc.).
Ultimately, no system may be totally environmentally
sound and politically acceptable to impacted communities. Further, hasteful resolution of the nuclear
waste disposal problem is not required to protect the
public health and safety. But progress on this problem
must continue as the volume of nuclear waste continues
to accumulate in spent fuel pools at existing power
plant sites.

The risk of accident during transport and the
potential impacts. As the nuclear waste volume
increases, so will the risk of accident in relation to
the increase of the nuclear waste volume. The trade-
off between minimizing transport costs and avoiding
major population centers during nuclear waste shipments
to radioactive waste repositories needs to be carefully
considered by the DOT.

The safety of nuclear waste transport needs more
attention. As the number of shipments increases future
decommissionings, more communities will be confronted
with this problem. Although the DOT and NRC are best
qualified to determine nuclear waste shipment routes,
more consultation with state and local governments is
necessary. Specifically, a more vigorous training
program for state and local response to transport
accidents is required, and local governments need to be
better informed about nuclear waste shipment routes.
The implication of doing otherwise is to needlessly
increase the safety risks to the general public in the
affected communities.

NOTES

1. Only one commercial nuclear fuel reprocessing
plant has ever operated in the U.S.--the abandoned
facility at West Valley, New York. West Valley shut
down after six years of operation (1966-1972) because
of financial difficulties and contamination problems.
The steel tanks at West Valley have leaked radioactive
acids into the surrounding soils. Current estimates
for the cost of cleanup and decontamination are $600
million to $1 billion. For details, see Cohen, (1979).

2.   For a more technical description of mothballing and entombment, see Manion and LaGuardia (1976, 1980).

3.   The NRC is preparing a generic EIS for decommissioning a variety of nuclear facilities, in anticipation of regulatory changes that might follow its recent reevaluation of decommissioning policy (U.S. Nuclear Regulatory Commission, 1981). Plans for decommissioning policy reevaluation are discussed in Calkins, (1980).

4.   The depreciation-accounts-approach to recovering decommissioning costs increases the plant depreciation base to be collected on, as part of the normal operating revenues from ratepayers during the plant's operating lifetime. Since these funds would normally be invested in new capital facilities in the utility system until needed for decommissioning, they will not represent liquid assets. Thus, the decommissioning funds might exist only in the utility company's accounting record books and financial reports.

5.   For an interesting account of radiation releases and health effects among the plant operators and general public that have occurred at the Shippingport nuclear power plant site, see the discussion in Sternglass (1981, Ch. 15).

6.   Any nondismantlement decommissioning option requires an extremely long-term, although indefinite, commitment of land resources at a nuclear power plant site. Since this commitment can be viewed as a negative environmental impact, it would most likely require justification on such bases as minimization of occupational radiation exposure, optimum cost, and maintenance of a hazard-free potential to public, health and safety (Manion and LaGuardia, 1980).

7.   The decommissioning of a federally owned nuclear power plant will have no impact on local government property tax receipts, since federal property is not subject to local taxation. A few nuclear facilities, such as the Shippingport nuclear power plant, have been jointly owned by an electric utility and the federal government (U.S. Department of Energy, 1982a).

8.   The citizens of Washington approved an initiative in November, 1980 prohibiting the import to the state of nuclear waste after July, 1981, unless approved by interstate compact and majority vote of both Houses of Congress. The implementation of this measure is uncertain, since it is currently being challenged in the courts. The outcome will have major implications for operations at Hanford and other nuclear waste repositories.

REFERENCES

Bupp, I.C. and J.C. Derian. 1978. Light Water: How the Nuclear Dream Dissolved. New York: Basic Books.

Calkins, G.D. 1980. Plan for Re-evaluation of NRC Policy on Decommissioning of Nuclear Facilities. NUREG-0436. Washington: U.S. Nuclear Regulatory Commission, Revision 1.

Cameron, D.M. 1981. Hot Hoosier Highways: A Report on the Preparedness of State and Local Officials in Indiana to Respond to Accidents in the Transportation of Radioactive Material. Bloomington: Indiana Public Interest Research Group.

Cohen, B.L. 1979. "The Situation at West Valley," Public Utilities Fortnightly, 104:26-32.

Comptroller General of the U.S. 1977. Cleaning up the Remains of Nuclear Facilities - a Multibillion Dollar Problem. EMD-77-46. Washington: U.S. General Accounting Office.

Fako, J.R. and R.J. Dickson. 1981. "Nuclear Power Plant Decommissioning as a Cost of Service," Public Utilities Fortnightly, 108:31-34.

Ferguson, J.S. 1980. "The Capital Recovery Aspects of Decommissioning Power Reactors," Public Utilities Fortnightly, 106:34-42.

Friedlander, G.D. 1978. "Decommissioning Commercial Reactors," Electrical World, 189: 44-48.

Fuller, J.G. 1976. We Almost lost Detroit. New York: Ballantine Books.

Goffman, J. 1981. Radiation and Human Health. San Francisco: Sierra Club Books.

Haber, S. 1982. Year-end Nuclear Legislative Report. Washington: Atomic Industrial Forum.

Hill, D., B.L. Pierce, W.C. Metz, M.D. Rowe, E.T. Haefele, F.C. Bryant, and E.J. Tuthill. 1982. "Management of High-level Waste Repository Siting," Science, 218:859-64.

Iwler, L. 1978. "Regulatory Bodies Ready for N-plant Retirements," Electrical World, 190: 19-20.

Kasperson, J.S., R.E. Kasperson, C. Hohenemser, and R.W. Kates. 1979. "Institutional Responses to Three Mile Island," Bulletin of the Atomic Scientists, 35:20-24.

Lear, G. and P.B. Erickson. 1975. "Decommissioning and Decontamination of Licensed Reactor Facilities and Demonstration Nuclear Power Plants." In Proceedings of the First Conference on Decontamination and Decommissioning (D&D) of ERDA Facilities. Idaho Falls, Idaho: U.S. Energy Research and Development Administration, pp. 31-45.

Lovins, A.B. and L.H. Lovins. 1981. Energy/war: Breaking the Nuclear Link. New York: Harper Colophon Books.

Mandahl, B., et al. 1979. _Technology and Costs for Dismantling Swedish Nuclear Plants_. KBS Technical Report 79-21. Stockholm: International Institute for Energy and Human Ecology, Royal Swedish Academy of Sciences.

Manion, W.J. and T.S. La Guardia. 1976. _An Engineering Evaluation of Nuclear Power Reactor Decommissioning Alternatives_. Washington: Atomic Industrial Forum.

Manion, W.J. 1980. _Decommissioning handbook_. Washington: U.S. Department of Energy.

National Academy of Sciences. 1980. _The Effects on Populations of Exposure to Low Levels of Ionizing Radiation: BEIR III_. Washington: National Research Council.

Oak, H.D., et al. 1980. _Technology, Safety and Costs of Decommissioning a Reference Boiling Water Reactor Power Station_. NUREG/CR-0672. Washington: U.S. Nuclear Regulatory Commission, Vol. 1.

Peters, J. 1983. Personal correspondence, Pittsburgh: Nuclear Operating Division, Duquesne Light Company.

Radioactive Waste Management Committee. 1982. _International Decommissioning Technology Exchange Questionnaire: Draft Summary Report_. Paris: OECD Nuclear Energy Agency.

Robbins, J.C. 1981. _Routing Hazardous Materials Shipments to Reduce Population Exposure Risk_. Ph.D. dissertation, Department of Geography, Indiana University.

Schneider, K.J. and C.E. Jenkins. 1977. _Technology, Safety and Costs of Decommissioning a Reference Nuclear Fuel Reprocessing Plant_. NUREG-0278. Washington: U.S. Nuclear Regulatory Commission, Vol. 1.

Schwent, V.L. 1978. Costs and Financing of Reactor Decommissioning: Some Considerations. Paper presented to a meeting of the National Association of Regulatory Utility Commissioners, Washington.

Smith, R.I., G.J. Konzek, and W.E. Kennedy. 1978. _Technology, Safety and Costs of Decommissioning a Reference Pressurized Water Reactor Power Station_. NUREG/CR-0130. Washington: U.S. Nuclear Regulatory Commission, Vol. 1.

Solomon, B.D. 1982. "U.S. Nuclear Energy Policy: Provision of Funds for Decommissioning," _Energy Policy_, 10:109-19.

Sternglass, E. 1981. _Secret Fallout: Low-level Radiation from Hiroshima to Three Mile Island_. New York: McGraw-Hill.

U.S. Department of Energy. 1982a. _Decommissioning of the Shippingport Atomic Power Station_. EIS-0080F. Washington: U.S. Government Printing Office.

U.S. Department of energy. 1982b. _Nuclear Reactors Built, Being Built or Planned_. DOE/TIC-8200-R40. Washington: U.S. Government Printing Office.

412

U.S. Nuclear Regulatory Commission. 1980. _Public Information Circular for Shipments of Irradiated Reactor Fuel_. NUREG-0725. Washington: U.S. Government Printing Office.

U.S. Nuclear Regulatory Commission. 1981. _Draft Generic Environmental Impact Statement on Decommissioning of Nuclear Facilities_. NUREG-0586. Washington: U.S. Government Printing Office.

Zeigler, D.J., S.D. Brunn, and J.H. Johnson. 1981. "Evacuation from a Nuclear Technological Disaster," _The Geographical Review_, 71:1-16.

Index

413

Contributors

A. M. Aikin (Ph.D. McGill University), chemist and chemical engineer, worked for Atomic Energy of Canada Limited from 1949 to 1976, for several years as Vice-President in charge of Whiteshell Nuclear Research Establishment in Manitoba. From 1976 to 1981, he was a consultant in nuclear energy matters. He is now retired.

Earl Cook (Ph.D. University of Washington) is late Distinguished Professor of Geography and Geology and Harris Professor of Geosciences, Texas A&M University. He published extensively on energy matters and has published a widely acclaimed book <u>Energy, Society and Man</u>. He was formerly Dean of the College of Geosciences (1971-1981) at Texas A&M University, and was one-time President of the Association of American State Geologists.

Susan L. Cutter (Ph.D. University of Chicago) is Associate Professor, Department of Geography, School of Urban and Regional Policy, Rutgers University. Her research interests are in behavioral geography and environmental management. Her most recent publications have focussed on the human responses to the accident and clean-up operations at the Three Mile Island nuclear power plant. She is a member of the Society of Risk Analysis.

Mary L. Daum (Ph.D. University of Wisconsin-Madison) is Research Collaborator, Department of Applied Science, Brookhaven National Laboratory. Formerly she was Instructor, Department of Geography, Northern Illinois University and the Department of Geography, University of Wisconsin-Rock Co. She is currently involved in several research projects dealing with acid precipitation.

417

Patrick G. Derr (Ph.D. University of Notre Dame) is Associate Professor of Philosophy, Clark University. He is Consultant in Clinical Ethics, Worcester City Hospital and Research Affiliate, Center for Environment, Technology and Development, Clark University. His research has focused on such topics as equity problems in technological hazards management. He is currently Associate Director of the New England Center for Philosophy and Public Affairs.

Jerome E. Dobson (Ph.D. University of Tennessee) is Leader of the Resource Analysis Group in the Energy Division of Oak Ridge National Laboratory. His specialties include the use of automated geographic analysis techniques to address policy issues regarding energy resource development, energy planning, and emergency management. His publications include works on automated geography, siting analysis, technology assessment, and water resources. He is a former chairman and current director of the Energy Specialty Group of the Association of American Geographers.

John Fernie (Ph.D. Edinburgh University, Scotland) is Senior Lecturer of Geography, Huddersfield Polytechnic, England. His research has focused on energy matters in Europe, most recently including nuclear power. Previous teaching appointments have included the University of Victoria and Arizona State University. He is author of <u>A Geography of Energy in the United Kingdom</u>.

Michael Greenberg (Ph.D. Columbia University) is Professor of Environmental Planning and Geography, and Co-director of the Graduate Program in Public Health, Rutgers University and the University of Medicine and Dentistry, New Brunswick. His current research foci are the relationship between urbanization and chronic diseases and hazardous waste sites. He is recent author of <u>Urbanization and Cancer Mortality</u>. He is recipient of a special award of merit from the U.S. Environmental Protection Agency.

F. Kenneth Hare (Ph.D. Universite de Montreal) is Provost of Trinity College and University Professor of Geography and Physics at the University of Toronto. He is recipient of six honorary degrees, and is a fellow of King's College (1967), Fellow of the Royal Society of Canada (1968), and is Officer of the Order of Canada (1978). He was Chairman of the Federal Study Group on Nuclear Waste Management (1977), Co-Chairman of the National Academy of Sciences/Royal Society of Canada Committee on Acid Precipitation (1980-1982), member of the Advisory Council of the Electric Power Research Institute (1979-1980), and a member of the Scientific Advisory Committee on Climatic Impact of the United Nation World Climate Impact Program (1980-1983).

James H. Johnson, Jr. (Ph.D. Michigan State University) is Assistant Professor of Geography at the University of California at Los Angeles. His research is concerned with human responses to technological hazards, especially nuclear reactor accidents, and he has testified before the Nuclear Regulatory Commission's Atomic Safety and Licensing Board in emergency planning proceedings for the Diablo Canyon (California) and Shoreham (New York) nuclear power plants.

Michael Kaltman (M.A. University of Pennsylvania) is Regional Planning Analyst with the U.S. Nuclear Regulatory Commission's Office of Nuclear Reactor Regulation in Washington, D.C. Prior positions include Chief of the Research Section in the Baltimore City Department of Planning, and Housing Planner for New York State. He has sponsored research on such topics as land use, population, socioeconomic impacts of constructing and operating facilities, and construction labor force impacts.

Roger E. Kasperson (Ph.D. University of Chicago) is former director and member of the Center for Technology, Environment, and Development and Professor of Government and Geography at Clark University. He has published widely on issues related to risk assessment and risk management, nuclear energy policy, and citizen participation. He is editor of the recent book Equity Issues in Radioactive Waste Management. Professor Kasperson has served as consultant to the Presidential Commission on the Accident at Three Mile Island, the Council on Environmental Quality, the California Energy Commission, among others. From 1977 to 1983, he served as a member of the National Research Council's Board of Radioactive Waste Management and chaired its panel on Social and Economic Issues in Siting Nuclear Waste Repositories. He is presently on the editorial boards of Environment and Risk Analysis.

Robert W. Kates (Ph.D. University of Chicago) is Research Professor, is a member of the Center for Technology, Environment, and Development, Clark University, and Professor of Geography. He is recognized internationally for his contributions to risk assessment and management of environmental hazards, and theory of the human environment. He has been a visiting scholar at several institutions and participated in many collaborative research projects including the Collaborative Research on Natural Hazards project with the Universities of Chicago, Colorado and Toronto. He is co-author of the Environment as Hazard and has authored other books on hazard management and policy. He has received the Honors Award of the Association of American Geography, honorary membership in Phi Beta Kappa, and a

420

MacArthur Prize Fellowship. He is a member of the National Academy of Sciences.

Donald A. Krueckeberg (Ph.D. University of Pennsylvania) is Professor of Urban Planning and Policy Development at Rutgers University. He is former editor of the <u>Journal of the American Planning Association</u>, and is author of numerous articles and several important books related to demographic analysis and public policy.

William C. Metz (Ph.D. University of Pittsburgh) is a Program Manager at Brookhaven National Laboratory. His research and publications have concentrated on energy facility siting methodologies, socioeconomic assessment and mitigation strategy development, barriers to high-level radioactive waste disposal, land use and population changes around nuclear power stations, and energy policy analyses. Previously, he worked with the Westinghouse Electric Corporation in the Environmental Systems Department.

Stanley Openshaw (Ph.D. University of New Castle) is Lecturer in the Department of Geography, Newcastle University, England. His research and publications have been in the area of spatial analysis of reactor siting policies, the geographical effects of hypothetical reactor accidents, and associated casualty predictions. His most recent work has been in reactor accident prediction models, in the comparative evaluation of sites using US and UK criteria, and in computer-based searches for sites. He is author of a recent book on the effects of nuclear war on Britain.

Timothy O'Riordan (Ph.D. University of Cambridge) is Professor of Environmental Sciences at the University of East Anglia, Norwich, England. He is internationally known for his research publications on environmental issues and is author of the widely used <u>Environmentalism</u>. He is holder of the Royal Geographical Society's Gill Memorial Award, and is Vice Chairman of the United Kingdom's Social Science Research Council Committee on Environmental Planning. He is one of the European Editors of <u>Risk Analysis: An International Journal</u>, and he presently holds a research grant to study the public inquiry into Britain's first pressurized water nuclear reactor at Sizewell in Suffolk.

Martin J. Pasqualetti (Ph.D. University of California at Riverside) is Associate Professor of Geography and Honors Research Professor at Arizona State University. His research and publications have addressed land use/energy problems of the development of geothermal energy, solar energy, and coal. Dr. Pasqualetti's most recent research has addressed environmental impacts of the nuclear fuel cycle, and the land- oriented policy issues attendant to decommissioning nuclear power plants. He has been Project Director on energy education projects funded by the U.S. Department of Energy and the National Science Foundation, and is a member of the Arizona Council on Energy Education. He is co-founder and current Chairman of the Energy Specialty Group of the Association of American Geographers.

Kenneth T. Pearlman (J.D. Columbia Law School, M.C.P. University of Pennsylvania) is Associate Professor of City and Regional Planning, Department of City and Regional Planning, The Ohio State University, with adjunct appointment to the Ohio State University Law School. He is current Editor of the Journal of the American Planning Association. His current research include land use controls, land use policy, environmental planning law, and social policy law.

K. David Pijawka (Ph.D. Clark University) is Assistant Director of the Center for Environmental Studies, and Assistant Professor in the Center for Public Affairs, Arizona State University. His research and publications have focused on risk assessment and technology policies, social impacts of energy development and environmental planning. He recently published a major study on the socioeconomic impacts of nuclear power plants for the U.S. Nuclear Regulatory Commission. Other research has included pesticide policy and regulation, decommissioning energy facilities, and hazardous waste planning.

Clark Prichard (Ph.D. Michigan State University) is an Economist and Regulatory Analyst with the U.S. Nuclear Regulatory Commission. He has also had positions with the International Monetary Fund and the Federal Energy Administration. Dr. Prichard was also honored as a National Science Foundation Fellow. He has extensive experience in areas of energy economics and regulatory policy.

Jeffrey P. Richetto (Ph.D. The Ohio State University) is Assistant Professor in the Department of Geography, Division of Urban and Regional Planning at the University of Alabama. He has published internationally on environmental and planning aspects of nuclear siting, and he has served as research analyst with the Ohio Power Siting Commission. His current research interests include energy resource development and energy facility modeling.

John E. Seley (Ph.D. University of Pennsylvania) is Associate Professor of Urban Studies at Queens College,

The City University of New York, Flushing. He has taught at the University of Minnesota, Cornell University, and Clark University. He has been a visiting fellow at the School of Architecture and Urban Planning at Princeton University. He has served on two committees of the National Academy of Sciences/National Research Council, including the Panel on Social and Economic Aspects of Radioactive Waste Management. He is recent author of <u>The Politics of Public-Facility Planning</u>.

Barry D. Solomon (Ph.D. Indiana University) is Visiting Assistant Professor of Geography and Mineral Resource Economics, and Research Associate in the Regional Research Institute, West Virginia University. His research and publications have emphasized power plant siting, coal development modeling, biomass fuels, and nuclear energy policy issues. He is current chairperson of the Public Policy Committee of the Association of American Geographer's Energy Specialty Group, and is co-investigator in a study of the state of the art in socioeconomic impact analysis modeling of energy resource development, funded by the U.S. Department of Energy.

John H. Sorensen (Ph.D. University of Colorado) is a Research Associate in the Resource Analysis Group at Oak Ridge National Laboratory. His research deals primarily with human response to natural and technological hazards and emergency planning. He is currently directing a major study of the social and psychological impacts of restarting Three Mile Island Unit 1 and working on projects dealing with emergency planning for nuclear power plants and siting hazardous facilities. He was formerly research associate at the Institute of Behavioral Science at the University of Colorado.

Nancy D. Waite (M.S. candidate in City and Regional Planning, The Ohio State University) is a neighborhood planner for the City of Columbus, Ohio. She has written on the topic of land use control techniques available to maintain low population zone around nuclear power plants, and is currently conducting thesis research on conflict resolution in communities that are sites for hazardous waste facilities.

Thomas J. Wilbanks (Ph.D. Syracuse University) is Associate Director and Head of Programs and Planning for the Energy Division of Oak Ridge National Laboratory. He served formerly on the faculties of Syracuse University and the University of Oklahoma (where he was chairman of the Department of Geography). He has published extensively on energy policy, regional development and other issues and has provided advice to international, federal, state, and local government agencies. He has served as Treasurer of the Association of American Geographers, and is currently a member of the Committee on Behavioral and Social Aspects of Energy Consumption and Production, National Academy of Sciences/National Research Council.

Donald J. Zeigler (Ph.D. Michigan State University) is Assistant Professor and Director of the Geography Program at Old Dominion University in Norfolk, Virginia. His primary interest is in human responses to technological hazards, particularly nuclear power, and he has done extensive research on the Three Mile Island and Shoreham nuclear power plants.

Wilbur Zelinsky (Ph.D. University of California at Berkeley) is Professor of Geography at The Pennsylvania State University. He was recipient of the Award for Meritorious Contribution to the Field of Geography by the Association of American Geographers, and President of the same association (1972-1973). His work has focused on the cultural and social aspects of Geography, and he currently serves on the editorial boards of two scholarly journals. He was a Guggenheim Fellow 1981-1982.